Proceedings of ICIPE 2011

The 7th International Conference on Inverse Problems in Engineering

Edited by

Alain J. Kassab

Eduardo A. Divo

Centecorp Publishing

Orlando, Florida, USA

ISBN-13: 978-0615471006 (Centecorp Publishing)
ISBN-10: 0615471005

First Printing

Proceedings of ICIPE 2011

***Alain J. Kassab* - Chair**
***Eduardo A. Divo* - Co-Chair**

Local organizing committee

John Cannon (University of Central Florida, USA)
Eduardo Divo (Daytona State College, USA)
Kirk Dolan (Michigan State University, USA)
Jay Frankel (University of Tennessee, USA)
Alain Kassab (University of Central Florida, USA)
Faissal Moslehy (University of Central Florida, USA)
Zuhair Nashed (University of Central Florida, USA)
Keith Woodbury (University of Alabama, USA)

International scientific committee

Jean-Luc Battaglia (University of Bordeaux, France)
James Beck (Michigan State University, USA)
Ryzsard Bialecki (Silesian Technical University, Poland)
Marc Bonnet (Ecole Polytechnique, France)
Cara Brooks (Rose-Hulman Institute of Tech., USA)
Tadeusz Burczynski (Silesian Technical Univ., Poland)
John Cannon (University of Central Florida, USA)
Miguel Cerrolaza (Central University, Venezuela)
Jongeun Choi (Michigan State University, USA)
Marcelo Colaco (Federal Univ. of Rio de Janeiro, Brazil)
Andrei Constantinescu (Ecole Polytechnique, France)
Renato Cotta (Federal Univ. of Rio de Janeiro, Brazil)
Kirk Dolan (Michigan State University, USA)
Kevin Dowding (Sandia National Labs, USA)
George Dulikravich (Florida International Univ., USA)
Ashley Emery (University of Washington, USA)
Colin Fox (University of Otago, New Zealand)
Bojan Guzina (Univ. of Minnesota, USA)
Jay Frankel (University of Tennessee, USA)
Brizeida Gamez (University of Carabobo, Venezuela)
Alemdar Hassanov (Izmir University, Turkey)
Cheng-Hung Huang (Nat. Cheng Kung Univ., Taiwan)
Derek Ingham (University of Leeds, UK)
Alexander Katsevich (Univ. of Central Florida, USA)
Patricia Lamm (Michigan State University, USA)
German Larrazabal (Univ. of Carabobo, Venezuela)
Antonio Leitao (Federal Univ. of Santa Catarina, Brazil)
Daniel Lesnic (University of Leeds, UK)
Toshiro Matsumoto (University of Nagoya, Japan)
Denis Maillet (Insitut Nat. de Poly. de Lorraine, France)
Faissal Moslehy (University of Central Florida, USA)
Zuhair Nashed (University of Central Florida, USA)
Antonio Silva Neto (Instituto Politécnico, UERJ, Brazil)
Andrew Nowak (Silesian Technical University, Poland)
David Ojeda (University of Carabobo, Venezuela)
Helcio Orlande (Federal Univ. of Rio de Janeiro, Brazil)
Darrell Pepper (University of Las Vegas, USA)
Keith Woodbury (University of Alabama, USA)
Luiz Wrobel (Brunel University, UK)
Nicholas Zabaras (Cornell University, USA)

Hosted at

University of Central Florida, Orlando, Florida
May 4-6, 2011, Orlando, Florida, USA

Conference Sponsors

United States National Science Foundation
Inverse Problems in Science and Engineering

History of the Conferences

During the years that the international conference (International Conference on Inverse Problems in Engineering - ICIPE) does not take place, informal two-day seminars (Inverse Problem Symposia - IPS) have been held at various universities in the USA. The sites and associated organizers are listed for purposes of archiving the historical continuity of these meetings. Previous international meetings (International Conferences on Inverse Problems in Science and Engineering- ICIPE) were held at Palm Coast, FL, in 1993, at Le Croisic, France, in 1996, at Port Ludlow, WA, in 1999, at Angra dos Reis, Brazil, in 2002, at Claire College, Cambridge, England, in 2005, and at Dourdan, France, in 2008. These international conferences grew out of several informal meetings held in the years prior to the Palm Coast conference, which focused mainly on inverse problems in heat transfer. Nicholas Zabaras broadened the scope of the Conference in the first international conference in 1993, especially into the field of solid mechanics. This conference was successful in attracting participation from the mechanics group.

Year	Venue	Chair
1988 – 91	Michigan State University, East Lansing, Michigan	James V. Beck
1992	Michigan State University, East Lansing, Michigan	Keith A. Woodbury
1993	*Palm Coast, Florida, USA*	*Nicholas Zabaras*
1994	University of Cincinnati, Cincinnati, Ohio	Diego Murio
1995	Ohio State University, Columbus, Ohio	Hank Busby
1996	*Le Croisic, France*	*Didier Delaunay*
1997	Rensselaer Polytechnic Institute, Troy, New York	Antoinette Maniatty
1998	Ball State University, Muncie, Indiana	Lija Guo
1999	*Port Ludlow, Washington, USA*	*Ashley Emery*
2000	Texas A&M University, College Station, Texas	Ted Watson
2001	University of Texas – Arlington, Arlington, Texas	A. Hadji-Sheikh
2002	*Angra dos Reis, Brazil*	*Helcio R.B. Orlande*
2003	The University of Alabama, Tuscaloosa, AL	Keith A. Woodbury
2004	University of Cincinnati, Cincinnati, Ohio	Diego Murio
2005	*Claire College, Cambridge, England*	*Daniel Lesnic*
2006	Iowa State University, Ames, Iowa	Mark Bryden
2007	Michigan State University, East Lansing, Michigan	Neil T. Wright
2008	*Dourdan, France*	*Denis Maillet*
2009	Michigan State University, East Lansing, Michigan	Kirk Dolan
2011	*University of Central Florida*	*Alain Kassab*

Sites and organizers for the International Conferences and symposia (International Conference on Inverse Problems in Engineering: Theory and Practice, ICIPE are italicized).

Preface

This is the 24th in the series of national and international meetings on *Inverse Problems* that were initiated at Michigan State University in 1988. In particular, it is the 7th of the international conferences in this series. Papers were solicited from graduate students, faculty, and scientists both nationally and internationally. Indeed, the international scientific committee and conference participants are from a variety of countries, including the USA, France, Poland, England, Russia, Japan, Brazil, among others. The topics covered in the conference are broad spanning from applications to thermal sciences, including estimation of thermal properties, estimation of heat flux functions, determination of thermocouple error, and other inverse heat transfer problems to mechanics, including such topics as damage identification and NDE to medical application such as LVAD controller design. The primary objective of the ICIPE 2011 conference is to bring together researchers and graduate students in sciences and engineering to present recent research results in inverse problems and to provide a forum in which to discuss their findings and to foster interdisciplinary interaction. Indeed, a noteworthy feature of all ICIPE meetings is the balanced focus on theory, applications, as well as the combination of both.

The meeting consisted of three days of technical presentations enriched by a workshop on *Parameter Estimation* and the *Inverse Heat Conduction Problems* (IHCP) which was delivered over the course of two days by Professor James Beck, Professor Robert McMasters and Professor Kirk Dolan. The workshop was open to all conference attendees, faculty and graduate students. The organizers gratefully acknowledge the financial support from the *United States National Science Foundation* which made possible the organization and delivery of the Workshop as well as the support of selected graduate students to attend the meeting and participate in the workshop. The contents of the workshop, slide presentations and associated materials as well as a recorded video of the parameter estimation lectures, are available for download from the conference website: http://www.icipe2011.org

A business meeting whose purpose was the planning of the next conference was held immediately following the close of the conference. The meeting was facilitated by Alain Kassab and proposals for hosting the next meeting were solicited and discussed.

As with previous meetings in this series, has attracted outstanding contributions collected and published in this book. This reflects the excellent work of all contributing authors and the care taken by the Scientific Advisory Committee and other colleagues in reviewing the presentations. As conference organizers, we are grateful to our colleagues who have made this conference a great success, and we hope we have achieved our aim to provide a friendly atmosphere that maximizes opportunities for interactions between participants.

We look forward and with anticipation to the next inverse problems meeting in this series.

Alain J. Kassab
Eduardo A. Divo
Orlando, Florida, USA, May 2011

Table of Contents

1. Mathematical Foundations and Algorithms

2. Inverse Problem in Materials, Bioengineering, and Chemistry

3. Inverse Problems in Machining

4. Inverse Problems in Heat Transfer: source term reconstruction

5. Inverse Problems in Heat Transfer: boundary condition and geometric reconstruction

6. Inverse Problems in Heat Transfer: property evaluation

7. Inverse Problems in Fluid Mechanics and Porous Media

8. Inverse Problems in Solid Mechanics

9. Stochastic Methods

10. Optimization and Inverse Design

COMPUTING SENSITIVITIES OF FINITE ELEMENT RESPONSES USING THE COMPLEX VARIABLE SEMI-ANALYTIC METHOD

Weiya Jin[1], Brian H. Dennis[2], and Bo Ping Wang[2]

[1]College of Mechanical Engineering
Zhejiang University of Technology, ZJUT
Hangzhou, Zhejiang, China
Email: jinweiya@zjut.edu.cn

[2]Department of Mechanical and Aerospace Engineering
University of Texas at Arlington
Arlington, TX, U.S.A.
Email: dennisb@uta.edu

Abstract

In many inverse and optimization applications, the semi-analytical method (SAM) is typically used in large scale finite element programs to compute the sensitivity of the objective function and constraints with respect to an input parameter since its computational efficiency. However, the SAM can lead to inaccurate sensitivities if the step size is not chosen carefully. The paper proposes the Complex Variable Semi-analytical Method (CVSAM), which combines the computational efficiency of the SAM and the accuracy of the complex variable method to compute sensitivities of finite element response with respect to input variables. The comparison of sensitivities computed by the CVSAM with an analytical solution for a non-linear heat conduction problem demonstrates the accuracy of the CVSAM. Application to the inverse heat conduction problem of shape determination demonstrates the benefits of accurate sensitivities provided by the CVSAM.

1. Introduction

The finite element method (FEM) is one of the most popular numerical methods in use today for stress and heat transfer analysis. The method has recently been extended beyond simple analysis and has been applied to various optimization and inverse problems. For many inverse and optimization applications, the sensitivity of the objective function and constraints with respect to an input parameter is required. If the objective function depends on a finite element calculation, the sensitivity of the response is ultimately required. Methods that can compute these responses accurately and efficiently are therefore needed. Many sophisticated approaches have been proposed including automatic differentiation [1, 2] and the adjoint variable approach [3]. However, the simple finite difference approach remains popular due to its programming simplicity and generality since very few modifications to the analysis code are required.

In practice, the semi-analytical method (SAM) is typically used in large scale finite element programs. This method combines analytical differentiation with finite differencing to create an approach that is more computationally efficient for large numbers of input parameters. However, the semi-analytical approach can lead to inaccurate sensitivities if the step size is not chosen carefully [4, 5, 6, 7]. For many problems there is an optimal step size but it is not known *a priori*.

It is known that dependence on step size choice can be eliminated if finite differencing in the complex plane is used. However, this requires the use of complex variables for all floating point numbers in the analysis code, which doubles the memory requirements and may increase run times by a factor of three. To overcome these computational efficiency drawbacks, we combine the semi-analytical method with the complex variable method to efficiently compute the first derivatives of a finite element response with respect to an input variable. With this semi-analytic complex variable method (CVSAM) the accuracy of the complex variable method is achieved with the computational efficiency of the semi-analytical method.

The paper proposes the Semi-analytical Complex Variable Method (CVSAM) [8] to compute the sensitivities of finite element response with respect to input variables. It combines the advantages of two methods and eliminates the drawbacks of both methods. With the CVSAM, the accuracy of the complex variable method is achieved with the computational efficiency of the SAM. The application of the CVSAM to non-linear heat conduction problem and inverse problem of heat conduction demonstrates its accuracy, consistency, and computationally efficiency.

2. The Complex Variable Semi-Analytical Method

In this section we introduce the CVSAM applied to systems of equations resulting from discretization by FEM. The finite element global equilibrium equation is

$$Ku = f \tag{1}$$

where the stiffness matrix K, displacement u and load vector f are functions of input random variables $X = (X_1, X_2, \cdots, X_n)$.

The SAM combines the efficiency of the analytical method with the ease of use and general nature of the finite difference method. Differentiate Equation (2) on both sides, we obtain

$$\frac{\partial K}{\partial X}u + K\frac{\partial u}{\partial X} = \frac{\partial f}{\partial X} \tag{2}$$

Then the derivative of displacement with respect to random variables X can be written as

$$\frac{\partial u}{\partial X} \approx \frac{\Delta u}{\Delta X} = K^{-1}\frac{\Delta f}{\Delta X} - K^{-1}\frac{\Delta K}{\Delta X}u \tag{3}$$

In Equation (3), the same stiffness matrix K is used for computing displacement u and displacement sensitivity with respect to all the random variables $\partial u / \partial X$ .Solving K^{-1}, the most time consuming computation, is required to be calculated only once for both Equation (1) and Equation (3). This advantage makes the SAM programming convenient and computationally efficient. However, since the finite difference method (FDM) is used in Equation (3) to compute $\Delta K / \Delta X$ and $\Delta f / \Delta X$, the SAM inherits the drawback of the FDM. The sensitivity $\partial u / \partial X$ is sensitive to the step size. An optimal step size exists that either small enough to minimize the truncation error or large enough to avoid subtractive cancellation error. The choice of proper step size in advance becomes very difficult for a complex problem to compute sensitivities with respect to various variables.

The complex variable method (CVM) is applied to computational fluid dynamics field to obtain sensitivities [9, 10, 11, 12]. It is based on a Taylor series expansion that takes a complex step in the imaginary dimension.

$$\begin{aligned} Ku = f(x+i\Delta x) = f(x) + i\Delta x f'(x) \\ - \frac{\Delta x^2 f''(x)}{2!} - \frac{\Delta x^3 f'''(x)}{3!} + \cdots \end{aligned} \tag{4}$$

The real and imaginary parts of Equation (4) are grouped and the first order derivative can be obtained as

$$f'(X) = \frac{\mathrm{Im}[f(x+i\Delta x)]}{\Delta x} + O(\Delta x^2). \tag{5}$$

We note that the calculation of the first order derivatives in Equation (5) does not involve the subtraction of two numbers. Therefore, the CVM avoids the subtractive cancellation errors that plague the finite difference method. The step size can be as small as possible with no loss of accuracy. However, the more accurate first derivatives come at a price, namely more computation time and memory. The factorization of a global stiffness matrix with all values declared as complex variable is the main cause of extra computational time and space.

The CVSAM combines the CVM with the SAM, and takes advantage of efficiency of the SAM and accuracy of the CVM to obtain the sensitivity of finite element response. The stiffness matrix in Equation (1) is factored once with all the variables declared as real variables. The factored K and u obtained are then substituted into Equation (3), where the numerical differentiation of stiffness matrix and load vector $\partial K/\partial X$ and $\partial f/\partial X$ are computed by the CVM. Therefore, the sensitivity can be computed as

$$\frac{\partial u}{\partial X} \approx K^{-1}\left\{\frac{\mathrm{Im}[f(X+i\Delta X)]}{\Delta X} - \frac{\mathrm{Im}[K(X+i\Delta X)]}{\Delta X}u\right\} \tag{5}$$

The computation of $\mathrm{Im}[f(X+i\Delta X)]/\Delta X$ and $\mathrm{Im}[K(X+i\Delta X)]/\Delta X$ can be done in an element-by-element fashion so as to avoid the creation of a global complex vector or matrix. This saves tremendously on required memory. Since the CVM is not sensitive to the choice of step size, the CVSAM can always obtain accurate sensitivities without extra computational cost.

3. Application To Non-Linear Heat Conduction

In this section we demonstrate the accuracy of the SACVM for a non-linear heat conduction problem within a slab. This case has an exact solution that can be used for comparison. The slab has unit thickness $L=1$ m, with temperature-dependent thermal conductivity. The boundary temperatures are maintained at $T_{Left}=0^{\circ}\mathrm{C}$ and $T_{Right}=100^{\circ}\mathrm{C}$. The conductivity is represented by two piecewise linear segments that interpolate the conductivity values between three temperatures: $k_1=1.0\mathrm{W/m{\cdot}^{\circ}C}$ at $T_1=0^{\circ}\mathrm{C}$, $k_2=2.0\mathrm{W/m{\cdot}^{\circ}C}$ at $T_2=50^{\circ}\mathrm{C}$, and $k_3=6.0\mathrm{W/m{\cdot}^{\circ}C}$ at $T_2=100^{\circ}\mathrm{C}$.

The two piecewise linear segment is

$$k(T)=\begin{cases} k_1\left(1-\dfrac{T-T_1}{T_2-T_1}\right)+k_2\left(\dfrac{T-T_1}{T_2-T_1}\right), & T_1\le T\le T_2 \\ k_2\left(1-\dfrac{T-T_2}{T_3-T_2}\right)+k_3\left(\dfrac{T-T_2}{T_3-T_2}\right), & T_2\le T\le T_3 \end{cases} \tag{6}$$

The analytical solution of this problem is [13]

$$T=\begin{cases} \dfrac{\beta_1}{2}T^2+(k_1-\beta_1T_1)T+C_1 & T_1\le T\le T_2 \\ \dfrac{\beta_2}{2}T^2+(k_2-\beta_2T_2)T+C_2 & T_2\le T\le T_3 \end{cases} \tag{7}$$

$$\text{where } \beta_i=\left(\frac{k_{i+1}-k_i}{T_{i+1}-T_i}\right), \qquad i=1,2 \tag{8}$$

$$\begin{aligned} C_1=&-\left[\frac{\beta_1}{2}T_L^2+(k_1-\beta_1T_1)T_L\right]\left(1-\frac{x}{L}\right) \\ &+\left\{\frac{(\beta_2-\beta_1)}{2}T_2^2+\left[(k_2-\beta_2T_2)-(k_1-\beta_1T_1)T_2\right]\right\}\frac{x}{L} \\ &-\left[\frac{\beta_2}{2}T_R^2+(k_2-\beta_2T_2)T_R\right]\frac{x}{L} \end{aligned} \tag{9}$$

$$C_2 = -\left[\frac{\beta_1}{2}T_L^2 + (k_1 - \beta_1 T_1)T_L\right]\left(1-\frac{x}{L}\right) + \left\{\frac{(\beta_2-\beta_1)}{2}T_2^2 + [(k_2-\beta_2 T_2)-(k_1-\beta_1 T_1)T_2]\right\}\left(\frac{x}{L}-1\right) - \left[\frac{\beta_2}{2}T_R^2 + (k_2-\beta_2 T_2)T_R\right]\frac{x}{L} \quad (10)$$

The mesh of the finite element model is shown in Fig. 1. There are total 1212 nodes and 2202 elements.

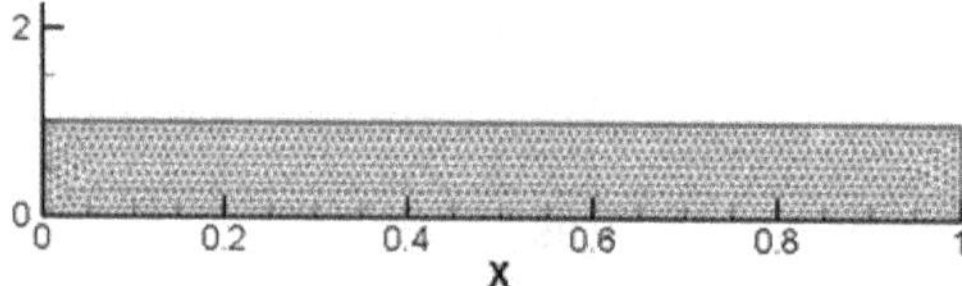

Figure 1: Triangular mesh of the slab

The CVSAM is used to compute the sensitivities of temperatures with respect to thermal conductivities k_1, k_2 and k_3 along the length of slab. The comparison of sensitivities obtained from the CVSAM with analytical formula is shown in Figure 2. The step size used in the CVSAM is 10^{-10}.

Figure 2 shows the sensitivities computed from the CVSAM agree well with the analytical solutions. Furthermore, the same sensitivities are obtained even for the step size smaller than 10^{-200}. Therefore, the CVSAM can always obtain accurate derivatives without being sensitive to the step size. The procedure and codes for the CVSAM are applicable to sensitivity analysis in linear or nonlinear heat conduction problems.

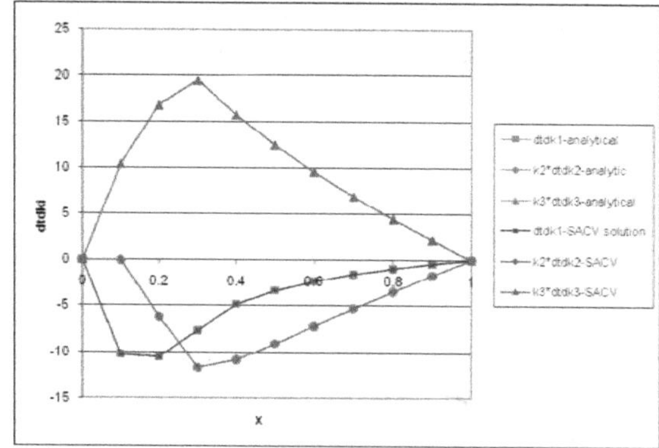

Figure 2: Sensitivities obtained from CVSVM and analytical solution

4. APPLICATION TO INVERSE HEAT CONDUCTION

In this section we demonstrate the use of CVSAM for the inverse determination of an unknown shape based on temperature measurements. In this example, we consider heat conduction across an infinitely long hollow cylinder has outer radius $r_o = 10$ m. The temperatures at inner and outer boundary are maintained at $T_o = 10°\text{C}$ and $T_i = 100°\text{C}$, respectively. The conductivity of the cylinder is $k_1 = 10\text{W/m}\cdot°\text{C}$. No heat source is introduced. We consider the steady heat conduction on the cross section of cylinder. The temperatures of 5 points on the cross section are known measurements. The interest of this inverse problem is to find the optimal inner radius of the cylinder. The model is shown in Figure 3, and the measured temperatures of 5 points are listed in Table 1.

This problem is solved using unconstrained numerical optimization based on a gradient search technique. The objective function is given in Equation 11. The goal is to minimize the square of the error between the temperature predicted by the FEM model and the measured temperatures. The objective can be stated as

$$\text{Minimize } \boldsymbol{Error(r)} = \sum_{i=1}^{5} \left(\boldsymbol{T}_i - \boldsymbol{T}_{i(measured)}\right)^2 \qquad (11)$$

where T_i is the temperature at 5 points predicted by the model, while $T_{i(measured)}$ is the temperature measured at the same point. The mesh used and the temperature distribution are shown in Figure 4. It has 592 elements and 348 nodes. The sensitivities were calculated using SAM and CVSAM. The abbreviated simulated measured temperatures used are given in Table 1. The double precision values were used for all computations.

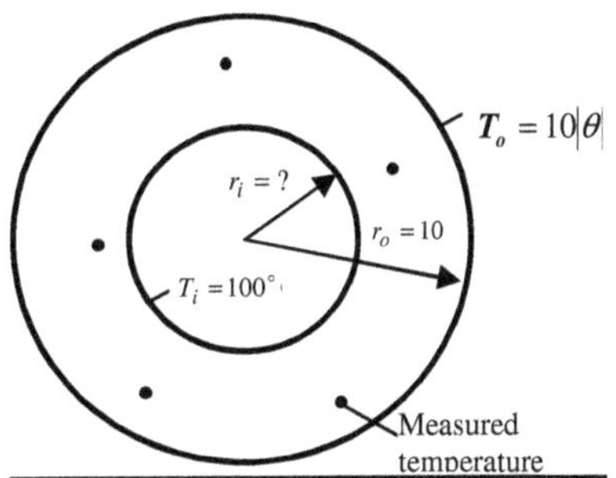

Figure 3: Inverse problem model

Table1: Measured temperatures of 5 points

Points	Temperatures
1	82.80
2	15.44
3	65.75
4	33.49
5	45.52

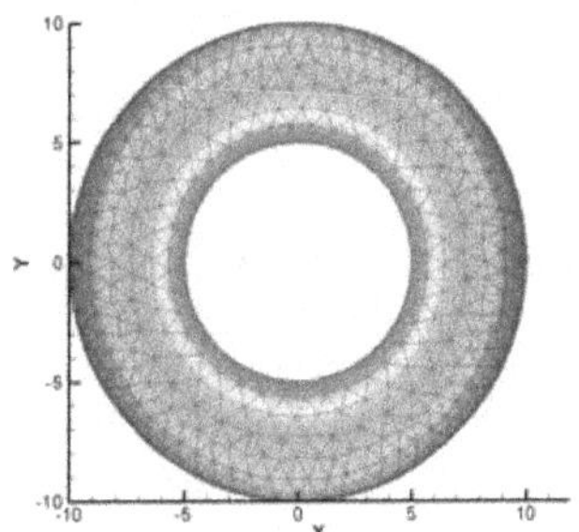

Figure 4: Mesh and forward solution

To minimize the error, the computation of the gradient of the objective is required.

$$\frac{\boldsymbol{dError(r)}}{\boldsymbol{dr}} = 2\sum_{i=1}^{5} \left(\boldsymbol{T}_i - \boldsymbol{T}_{i(measured)}\right)\frac{\partial \boldsymbol{T}_i}{\partial \boldsymbol{r}} \qquad (12)$$

We consider two cases. In the first case, we solved the inverse problem using sensitivities computed using SAM. For the second case we used sensitivities computed with CVSAM. For both cases a gradient-based unconstrained non-linear optimization algorithm [14] was used to minimize the objective function. For both cases we varied the initial guess for the unknown inner radius and the perturbation step size and noted the effect on the solution and the number of iterations required. The results are shown in Tables 2 and 3 for cases one and two, respectively.

Table 2: SAM Results

Initial guess (m)	Step size (m)	Iterations	Predicted inner radius (m)
4.9	1.0E-01	27	4.987451
	1.0E-02	27	4.967940
	1.0E-06	26	4.964337
5.05	1.0E-01	10	5.000000
	1.0E-02	30	5.050000
	1.0E-06	30	5.050000
5.1	1.0E-01	34	4.998439
	1.0E-02	30	5.100000
	1.0E-06	30	5.100000

Table 3: CVSAM Results

Initial guess (m)	Step size (m)	Iterations	Predicted inner radius (m)
4.7	1.0E-01	12	5.000000
	1.0E-06	12	5.000000
	1.0E-100	12	5.000000
5.5	1.0E-01	8	5.000000
	1.0E-06	10	5.000000
	1.0E-100	8	5.000000
6.7	1.0E-01	9	5.000000
	1.0E-06	9	5.000000
	1.0E-100	9	5.000000

The results show that subtractive error present in the SAM due to the inherent finite differencing has a significant effect. The SAM results in more iteration to reach a solution close to, but not the same as the exact solution. Furthermore, an initial guess close to the exact solution was required. In contrast, using CVSAM resulted in the exact solution for a wide range of initial guesses perturbation sizes. In addition, use of CVSAM sensitivities reduced the number of optimizer iterations by a factor of three compared to the SAM case.

5. Conclusions

The paper introduces the complex variable semi-analytical method (CVSAM) to compute sensitivities of finite element response with respect to input variables. The method retains the computational efficiency of the semi-analytical method (SAM) while achieving high accuracy for small perturbation step sizes. The CVSAM was applied to a non-linear heat conduction problem with excellent agreement with the analytical solution for all step sizes tried. The method was then applied to an inverse shape detection problem where it was compared with SAM. The CVSAM results were more accurate and were achieved in one third the iterations compared to the SAM case. The CVSAM appears to be a promising approach for computing accurate sensitivities for optimization-based inverse problems in a computationally efficient manner.

6. REFERENCES

[1] Liu, P. L., and Der Kiureghian, A., “Finite element reliability methods for geometrically non linear stochastic structures,” Tech. Rep. No. UCB/SEMM/89-05, University of California, Berkeley, 1989.

[2] Zhang, Y. and Der Kiureghian, A., “Dynamic response sensitivity of inelastic structures,” Comp. Meth. Appl. Mech. Eng., Vol. 108, pp. 23-26, 1993.

[3] T. H. Lee, “An adjoint variable method for design sensitivity analysis of elastoplastic structures”, J. Mech. Scie. & Tech., Vol. 13, pp. 246-252, 1999.

[4] Barthelemy, B., Chon, C. T. and Haftka, R. T., "Accuracy problems associated with semi-analytical derivatives of static response," Finite Elements in Analysis and Design, Vol. 4, pp. 249-265, 1988.

[5] Cheng, G., Gu, Y., Zhou., “Accuracy of semi-analytical sensitivity analysis,” Finite Elements in Analysis and Design, Vol. 6, pp. 113-128, 1989.

[6] Barthelemy, B., and Haftka, R. T., "Accuracy analysis of the semi-analytical method for shape sensitivity calculation," Mech. Struc. & Mach., Vol. 18, No. 3, pp. 407-432, 1990.

[7] Olhoff, N., Rasmussen, J., “Study of inaccuracy in semi-analytical sensitivity analysis-a model problem.” Struc. Optim., Vol. 3, pp 203-213, 1991.

[8] Jin W., Dennis H. B., and Wang B. P., “Improved Sensitivity Analysis using a Complex Variable Semi-analytical Method”, Struc. Multidisciplinary Optim., Vol. 41, No. 3, pp. 433-439, 2010.

[9] Martins, J. R .R. A., Sturdza P., and Alonso, J. J., “The complex-step derivative approximation,” ACM Transa. on Mathe. Software, Vol. 29, No. 3, pp. 245-262, 2003.

[10] Martins, J. R. R. A., Kroo, I. M., and Alonso, J. J., “An automated method for sensitivity analysis using complex variables,” 38th Aerospace Science Meeting and Exhibit, AIAA paper, AIAA-2000-0689, pp. 1-12, 2000.

[11] Anderson, W. K., Newman, J. C., Whitfield D. L., and Nielsen E. J., “Sensitivity analysis for the Navier-Stokes equations on unstructured meshes using complex variables,” AIAA paper, AIAA-99-3294, pp. 381-389, 1999.

[12] Rodriguez, D. L., “A multidisciplinary optimization method for designing inlets using complex variables,” AIAA paper, AIAA-2000-4875, pp. 1-10, 2000.

[13] Dowding K. J. and Blackwell B. F., “Sensitivity analysis for nonlinear heat conduction,” Journal of Heat Transfer, Vol. 123, 2001.

[14] Optimization Toolbox, Matlab Documentation, 2008.

General Ray Method for Solution of Inverse Coefficient Problems for Laplace Equation

A. Grebennikov

Facultad de Ciencias Físico Matemáticas, Benemérita Universidad Autónoma de Puebla, Av. San Claudio y Río verde, Ciudad Universitaria, CP 72570, Puebla, Pue., México, Email: agrebe50@yahoo.com.mx

Abstract

A new approach for solution of inverse coefficient problem for Laplace equation is considered. This approach is based on proposed by author General Ray Principle and leads to the new *GR*-Method, which consists in reduction of the partial differential equation to the family of ordinary differential equations, using local traces for considered functions and operators. New method presents the solution of considered problems by explicit analytical formulas that use the direct and inverse Radon transform. In the case of noised input data the regularization with Recursive Spline Smoothing is used. Proposed method is realized by fast and stable algorithms and MATLAB software. The quality of algorithms is demonstrated by numerical experiments. Applications to electric tomography and mathematical simulation in creation of nano-composite materials with special heat-conductive properties are considered.

Key Words: inverse problems, PDE coefficients, electric tomography

1. Introduction

Mathematical and computer simulation is very important for modern investigations in many applied areas. This simulation leads to solving direct and inverse problems [1]. Basic mathematical models that describe processes include traditional differentials equations. In this paper we will consider problems that are described by the two-dimensional Laplace type equation

$$\nabla(\varepsilon(x,y)\nabla u(x,y))=0, \quad (x,y)\in\Omega \tag{1}$$

where Ω – some limited open region on a plane. In the case of modelling the electric field, function $u(x,y)$ is potential, the function $\varepsilon=\varepsilon(x,y)$ characterize the electro-conductivity or permittivity of a media. In the case of modelling the stationary heat conductive field, function $u(x,y)$ describes distribution of the temperature in two-dimensional domain Ω with thermo-conductivity $\varepsilon(x,y)$.

If the function $\varepsilon(x,y)$ is known and some boundary conditions are given, we have the direct problem for reconstruction the function $u(x,y)$ only. If bought functions $u(x,y)$ and $\varepsilon(x,y)$ are unknown, we have the inverse coefficient problem [2]. In traditional statement of inverse problems [2], [3] it is supposed also that some family of functions $J_n(x,y)$, $u^0(x,y)$ are known on the boundary curve Γ and the next boundary conditions are satisfied:

$$\varepsilon(x,y)\frac{\partial u(x,y)}{\partial n}=J_n(x,y), \quad (x,y)\in\Gamma, \tag{2}$$

$$u(x,y)=u^0(x,y), \quad (x,y)\in\Gamma, \tag{3}$$

where $\frac{\partial}{\partial n}$ is the normal derivative in the points of the boundary curve Γ. Mentioned families of functions in the boundary conditions, as the rule, correspond to some scanning scheme [3].

Solution of such inverse problems is very important in mathematical simulation of electrical tomography [3] and investigation of heat-conductive properties of materials [4]. But all mathematical statements and known methods for solution of considering inverse problems are non-linear [2], [3] and require a lot of time and memory in its computer realization, that is not appropriated in modern investigations in mentioned applied areas.

We propose here another approach for the mathematical modelling of the distribution of the external physical field and measurement of the external data. Proposed modelling is based on General Ray Principle (GRP) [5] and use the classic Radon transformation [6]. GRP leads to the new explicit General Ray (GR) Method and a fast linear algorithm for numerical solution of inverse problems. This approach and algorithms are justified with the numerical experiments on the simulated model examples.

2. General Ray Method

We will use the General Ray Principle, i.e. to consider the physical field as the stream flow of "general rays" [5]. We will consider the distribution of the field as the parallel scanning beam with lines l, which have the parametric presentation: $x = p\cos\varphi - t\sin\varphi$, $y = p\sin\varphi + t\cos\varphi$, where $|p|$ is a length of the perpendicular, passed from the centre of coordinates to the line l, φ is the angle between the axis x and this perpendicular. We will consider Ω as a convex domain with smooth boundary curve Γ.

Let the external field in scanning scheme be homogeneous in the direction orthogonal to the line l. Using parameterization for the line l, we shall consider the potential $u(x,y)$ and function $\varepsilon(x,y)$ for $(x,y)\in l$ as functions (traces) $\bar{u}(t)$ and $\bar{\varepsilon}(t)$ of variable t, which depend on parameters p, φ. Following to the GRP, we change the equation (1) on the family of ordinary differential equations, every of such can be written on the line l for every fixed p and φ in the form:

$$\left(\bar{\varepsilon}(t)\bar{u}_t'(t)\right)_t' = 0, \quad t\in[t_0,t_1], \tag{4}$$

where coordinates t_0, t_1 correspond to points of intersection of the straight line l and the boundary curve. We suppose that functions $v(p,\varphi)$ and $J(p,\varphi)$ are given and we can write boundary conditions

$$\bar{\varepsilon}(t_0)\bar{u}_t'(t_0) = J(p,\varphi), \tag{5}$$

$$\bar{u}(t_1) - \bar{u}(t_0) = v(p,\varphi) \tag{6}$$

The family of equations (4) – (6) we consider as the basic mathematical model in application of GRP for considering type of inverse problems.

We suppose that different components in the considered structure have the smooth distribution, such as functions $\bar{\varepsilon}(t)\bar{u}_t'(t)$ and $\bar{u}_t'(t)$ are continuous. We suppose also that $J(p,\phi)\neq 0$ for every p, φ. Integrating twice the equation (4) on t and using boundary conditions (5) - (6), we obtain for $\varepsilon(x,y)$ the next basic formula of the General Ray Method:

$$\varepsilon(x,y) = 1/R^{-1}\left[\frac{v(p,\varphi)}{J(p,\varphi)}\right], \tag{7}$$

where R^{-1} is the inverse Radon transform operator that can be realized by known explicit analytical formula [6]. So, *GR*-Method gives the explicit solution of the inverse coefficient problem for considering case.

General Ray Method can be also applied to other types of equations describing no stationary problems, so as generalised for multidimensional case and for structures with piecewise constant characteristics.

3. Regularization of the *GR*-Method

Analysis of formula for inverse Radon transformation shows that its instability for discrete noised data is equivalent to the instability of the problem of the numerical differentiation with respect to the variable p of the noised function $\tilde{v}(p,\varphi)$. The regularization of the inversion of Radon transform was constructed by author in [7] on the base of the Recursive Smoothing by splines (RSS), which was used in [8] for post-processing to improve the electrical capacitance tomography image reconstruction.

RSS uses the explicit formulas for two-dimensional spline on the regular uniform grid $\{p_i,\varphi_j\}$, $i=-2,\dots,n+2$; $j=-2,\dots,n+2$. Let $s_i(u)$ be a local basic cubic spline, constructed on the units $w_{i-2},\dots w_{i+2}$; $i=0,\dots,n+1$; where w is p, or w is φ. Mentioned formulas are the next ones:

$$s_k(p,\varphi)=\sum_{i=1}^{n}\sum_{j=1}^{n} s_{k-1}(p_i,\varphi_i)s_i(p)s_j(\varphi), \tag{8}$$

$$k=1,\ 2,\dots,\overline{K};\quad S_0(p_i,\varphi_j)=\overline{v}(p_i,\varphi_j)\ . \tag{9}$$

The number of smoothes $\overline{K}$ is the regularization parameter, which can be chosen here in accordance with residual (discrepancy) principle, using the discrete estimation δ of the errors. It means, if the values of the exact function $v(p,\varphi)$ and the noised function $\overline{v}(p,\varphi)$ satisfy to the conditions

$$\left|v(p_i,\varphi_j)-\overline{v}(p_i,\varphi_j)\right|\le\delta, i,j=1,\dots,n\,,$$

then $\overline{K}$ is chosen as maximum among all k, for which the inequality is fulfilled:

$$\sum_{i=1}^{n}\sum_{j=1}^{n}\left|S_k(p_i,\varphi_j)-\overline{v}(p_i,\varphi_j)\right|^2\le c\delta^2n^2\,, \tag{10}$$

where $c=\text{conts}>1$.Theoretical and numerical justifications of the regularization properties of this type of smoothing are presented in [9], [10].

If for structures with piecewise constant characteristics the set $\hat{\varepsilon}=\{\varepsilon_0,\varepsilon_1,\varepsilon_2\}$ of the known values ε_i of the function $\varepsilon(x,y)$ is given, then the algorithm includes also the projection of the pre-reconstructed data to the set $\hat{\varepsilon}$ with respect to the absolute or relative criterions [7].

We underline that presented algorithms are based on fast numerical realization of the inverse Radon transform and on explicit approximation formulas (8) – (10) that do not require solving any equations. It guarantees the fast property in a hole of proposed algorithms for solution of considered inverse problem in convex domain Ω.

4. Application to the Electric Tomography

Computer Tomography consists in the image reconstruction of an interior of a body using the measurements on its surface of characteristics of some external field. It can be state mathematically as a coefficient inverse problem for a differential equation describing the distribution of the field in considered region. Coefficients are the functions of the space variables and characterize properties of a media.

Electrical Impedance Tomography (EIT) is the most developed approach for electric tomography that includes the electric resistance (ERT) or capacitance tomography (ECT) schemes [3].

In ERT the function $J_n(x,y)$ is given, function $u^0(x,y)$ is measured. In ECT the function $u^0(x,y)$ is given, the value of the normal component of electric induction $J_n(x,y)$ is related with measured mutual capacitances [3]. In both ERT and ECT schemes the electric field is produced by the same electrodes that serve as measuring elements i.e. the electrodes are active. May be this activity of electrodes, which provokes its mutual influence, is the cause of impossibility to use in real experiments a great number of electrodes and obtain the sufficiently large number of high-quality measurements.

We propose here another variant of the Electrical Tomography, when the external electromagnetic field $\vec{V}(l)$, is produced by active electrodes, located outside of the Ω, initiates some distribution of the electric potential inside the domain Ω. At that, we propose that measurements of necessary values would be realized on the boundary curve Γ with another, no active electrodes.

We put bellow Ω as a circle of radius *r*, so that $t_0 = -\bar{t}$, $t_1=\bar{t}=\sqrt{r^2-p^2}$. The corresponding scheme is presented at Fig. 1, where active electrodes are marked with arcs on the external circle $\mathcal{A}$ of radius *R* and inactive electrodes are marked as arcs on the internal circle $\mathcal{B}$, which is the board of the domain.

In proposing new scheme we can calculate generated potential $\bar{u}\left(-\bar{t}\right)$, and induction function $J(p,\varphi)$ on direction *l*, so we need to make measurements only of values of the potential $\bar{u}\left(\bar{t}\right)$.

It is very important that electrodes on the boundary Γ do not produce the external electric field (are not active) and serve only for measuring data. Therefore, the proposed approach gives in principal the possibility to use a large number of electrodes and measurements of the input values of functions $\bar{u}\left(\bar{t}\right)$ and reconstruct the desired image more perfectly.

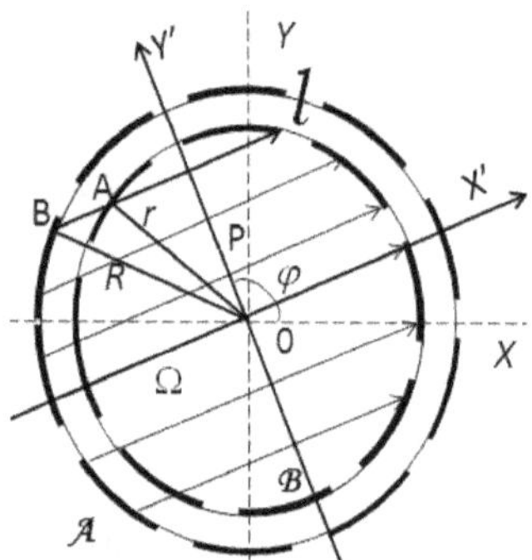

Figure 1: The measurement scheme of the external data.

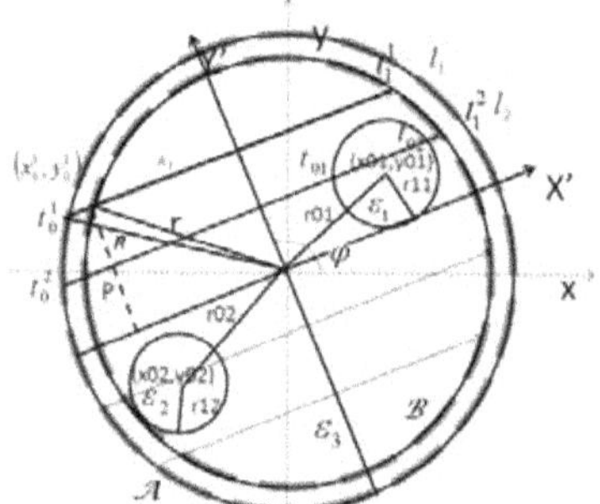

Figure 2: Illustration for synthetic example.

5. Application to Heat Transfer

Investigation of heat-conductive properties of materials is very important for many applied areas. Mathematical and computer simulation is one of the important steps in this investigation. This simulation leads to solving direct and inverse heat transfer problems. Modern investigations are characterized by a penetration with more substantially detailed in the structure of investigated objects and phenomena. Basic mathematical models that describe the heat-conductive processes include traditional differentials equations, nevertheless frequently with specific elements. This requires elaboration of the new analytical and numerical methods of its study, adapted to the modern requirements. One of the most important of these requirements is the possibility to obtain a sufficient increase of the exactitude at the solution of the problems in the real time allowed, or, that is equivalent, to resolve the problem with the appropriate exactitude by the fastest manna. The mathematical models and the known numerical methods often do not satisfy to these requirements at their computer realization, particularly for solution of desired heat transfer problems.

6. Numerical Experiments

We have constructed the numerical realization of formula (7) that we call "*GR*–algorithm". This algorithm does not require solving any equation, because the Radon transform can be inversed by fast algorithm using discrete FFT algorithm.

We tested scanning GR-algorithm on mathematically simulated model examples. The first and second presented experiments correspond to piecewise constant structure. We considered inside the unit circle Ω two different internal elements Ω_1, Ω_2 of different permittivity, as it is shown at Fig. 2.

Simulation consists in the next steps:

1) analytic solution the direct Cauchy problem for equation (4) with known values of functions $J(p,\varphi)$ and $\bar{u}\left(-\bar{t}\right)$ for every fixed angle φ and parameter *p*,

2) calculation the value $\bar{u}\left(\bar{t}\right)$ and explicit form of $v(p,\varphi)$,

3) numerical realization of formula (7).

The deduction of the corresponding formulas at steps 1) and 2) are presented in details at [11]. These formulas give us the important result of this simulation: the relation $v(p,\varphi)/J(p,\varphi)$ dos not depend on function $J(p,\varphi)$ so as on the value of the generated potential at the border, so on radius of the external circle $\mathcal{A}$, that confirms the validity of proposed scheme and constructed algorithm. The step 3) was realized on discrete simulated data with *n* nodes for every variable.

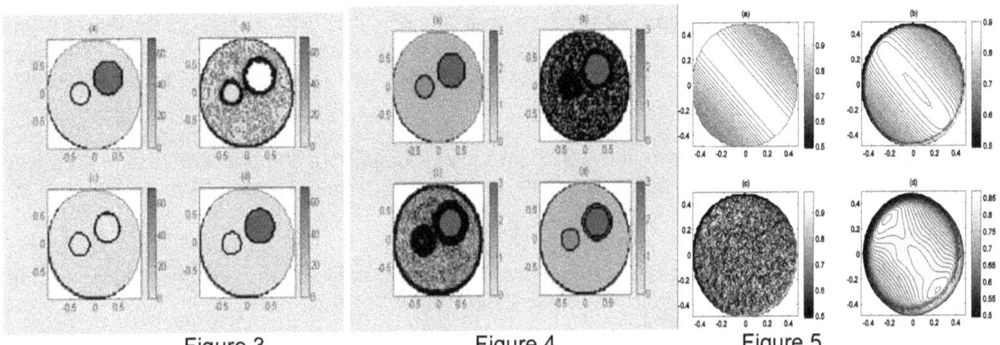

Figure 3. Figure 4. Figure 5.

In the first example we use exact values in *n*=31 discrete points for the case $\varepsilon_0(x,y)=1$, $\varepsilon_1(x,y)=2$, $\varepsilon_2(x,y)=70$. It is difficult case for the reconstruction, because it corresponds to the greater scale of values $\hat{\varepsilon}=\{\varepsilon_0,\varepsilon_1,\varepsilon_2\}$, when the post-processing (projection) is required even for the pre-reconstruction that used exact data. In Fig. 3 there are presented reconstructions of the structure image by *GR*–algorithm: graph (a) – exact distribution; graph (b) – reconstruction without post-processing, graph (c) - reconstruction with post-processing using the absolute criterion projection; (d) – reconstructions with post-processing using the relative criterion projection.

The second presented numerical experiment corresponds to the reconstruction of the structure for the case $\varepsilon_0(x,y)=1$, $\varepsilon_1(x,y)=2$, $\varepsilon_2(x,y)=3$, using simulated noised input data, i.e. values of a function $\bar{v}(p,\varphi)=v(p,\varphi)(1+\delta(p,\varphi))$, where $\delta(p,\varphi)$ is the randomized function with estimation: $\|\delta(p,\varphi)\|_{C[\Omega]}\le\delta$. Results of the regularized reconstruction for *n*=31, $\delta=0.05$ are presented on Fig. 4: graph (a) - exact $\varepsilon(x,y)$; graph (b) - reconstruction with noised $\bar{v}(p,\varphi)$ without regularization; graph (c) - reconstruction with noised $\bar{v}(p,\varphi)$ by regularised *GR*-algorithm with RSS only, without post-processing projections; graph (d) - reconstruction with noised $\bar{v}(p,\varphi)$ by regularised *GR*-algorithm with RSS and the post-processing, using absolute criterion projection of pre-reconstructed image.

The third experiment corresponds to the continue characteristic function $\varepsilon(x,y)=1/\cos(x+y)$ in the domain Ω as the circle of the radius $r=0.5$. The external stationary field produces in Ω the potential $u(x,y)=\sin(x+y)$, such as $J(p,\varphi)=\cos\varphi-\sin\varphi$, exact function $v(p,\varphi)=2\cos(p(\sin\varphi+\cos\varphi))\sin(\bar{t}(\sin\varphi-\cos\varphi))$, $\bar{t}=-\sqrt{0.25-p^2}$. A noised function $\bar{v}(p,\varphi)$ we constructed in the same form as for the second example. Results of the numerical experiments for n=51, estimation $\delta=0.03$ are presented on Fig. 5: graph (a) - exact $\varepsilon(x,y)$; graph (b) - reconstruction with exact $v(p,\varphi)$; graph (c) - reconstruction with noised $\bar{v}(p,\varphi)$ without regularization; graph (d) - reconstruction with noised $\bar{v}(p,\varphi)$ by regularised *GR*-algorithm with RSS.

The author acknowledges to VIEP of Merited Autonomous University of Puebla, Mexico, for the support of the part of this investigation in the frame of the Project No GRA-EXC11-I.

7. References

[1] Tikhonov, A.N. and Arsenin V. Ya., Methods for Solving Ill-Posed Problems. V.H. Winston & Sons, Washington, D.C., 1977.

[2] Isakov, V., Inverse Problems for Partial differential Equations, Springer, New York, 1998.

[3] Williams R.A. and Beck M.S., Process Tomography: Principles, Techniques and Applications, Butterworth-Heinemann, Oxford, 1995.

[4] Alifanov, O. M., Inverse Heat Transfer Problems, Springer-Verlag, Berlin-Heidelberg, 1994.

[5] Grebennikov, A.I., A novel approach for the solution of direct and inverse problems of some equations of mathematical physics. *Proceedings of the 5-th International Conference on Inverse Problems in Engineering: Theory and Practice,* (ed. D. Lesnic), Vol. II, , Chapter G04, pp. 1-10, 2005, Leeds University Press, Leeds, UK.

[6] Radon, J., Uber Die Bestimmung von Funktionen Durch Ihre Integrawerte Langs Gewisser Mannigfaltigkeiten. Berichte Sachsische Academic der Wissenschaften, Leipzig, Math.-Phys. Kl. N 69, pp. 262-267, 1917.

[7] Grebennikov, A.I., Regularization algorithms for electric tomography images reconstruction, WSEAS Transaction on Systems J., Issue 2, Vol. 2, pp. 487 -492, 2003.

[8] Grebennikov, A. and Gamio, Carlos, Fast post-processing algorithm for improving electrical capacitance tomography image reconstruction. Inverse Problems in Science and Engineering, Vol. 14, No. 1, January, pp. 64 - 74, 2006.

[9] Grebennikov, A.I., Spline Approximation Method and Its Applications, MAX Press, Moscow, 2004.

[10] Morozov, V.A. and Grebennikov, A. I., Methods for Solution of Ill-Posed Problems: Algorithmic Aspects, Moscow University Press, Moscow, 1992 (English edition in 2005).

[11] Grebennikov, A. and Reyes, S., New fast algorithm for solution of electrical tomography problem, Superficies y Vacío J., **23(S)**, pp. 176-179, August 2010.

FAULT DIAGNOSIS BASED ON INVERSE PROBLEM SOLUTION

Lídice Camps[1], Orestes Llanes[2], Antonio J. da Silva [3] and Mirtha Irizar[2]

[1] Departamento de Matemática. Instituto Superior Politécnico José Antonio Echeverría (ISPJAE). Marianao, La Habana, Cuba.
Email: lidice@mecanica.cujae.edu.cu
[2] Departamento de Automática y Computación. ISPJAE.
Email: orestes,mirtha@electrica.cujae.edu.cu
[3] Departamento de Ingeniería Mecánica y Energía. Instituto Politécnico de la Universidad del Estado de Río de Janeiro, IPRJ-UERJ.
Nova Friburgo, Brasil.
Email: ajsneto@iprj.uerj.br

Abstract

This work presents the formulation of the fault diagnosis as an inverse problem and the application of stochastic algorithms and their cooperative strategy for obtaining the solution of the optimization problem. The principal aim is to establish a basis for the development of new and viable parameter-estimation Fault Diagnosis Methods which alleviates some difficulties that the current methods cannot avoid. These difficulties are related with fault sensitivity and robustness to external disturbances. The selected algorithms were Differential Evolution and Ant Colony Optimization as well a cooperative strategy. This proposal is illustrated using simulated data of the inverted-pendulum benchmark. With the purpose of analyzing the advantages of such approach, mainly with respect to robustness and sensitivity, some experiments with noisy data and incipient faults are considered. The results indicate the suitability of the approach adopted.

Key Words: Ant Colony Optimization, Differential Evolution Algorithm, Fault diagnosis, Inverse Problem, Parameter Estimation, Robustness, Sensitivity.

1 Introduction

The automatic early detection, isolation and localization of faults that have an effect on industrial systems are of high interest in order to improve the reliability and safety, see [1, 2]. This process is called fault diagnosis or FDI, see [3].

The approaches for FDI are based on the system input and output measurements. In addition it is possible to achieve a model of the system and possible faults can be associated with specific parameters and states of the mathematical model. In this principle is based the fault diagnosis via parameter estimation methods, see [4, 3, 5]. The limitations of the current FDI methods lead to the necessity of development new alternatives that can deal with an appropriate balance between fault sensitivity and robustness to external disturbances, see [2, 6].

Despite the fact that the structure of the FDI methods via parameter estimation is analogous to the inverse problem of parameter estimation only some few and recent works are reported in that sense, see [7, 8]. With the aim of development new and viable FDI methods and taking into account this similitude, this work presents the formulation of the fault diagnosis as an inverse problem and the application of stochastic algorithms and their cooperative strategy for obtaining the solution of the optimization problem.

This application is illustrated using simulated data of the inverted-pendulum system benchmark (IPS) and implementing the stochastic algorithms Differential Evolution (DE), Ant Colony Optimization (ACO) and their cooperative strategy. With the purpose of analyze the advantages of this approach, principally in the topic of robustness and sensitivity, some experiments with noisy data and incipient faults were considered.

The results indicate the suitability of this approach for development robust and sensitive FDI methods. This is the main contribution of this work.

2 Structure of the Model-based FDI methods via parameter estimation

FDI based on system parameters which are mostly not directly measurable, requires online parameter estimation methods.

Let be

$$\begin{aligned} \dot{\mathbf{x}}(t) &= f(\mathbf{x}(t), \mathbf{u}(t), \Theta) \\ \mathbf{y}(t) &= g(\mathbf{x}(t)) \end{aligned} \tag{1}$$

the mathematical model that represents as close as possible the physical laws which govern the system behavior. The state vector is represented by $\mathbf{x}(t) \in \mathbb{R}^m$. The measurable input signal $\mathbf{u}(t) \in \mathbb{R}^l$ and output signal $\mathbf{y}(t) \in \mathbb{R}^p$ can be directly obtained by the use of physical sensors; the components of the model parameters vector $\Theta \in \mathbb{R}^n$ are identified with physical process coefficients $\rho \in \mathbb{R}^j$ and in general $n \neq j$.

The estimations of the parameters θ_n will allow to diagnose the fault once the relationship between θ_n and ρ_j is established.

2.1 Inverse problem of FDI

In the inverse problem of FDI we are interested in the model parameters θ_l which are unknown while the mathematical model of the system and measured data of the output and the input are available. The vector of parameters Θ can be obtained from the solution of the inverse problem of parameter estimation that can be formulated as a minimization problem

$$\begin{array}{ll} \text{min} & F(\hat{\Theta}) = \sum_{t=1}^{S} [\mathbf{y}(t) - \hat{\mathbf{y}}(t)]^2 \\ \text{s.a} & \Theta_{min} \leq \hat{\Theta} \leq \Theta_{max} \end{array} \tag{2}$$

where S is the number of sampling of the input **u(t)** and the output **y(t)**; and $\hat{\mathbf{y}}(t)$ are the estimations of the output that are obtained from **u(t)** and the solution of the direct problem that was given in Eqn. (1).

For the solution of the optimization problem that was specified in Eqn. (2) even in a noisy environment, stochastic algorithms can be implemented.

3 Differential Evolution and Ant Colony Optimization

This section describes the basis of the two algorithms that will take part in our study.

3.1 Differential Evolution

The Differential Evolution (DE) was proposed around 1995 for optimization problems, see [9]. DE is an improved version of the Goldberg's Genetic Algorithm taking the basis of Simulated Annealing. Some of the most important advantages of DE are: simple structure, simple computational implementation, speed and robustness, see [9].

Basically, DE generates a new solution vector by adding the weighted difference between a pair population vectors to a certain vector (the number of pair can be changed). This configuration is summarized by the notation $DE/\mathbb{X}^t/\gamma/\lambda$ where $\mathbb{X}^t$ denotes the vector to disturb in the iteration t, γ the number of pair of vectors for disturbing $\mathbb{X}^t$ and λ indicates the type of crossover to be used. In this case we have considered $DE/\mathbb{X}^{t(best)}/1/bin$. Expressing it in a more formal fashion

$$\mathbb{X}^{t+1} = \mathbb{X}^{t(Best)} + F_s\left(\mathbb{X}^{t(\alpha)} - \mathbb{X}^{t(\beta)}\right) \tag{3}$$

where $\mathbb{X}^{t+1}, \mathbb{X}^{t(Best)}, \mathbb{X}^{t(\alpha)}, \mathbb{X}^{t(\beta)} \in \mathbb{R}^n$ and F_s is the weight applied to random differential or scaling factor. In order to complete the mutation operator the crossover operator is defined for each vector component $x_1, x_2 \cdots x_n$.

$$x_n^{t+1} = \begin{cases} x_n^{t+1} & \text{if } R \leq C_R \\ x_n^{t(Best)} & \text{in other case} \end{cases} \tag{4}$$

where $0 \leq C_R \leq 1$ is the called crossover constant, R is a random number which is generated by the distribution λ which in this case is the binomial distribution.

Finally we define the selection operator

$$\mathbb{X}^{t+1} = \begin{cases} \mathbb{X}^{t+1} & \text{if } F(\mathbb{X}^{t+1}) \leq F(\mathbb{X}^{t(Best)}) \\ \mathbb{X}^{t(Best)} & \text{in other case} \end{cases} \tag{5}$$

in order to finish the procedure. The key parameters of control in DE are the population size, Z, the crossover constant, C_R, and the scaling factor, F_s. In [9] some simple rules for choosing the parameters of DE for any application are given.

3.2 Ant Colony Optimization

Ant Colony Optimization (ACO) was initially proposed, see [10], for integer programming problems but recently it has been successfully extended and adapted to continuous optimization problems. ACO is inspired on the behavior of ants seeking a path between their colony and a source of food. This behavior is due to the deposit and evaporation of pheromone.

A good point of this algorithm is that its parameters can be manipulated in order to aims a more exploitation or exploration structure which allow an efficient hybridization with other algorithms. For the continuous case the idea of the ACO is to mimic the natural behavior of the ants with simulated ones which are identified with a feasible solution. The first step is to divide the feasible interval of each variable of the problem in k possible values $x_n^1, x_n^2, \ldots, x_n^k$. In each iteration of the algorithm a family of Z new ants are generated based on the information obtained from the previous ants and based on a selection mechanism. The information of the previous ants is saved on the pheromone accumulative probability matrix $\mathbb{PC}$ (the matrix has dimensions $n \times k$ where n is the number of variables in the problem) which is updated at each iteration based on an evaporation factor C_{evap} and an incremental factor C_{inc}:

$$pc_{ij}(t) = \frac{\sum_{l=1}^{j} f_{il}(t)}{\sum_{l=1}^{k} f_{il}(t)} \tag{6}$$

where f_{ij} are the elements of the pheromone matrix and express the pheromone level of the discrete value j^{th} of the variable i. This matrix is updated in each iteration:

$$f_{ij}(t+1) = (1 - C_{evap})f_{ij}(t) + \delta_{ij,best}C_{inc}f_{ij}(t) \tag{7}$$

The selection mechanism considers a parameter q_0 and a family of N random number $q_1^{rand}, q_2^{rand}, \ldots q_n^{rand}$ for the zth ant to be generated. For each variable $x_n^{(z)}$ that will be part of the ant z^{th} is set the following generation mechanism:

$$x_n^{(z)} = \begin{cases} x_n^{\tilde{k}} & \text{if} \quad q_n^{rand} < q_0 \\ x_n^{\check{k}} & \text{if} \quad q_n^{rand} \geq q_0 \end{cases} \tag{8}$$

where $\tilde{k}: \ f_{n\tilde{k}} \geq f_{nk} \, \forall k$ and $\check{k}: \ (pc_{n\check{k}} > q_n^{rand}) \wedge (pc_{n\check{k}} \leq pc_{nk}) \ \forall k \geq \check{k}$.

The control parameter q_0 allows to control the level of randomness during the ant generation. One important remark is that ACO makes possible the simultaneous exploration of different solutions by handling with an ant colony and the pheromone matrix.

4 Experimental Study

The application of the two stochastic strategies DE and ACO for the inverse problem of FDI when the simulated data of the IPS benchmark are considered is now experimentally illustrated.

4.1 Inverted-Pendulum system

The system is formed by an inverted pendulum mounted on a motor-driven car. Here we consider only the two - dimensional problem where the pendulum moves only in the plane of the paper, see Fig.1.

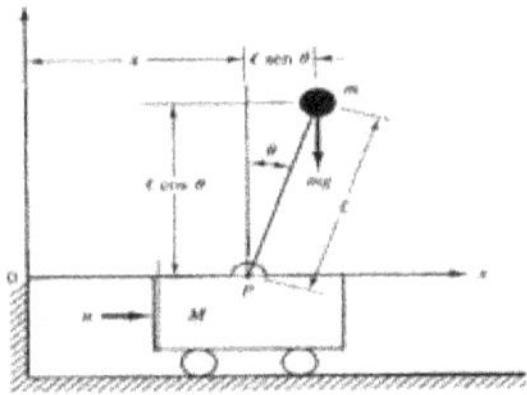

Figure 1: Inverted- Pendulum system

The mathematical model of the IPS has been widely studied, see [11, 12]. The system is described by a state-space representation of a linear time invariant system, affected by additive faults:

$$\begin{aligned} \dot{\mathbf{x}}(t) &= A\mathbf{x}(t) + B\mathbf{u}(t) + E_f f(t) \\ \mathbf{y}(t) &= C\mathbf{x}(t) + F_f f(t) \end{aligned} \tag{9}$$

The fault vector is denoted by $f(t)$. The matrices A, B, C, E_d, F_f are known and with appropiate dimensions:

$$A = \begin{pmatrix} 0 & 1 & 0 & 0 \\ 20.601 & 0 & 0 & 0 \\ 0 & 0 & 0 & 1 \\ -0.4905 & 0 & 0 & 0 \end{pmatrix} \quad B = \begin{pmatrix} 0 \\ -1 \\ 0 \\ 0.5 \end{pmatrix} \quad C = \begin{pmatrix} 1 & 0 & 0 & 0 \\ 0 & 0 & 1 & 0 \end{pmatrix} \quad E_f \begin{pmatrix} 0 & 0 & 0 \\ -1 & 0 & 0 \\ 0 & 0 & 0 \\ 0.5 & 0 & 0 \end{pmatrix} \quad F_f = \begin{pmatrix} 0 & 1 & 0 \\ 0 & 0 & 1 \end{pmatrix}$$

The state vector is $\mathbf{x} = [\theta \ \dot{\gamma} \ x \ \dot{x}]^t$, where γ and $\dot{\gamma}$ are the angle of the pendulum with respect to the vertical position and the angular velocity respectively; x and $\dot{x}$ are the position and the velocity of the car respectively. The outputs of

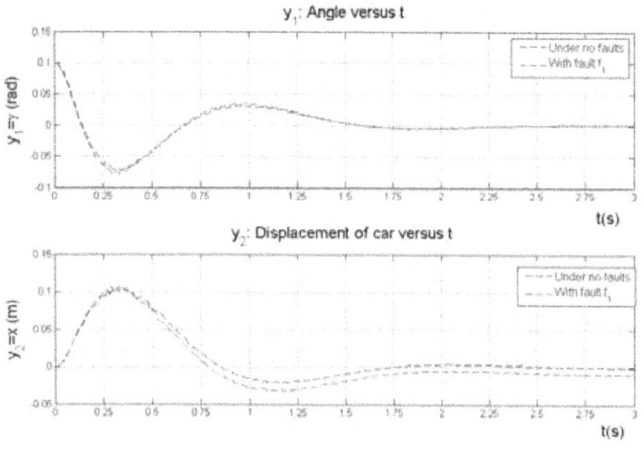

(a) Simulation with no faults (blue lines) and fault f_1 (red lines)

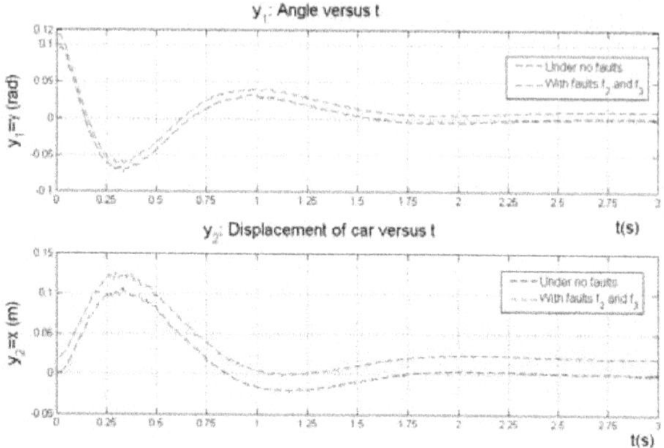

(b) Simulation with no faults (blue lines) and faults f_2, f_3 (red lines)

Figure 2: Behavior of the system under different situations, noise data 5 $\%$

the system are $\mathbf{y} = [\gamma\ x]^t$ and the input $\mathbf{u}(t) = F$ is the control force applied to the car. The relationship between each element of the fault vector $f(t) = [f_1\ f_2\ f_3]^t$ and the faults of the system is one-one: f_1 causes undesired movement of the car taking place in the actuator, this kind of fault is represented by an additive fault affecting the system input F; f_2 represents a fault in the sensor of γ and f_3 identifies faults in the sensor that measures x.

Let be $\Theta = f(t)$, and considering the nature of the faults and the properties of the IPS, then the elements of Θ have the following restrictions:

$$\begin{array}{lll} \theta_1 \in \mathbb{R}: & -1 \leq \theta_1 \leq 1 & \text{N} \\ \theta_2 \in \mathbb{R}: & 0 \leq \theta_2 \leq 0.01 & \text{rad} \\ \theta_3 \in \mathbb{R}: & 0 \leq \theta_3 \leq 0.02 & \text{m} \end{array} \tag{10}$$

In order to diagnose the system we can achieve estimations of $\Theta = f(t)$. In that sense the inverse problem of FDI that was formulated in Eqn. (2) should be solved.

4.1.1 Data simulation

The behavior of the system was simulated under no faults and under different faulty situations. The direct problem, Eqn. (9), was numerically solved by Runge Kutta 4. The Fig. 2 shows three different situations that were simulated.

4.2 Fault Diagnosis based on inverse problem: DE and ACO

In order to diagnose the faults, the minimization problem formulated in Eqn. (2) was implemented considering DE and ACO. All the implementations were made using MATLAB R2008a. The stopping criterion for both algorithms considered the number of iterations (1000) and the value of an objective function measurement ϵ_F $(\epsilon_F < 0.01)$:

$$\epsilon_F = \left\| F\left(\hat{\Theta}\right) \right\|_{+\infty} = max\left\{F_1, F_2\right\} \tag{11}$$

The function $\epsilon_F : \mathbb{R}^3 \to \mathbb{R}$ allows make comparations in a easier way. It is important to note that the computational effort of the ACO is greater than that of the DE : In each iteration the DE algorithm evaluates the objective function only once while the ACO makes Z evaluations.

4.2.1 DE implementation

The DE implementation was based on the description made in this paper with the following parameters values: population was considered to be $Z = 10$ and the mutation mechanism is $(DE/x_j^{best}/1/bin)$ with $C_R = 0.7$ and $F_s = 0.6$.

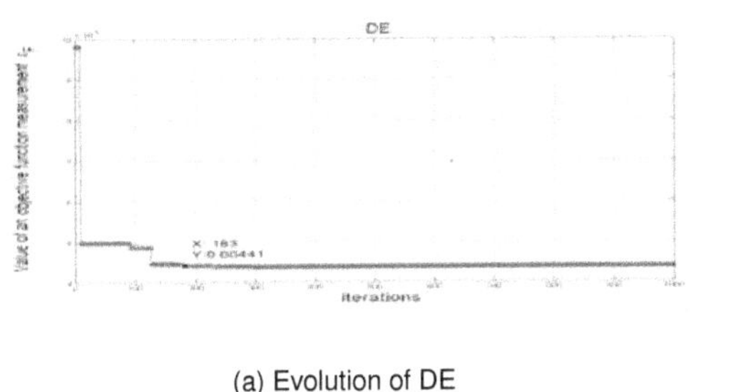

(a) Evolution of DE

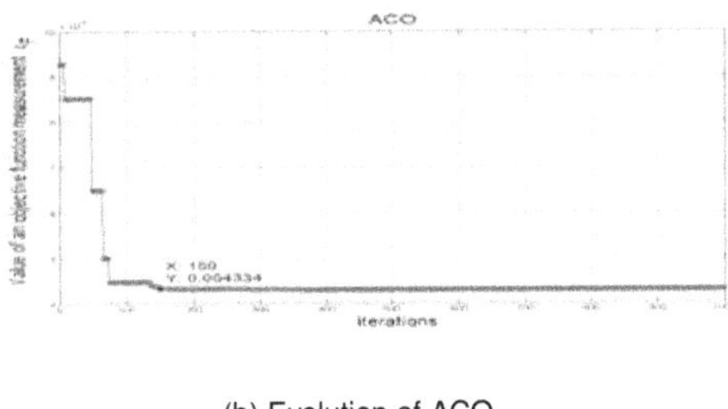

(b) Evolution of ACO

Figure 3: Evolution of the algorithms under th faulty situation $\Theta = [0\ 0.01\ 0.02]$, noise data 5 $\%$

4.2.2 ACO implementation

On the other hand the ACO implementation also considered the mechanism that was described before: the colony have 10 ants at each iteration, the value of k is the same for the three variables $k = 1001$, $q_0 = 0.75$, $C_{evap} = 0.30$ and $C_{inc} = 0.10$.

4.3 Results

The Table 1 shows the results of the diagnosis of different faulty situations by both algorithms. The number of experiments was 30 in each case. All cases considered data with 5 $\%$ of noise in order to simulate the effect of disturbances. The abbreviations that were used in the tables are: alg for the algorithm implemented and Mean Eval $F(\Theta)$ for the average of objective function evaluations that were achieved. The values of the vector parameter Θ indicate the magnitude of the faults that were introduced for the data simulation and the components of the vector Mean $\hat{\Theta}$ are the mean of the estimations of the magnitude of each fault. Both algorithms make a correct diagnosis of the faults . DE algorithm is more accurate in the determination of the fault magnitude.

Table 1: Diagnosis obtained in different faulty situation: data with 5 $\%$ of noise

alg	fault Θ	Mean $\hat{\Theta}$	Mean Eval $F(\Theta)$
DE	$\Theta = [0\ 0.01\ 0.02]$	$[0.054\ 0.01\ 0.0200]$	140
ACO	$\Theta = [0\ 0.01\ 0.02]$	$[-0.031\ 0.0085\ 0.0197]$	1200
DE	$\Theta = [0.5\ 0.01\ 0.02]$	$[0.4723\ 0.0094\ 0.02]$	128
ACO	$\Theta = [0.5\ 0.01\ 0.02]$	$[0.4219\ 0.0079\ 0.0199]$	1163
DE	$\Theta = [-0.5\ 0\ 0.004]$	$[-0.4109\ 0.0001\ 0.0038]$	154
ACO	$\Theta = [-0.5\ 0\ 0.004]$	$[-0.463\ 0.000014\ 0.00394]$	1201

The analysis of the behavior of the algorithms, see Fig. 3, revealed that after 200 iterations the objective function does not considerably decrease and the best solution of the population does not change significantly after the iteration 200. For that reason the maximum number of iterations was changed to 200.

With the purpose of analyze the robustness of the diagnosis via DE and ACO we make the same experiments that are exposed in the Table 1 but with more noise (15 $\%$). The results are in the Table 2 and revealed that the faults in the sensor, f_2 and f_3, are well detected by both algorithms but the diagnosis is better with DE, in fact the more accurate results are given by DE. The actuator fault f_1 is detected but its magnitude is not well established, the best approximation is given by ACO. In other words, the diagnosis via inverse problem with the use of DE and ACO is robust for sensor additive faults.

With the aim of improving the diagnosis of the actuator fault we propose a cooperative strategy between DE and ACO. This strategy is based on the fact that under 15 $\%$ of noise the ACO algorithm shown more robust diagnosis for the fault f_1 and DE did the same but for the rest of the faults. The computational cost of ACO is major than DE, and the adaptation of ACO to the continuous problem allows more accurate estimations without incrementing the computational cost and memory cost. That is the reason why we have decide to start with ACO for obtaining an initial estimation of Θ that ensure a robust diagnosis for f_1 and then use this solution as initial one for DE in order to improve the estimations for the sensors faults. Now we have considered the maximum number of iterations for ACO equal to $iter_{ACO} = 80$ and for DE the value $iter_{DE} = 120$. We reproduced the same experiments and the results are shown in the table 3.

Table 2: Diagnosis obtained in different faulty situation: data with 15 % of noise

alg	fault Θ	Mean $\hat{\Theta}$	Mean Eval $F(\Theta)$
DE	$\Theta = [0\ 0.01\ 0.02]$	[0.36 0.01 0.0200]	189
ACO	$\Theta = [0\ 0.01\ 0.02]$	[0.026 0.0074 0.0188]	2000
DE	$\Theta = [0.5\ 0.01\ 0.02]$	[0.6783 0.01 0.02]	160
ACO	$\Theta = [0.5\ 0.01\ 0.02]$	[0.519 0.00686 0.0079]	1782
DE	$\Theta = [-0.5\ 0\ 0.004]$	[-0.310 0.000023 0.004]	194
ACO	$\Theta = [-0.5\ 0\ 0.004]$	[-0.483 0.008 0.0033]	2000

Table 3: Diagnosis obtained in different faulty situation: data with 15 % of noise

alg	fault Θ	Mean $\hat{\Theta}$	Mean Eval $F(\Theta)$
ACO-DE	$\Theta = [0\ 0.01\ 0.02]$	[0.0018 0.01 0.02]	876
ACO-DE	$\Theta = [0.5\ 0.01\ 0.02]$	[0.5104 0.01 0.02]	740
ACO-DE	$\Theta = [-0.5\ 0\ 0.004]$	[-0.471 0.000009 0.004]	831

5 Conclusions

This study indicates that focusing the FDI as an inverse problem and implementing stochastic algorithms for solving the optimization problem is a viable way of development robust FDI methods based on a model.

The advantages observed in the application of the two algorithms to the FDI problem are: correct diagnosis, easy structure and robustness. The experiments show that DE is good for robust diagnoses of additive faults in the sensors and ACO for additive faults in the actuator. The cooperative strategy ACO-DE shows more robust and more sensitive (to actuators and sensors faults) diagnosis than pure DE or pure ACO.

In this sense the study of a more cooperative strategy between these two nature inspired algorithms will be done: considering the influence of the parameters of ACO on the estimations of the actuator faults as well the structure and parameters of DE algorithm in the estimations of sensors faults.

References

[1] R. Isermann, "Supervision, fault-detection and fault-diagnosis methods- an introduction," *Control Eng. Practice*, vol. 5, no. 5, pp. 639–652, 1997.

[2] S. Simani, C. Fantuzzil, and R. J. Patton, *Model-Based Fault Diagnosis in Dynamic systems Using Identification Techniques.* Springer-Verlag, 2002.

[3] R. Isermann, "Process fault detection based on modelling and estimation methods– a survey," *Automatica*, vol. 20, no. 4, pp. 387–404, 1984.

[4] M. Fuente, P. Vega, M. Zarrop, and M. Poch, "Fault detection in a real wastewater plant using parameter. estimation techniques," *Control Eng. Practice*, vol. 4, no. 8, pp. 1089–1098, 1996.

[5] R. Isermann, "Fault diagnosis of machines via parameter estimation and knowledge processing tutorial paper," *Automatica*, vol. 29, no. 4, pp. 815–835, 1993.

[6] S. Simani and R. J. Patton, "Fault diagnosis of an industrial gas turbine prototype using a system identification approach," *Control Engineering Practice*, vol. 16, pp. 769–786, 2008.

[7] M. Witczak, "Advances in model based fault diagnosis with evolutionary algorithms and neural networks," *Int. J. Appl. Math. Comput. Sci.*, vol. 16, no. 1, pp. 85–99, 2006.

[8] E. Yang, H. Xiang, and D. G. Z. Zhang, "A comparative study of genetic algorithm parameters for the inverse problem-based fault diagnosis of liquid rocket propulsion systems," *International Journal of Automation and Computing*, vol. 04, pp. 255–261, july 2007.

[9] S. R. and K. Price, "Differential evolution: A simple and efficient adaptive scheme for global optimization over continuous spaces," *International Computer Science Institute*, 1995.

[10] M. Dorigo, *Ottimizzazione, Apprendimento Automa¡tico, Ed Algoritmi Basati su Metafora Naturale.* PhD thesis, Politécnico di Milano, Italia, 1992.

[11] S. X. Ding, *Model- based Fault Diagnosis Techniques: Design Schemes, Algorithms, and Tools.* Springer, 2008.

[12] K. Ogata, *Modern Control Engineering.* Prentice- Hall, third edition ed., 1998.

On the multi-frequency reconstruction of penetrable obstacles via the linear sampling method

Bojan B. Guzina[1], *Fioralba Cakoni*[2] *and Cédric Bellis*[1,3]

[1] *Department of Civil Engineering, University of Minnesota, Minneapolis, MN*

[2] *Department of Mathematical Sciences, University of Delaware, Newark, DE*

[3] *Laboratoire de Mécanique des Solides, Ecole Polytechnique, Palaiseau, France*

Email: guzina@wave.ce.umn.edu, cakoni@math.udel.edu, bellis@lms.polytechnique.fr

Abstract

This paper deals with the multi-frequency reconstruction of penetrable scatterers by way of the linear sampling method. On introducing a suitable approximate solution to the linear sampling equation and making an assumption of continuous frequency sweep, two possible choices for a cumulative multi-frequency indicator function of the scatterer's support are proposed. The first alternative, termed the "serial" indicator, is taken as a natural extension of its monochromatic companion in the sense that its computation entails *space-frequency* (as opposed to space) L^2-norm of a solution to the linear sampling equation. Under a set of assumptions that include experimental observations down to zero frequency and compact frequency support of the wavelet used to illuminate the obstacle, this indicator function is further related to its time-domain counterpart. As a second possibility, the so-called "parallel" indicator is alternatively proposed as an L^2-norm, in the frequency domain, of the monochromatic indicator function. By way of a perturbation analysis which demonstrates that the monochromatic solution of the linear sampling equation behaves as $O(|k^2 - k_*^2|^{-m})$, $m \geqslant 1$ in the neighborhood of an isolated eigenvalue, k_*^2, of the associated interior transmission problem [1], it is found that the "serial" indicator is unable to distinguish the interior from the exterior of a scatterer in situations when the prescribed frequency band traverses at least one such eigenvalue. In contrast the "parallel" indicator is, due to its particular structure, shown to be insensitive to the presence of germane interior eigenvalues, and thus to be robust in a generic scattering configuration.

1 Introduction

In the context of inverse scattering, the past two decades have witnessed the inception and growth of a range of non-iterative techniques for obstacle reconstruction such as the linear sampling method [2]. Such point-probing algorithms commonly operate within the framework of monochromatic i.e. single-frequency obstacle illumination which postulates that the squared wave number, computed with reference to the background medium, is not an eigenvalue of the associated interior transmission problem. For common scattering configurations such eigenvalues form an at most countable set, with no accumulation points other than infinity [3, 4], which makes the featured restriction manageable if not desirable in the context of practical applications.

Besides (and before) the choice of an appropriate reconstruction technique, the critical issue for most inverse scattering problems is the richness of the observed data set. In general the latter can be extended either spatially, in terms of the aperture of experimental observations, or temporally, by considering multi-frequency or time-domain scattered waveforms. Notwithstanding the fact that the latter alternative is often far more tractable in terms of experimental implementation, the literature dealing with point-probing algorithms that transcend the customary monochromatic framework is relatively scarce. In particular, what largely remains unclear is the role of the eigenvalues of the germane interior problem (defined over the support of a hidden scatterer) toward the performance of point-probing methods in situations where the former are traversed by a given frequency sweep or the Fourier spectrum of a prescribed transient signal.

To help bridge the gap, this study focuses on the multi-frequency reconstruction of penetrable obstacles via the linear sampling method entailing far-field observations of the scattered field. On assuming that the (monochromatic) sampling equation is solved over a compact connected set of real-valued excitation frequencies ω, two possible choices for a cumulative, multi-frequency indicator function are proposed and scrutinized.

2 Preliminaries

Consider the time-harmonic scattering of scalar waves by a penetrable obstacle D in a homogeneous unbounded medium $\mathbb{R}^3$, endowed with sound speed c_o, due to an incident plane wave

$$u^i = e^{ikx\cdot d}, \qquad d \in \Omega. \tag{1}$$

Here $k=\omega/c_o$ is the wavenumber; ω denotes the frequency of excitation, and Ω is the unit sphere centered at the origin. The support of D is assumed to be such that $\mathbb{R}^3\setminus\overline{D}$ is connected, and that ∂D is of Lipschitz type. As a canonical example of the scattering by a penetrable obstacle, consider next the case where D is characterized by a spatially-varying sound speed $c(x)$ and associated index of refraction, $n(x)=(c_o/c)^2$, such that i) $c>c_D>0$ where c_D is a constant; ii) $n\in L_\infty(D)$, and iii) ∇n is sufficiently small so that it can be omitted from the field equation. For simplicity, an additional hypothesis is made that the mass density of the system, ρ, is constant throughout. In this case, the relevant scattering problem can be written as

$$\begin{aligned} \Delta u + k^2 u = 0 \qquad & \text{in } \mathbb{R}^3\setminus\overline{D}, \\ \Delta \varpi + k^2 n\, \varpi = 0 \qquad & \text{in } D, \\ \varpi - u = u^i, \quad \varpi_{,\nu} - u_{,\nu} = u^i_{,\nu} \qquad & \text{on } \partial D, \\ \lim_{|x|\to\infty} |x|\Big(\frac{\partial u}{\partial |x|} - iku\Big) = 0, & \end{aligned} \tag{2}$$

where $u_{,\nu}=\nabla u\cdot\nu$, and ν is the normal on ∂D (defined almost everywhere) oriented toward the exterior of D. Analogous to the case of scattering by a Dirichlet obstacle, it is known [5] that (2) permits a unique solution $(u,\varpi)\in H^1_{\text{loc}}(\mathbb{R}^3\setminus\overline{D})\times H^1(D)$.

By way of Green's theorem, it can be shown [5] that the scattered field u solving (2) permits integral representation

$$u(x,d) = \int_{\partial D}\Big(u(\xi,d)\Psi_{,\nu}(x,\xi,k) \,-\, u_{,\nu}(\xi,d)\Psi(x,\xi,k)\Big)\mathrm{d}s_\xi, \tag{3}$$

where Ψ is the radiating fundamental solution of the Helmholtz equation. Assuming illumination by plane waves, (3) can be used to demonstrate that

$$u(x,d) \;=\; \frac{e^{ik|x|}}{|x|}u_\infty(\hat{x},d) \,+\, O\left(|x|^{-2}\right) \qquad \text{as } |x|\to\infty, \tag{4}$$

where

$$u_\infty(\hat{x},d) \;=\; \int_{\partial D}\Big(u(\xi,d)(e^{-ik\hat{x}\cdot\xi})_{,\nu} \,-\, u_{,\nu}(\xi,d)e^{-ik\hat{x}\cdot\xi}\Big)\mathrm{d}s_\xi \tag{5}$$

is the so-called *far-field pattern* of the scattered field [5].

3 The linear sampling method

With reference to the direct scattering framework established in Section 2, the goal is to reconstruct the support D of a hidden obstacle on the basis of available information on the scattered field, synthesized via u_∞, for multiple incident fields. For the remainder of this section it is assumed, following the usual treatment [5], that the data are available at a single excitation frequency, ω, such that k^2 is not a *transmission eigenvalue* [4] for bounded domain D. Hereon, it is for simplicity assumed that the far-field pattern u_∞ is known for every direction of observation and every direction of plane-wave incidence, i.e. that the data are given by $u_\infty(\hat{x},d)$ for $\hat{x},d\in\Omega$. In this case, the linear sampling method revolves around solving the equation of the first kind

$$(Fg_z)(\hat{x}) \;=\; \Psi_\infty(\hat{x},z,k), \qquad \hat{x}\in\Omega, \tag{6}$$

where $F: L^2(\Omega) \to L^2(\Omega)$ is the so-called far-field operator given by

$$(Fg)(\hat{x}) := \int_{\Omega} u_\infty(\hat{x}, d)\, g(d)\, \mathrm{d}s_d; \tag{7}$$

g_z is the solution density used to construct an indicator function; z denotes the sampling point, and Ψ_∞ is the far-field pattern of Ψ as $|x| \to \infty$. With such premises, it can be shown [6] that:

- If $z \in D$ then for every $\epsilon > 0$, there exists a solution $g_z^\epsilon \in L^2(\Omega)$ of (6) such that
$$\| Fg_z^\epsilon(\cdot) - \Psi_\infty(\cdot, z, k) \|_{L^2(\Omega)} < \epsilon; \tag{8}$$
- When $z \in D$, one further has
$$\lim_{z \to \partial D} \| g_z^\epsilon \|_{L^2(\Omega)} \to \infty, \qquad \lim_{z \to \partial D} \| v_{g_z^\epsilon} \|_X \to \infty,$$
where
$$v_g(x) := \int_{\Omega} e^{ikx \cdot d}\, g(d)\, \mathrm{d}s_d \tag{9}$$
is the Herglotz wave function with kernel g, and
- When $z \in \mathbb{R}^3 \setminus \overline{D}$, then for every $\epsilon > 0$ there exists a solution $g_z^\epsilon \in L^2(\Omega)$ such that
$$\| Fg_z^\epsilon(\cdot) - \Psi_\infty(\cdot, z, k) \|_{L^2(\Omega)} < \epsilon$$
and
$$\lim_{\epsilon \to 0} \| g_z^\epsilon \|_{L^2(\Omega)} \to \infty, \qquad \lim_{\epsilon \to 0} \| v_{g_z^\epsilon} \|_{L^2(D)} \to \infty$$

With the above results in place, D can be reconstructed by employing a suitable regularization technique to solve the far-field equation $Fg_z = \Psi_\infty(\cdot, z, k)$ over an appropriate grid of sampling points, and using $\Pi(z) := 1/\| g_z \|_{L^2(\Omega)}$ as a characteristic function of the support of the scatterer.

3.1 Relationship with the solution to the interior problem

To shed light on the denseness claim (8), consider the so-called *interior transmission problem* [4]

$$\begin{aligned} \Delta v_z + k^2 v_z &= 0 && \text{in } D, \\ \Delta w_z + k^2 n\, w_z &= 0 && \text{in } D, \\ w_z - v_z &= \Psi(\cdot, z, k) && \text{on } \partial D, \\ (w_z)_{,\nu} - (v_z)_{,\nu} &= \Psi_{,\nu}(\cdot, z, k) && \text{on } \partial D \end{aligned} \tag{10}$$

which is, following earlier hypothesis, considered under the restriction that k^2 is not a transmission eigenvalue for D [7] – defined as the value of k^2 for which the homogeneous counterpart of (10) permits non-trivial solution.

Lemma 1. Assume that $z \in D$, and let k be such that $|k - k_0| \leqslant r$ for some $r > 0$ and $k_0 \in \mathbb{C}$. Under such restrictions there is a constant c_0 independent of k (but dependent on k_0 and r), such that any density $g_z^\epsilon \in L^2(\Omega)$ for which the affiliated Herglotz wave function (9) approximates component v_z of the unique solution (v_z, w_z) to (10) as $\| v_{g_z^\epsilon} - v_z \|_{L^2(D)} < c_0 \epsilon$, also satisfies the far-field inequality (8). Further, for any $c_0 \epsilon > 0$ there exists density $g_z^\epsilon \in L^2(\Omega)$ such that $v_{g_z^\epsilon}$ meets the postulated $L^2(D)$ inequality.

For the proof of Lemma 1, that makes use of the foregoing results, see [1]. Here it is particularly important to note that the ensuing estimates, while established in a time-harmonic setting, hold *uniformly* with respect to k over any closed region in the complex plane (hereon denoted by $\mathbb{C}$) – a result that provides a linchpin for the extension of the linear sampling to multi-frequency scattering configurations.

4 Multi-frequency reconstruction

As examined earlier, the linear sampling method considers inverse scattering at a single excitation frequency, ω, such that $k^2 = \omega^2/c_{\text{o}}^2$ is not an eigenvalue of the germane interior transmission problem for D. Assuming that ∂D is of Lipschitz type, it can be shown [3, 4] that the eigenspectrum of the interior transmission problem over D is at most *countable* with no finite accumulation points.

To examine the possibility and effectiveness of multi-frequency obstacle reconstruction, the ensuing study focuses on a generic situation where the scattered field due to multiple incident wavefields, synthesized via u_∞ or u, is monitored over a *frequency band*, $\omega \in F_\omega := [\omega_1, \omega_2] \subset \mathbb{R}^+$, $\omega_2 < \infty$. For clarity of exposition, all frequency-dependent quantities referred to in the sequel will have ω added to their list of arguments whereby $u(x,y)$ is superseded by $u(x,y,\omega)$, $g_z(y)$ by $g_z(y,\omega)$, and so on. In this setting, the multi-frequency counterpart of (6) can be postulated as

$$(Fg_z)(\hat{x},\omega) = \Psi_\infty(\hat{x}, z, \omega/c_{\text{o}}), \qquad \hat{x} \in \Omega, \qquad \omega \in F_\omega \tag{11}$$

where $F: L^2(\Omega) \times L^2(F_\omega) \to L^2(\Omega) \times L^2(F_\omega)$ is a bounded linear operator such that

$$(Fg)(\hat{x},\omega) := \int_\Omega u_\infty(\hat{x}, d, \omega)\, g(d,\omega)\, \mathrm{d}s_d. \tag{12}$$

For a systematic treatment of such extended inverse scattering problem, the key issues to be addressed pertain to: i) the choice of a "cumulative" indicator function that reflects the extended data set, and ii) the situation where the chosen frequency band traverses at least one interior eigenvalue, i.e. when

$$\Lambda \cap F_{k^2} \neq \emptyset, \quad F_{k^2} := \{k^2 : k = c_{\text{o}}^{-1}(\omega_1 + \eta(\omega_2 - \omega_1)),\, \eta \in [0,1]\}.$$

4.1 "Serial" and "parallel" indicator functions

Perhaps the most obvious extension of the monochromatic indicator function can be written as

$$\Pi_F^{(1)}(z) := \frac{1}{\| g_z \|_{L^2(\Omega)\times L^2(F_\omega)}} = \left(\int_{\omega_1}^{\omega_2} \| g_z(\cdot,\omega) \|^2_{L^2(\Omega)}\, \mathrm{d}\omega \right)^{-1/2}. \tag{13}$$

Assuming that $\Lambda \cap F_{k^2} = \emptyset$, one finds on the basis of the results highlighted in Section 3 that the "*serial*" indicator (13), similar to its monochromatic companion, becomes vanishingly small for $z \in \mathbb{R}^3 \setminus D$ which justifies its candidacy for a characteristic function of the support of the scatterer.

Another possible choice of a cumulative indicator function can be written as an L^2-norm of the "monochromatic" indicator over the featured frequency band, i.e.

$$\Pi_F^{(2)}(z) := \left\| \frac{1}{\| g_z \|_{L^2(\Omega)}} \right\|_{L^2(F_\omega)} = \left(\int_{\omega_1}^{\omega_2} \| g_z(\cdot,\omega) \|^{-2}_{L^2(\Omega)}\, \mathrm{d}\omega \right)^{1/2}, \tag{14}$$

where "Ω" signifies Ω and S_s in the case of incident plane waves and point sources, respectively. The reasoning behind the "*parallel*" indicator (14) is that of constructive interference where, again assuming that $\Lambda \cap F_{k^2} = \emptyset$, distributions $1/\| g_z(\cdot,\omega) \|_{L^2(S_s)}$, $\omega \in F_\omega$ reinforce each other in exposing the support of the scatterer by jointly vanishing when $z \in \mathbb{R}^3 \setminus D$.

To ensure the robustness of the multi-frequency reconstruction scheme, however, the critical issue with both (13) and (14) is their behavior and performance in situations when $\Lambda \cap F_{k^2} \neq \emptyset$ – a possibility that cannot be discarded beforehand for the logical value of the latter inequality is, for given F_{k^2}, dependent on the geometry and nature of a hidden scatterer. Given the fact that both $\Pi_F^{(1)}$ and $\Pi_F^{(2)}$ vanish when $z \notin D$ and $\Lambda \cap F_{k^2} = \emptyset$, of particular concern here is the situation when $z \in D$ and F_{k^2} contains at least one eigenvalue of the relevant interior problem. Indeed, if either candidate for a cumulative indicator function necessarily vanishes in this case, such behavior would preclude its utility as a characteristic function of the support of the obstacle in a generic scattering environment.

4.2 Behavior of the solution in a neighborhood of an eigenvalue

To expose the utility of (band-limited) cumulative indicator functions proposed in Section 4.1, it is critical to understand the behavior an approximate solution, g_z^ϵ, to the far-field equation (6) in the *neighborhood* of a "resonant" frequency ω_*, such that $\omega_*^2/c_o^2 = k_*^2 \in \Lambda$. In the context of far-field scattering, the first result in this direction was provided in [8] where it was shown that for $k^2 = k_*^2 \in \Lambda$ and almost every $z \in D$, Herglotz wave function $v_{g_{z,\delta}}^\epsilon$ (where $\epsilon = \epsilon(\delta)$ and $g_{z,\delta}^\epsilon$ is the Tikhonov-regularized solution of (6)) becomes unbounded, when $\delta \to 0$, in the $L^2(D)$-norm. In the context of multi-frequency indicator functions (13) and (14), however, it is necessary to examine the *blow-up rate* of the relevant solution $g_{z,\epsilon}$ in the neighborhood of an eigenvalue $k_*^2 \in \Lambda$. Specifically, one needs to know whether $\|g_z(\cdot,\omega)\|_{L^2(\Omega)}$ is square-integrable with respect to ω over a given interval $[\omega_1, \omega_2]$, containing "resonant" frequency ω_* that corresponds to an eigenvalue of the germane interior problem.

Theorem 1. Let k_*^2 be an isolated transmission eigenvalue, and consider $\alpha > 0$ such that the ball $\mathcal{B}_{k_*^2,\alpha} := \{k^2 : |k^2 - k_*^2| < \alpha,\ k^2 \neq k_*^2\}$ does not contain any eigenvalues other than k_*^2. Further, let g_z^ϵ be the approximate solution of either the far-field or the near-field equation, specified in Lemma 1. Then for sufficiently small $\epsilon > 0$ and $\alpha > 0$, all $k^2 \in \mathcal{B}_{k_*^2,\alpha}$, and almost every $z \in D$ one has

$$\|v_{g_z^\epsilon}\|_{L^2(D)} \geqslant \frac{C_1}{|k^2 - k_*^2|} \quad \text{and} \quad \|g_z^\epsilon\|_{L^2(\Omega)} \geqslant \frac{C_2}{|k^2 - k_*^2|}. \tag{15}$$

Here v_g is given by (9), while C_1 and C_2 are positive constants dependent on z, k_* and α, but not on k and ϵ.

For the proof of Theorem 1, see [1].

Remark 1. *From Theorem 1, it is clear that* $\|g_z^\epsilon\|_{L^2(\Omega)}$, $\Omega = \Omega, S_s$ *behaves as* $O(|\omega - \omega_*|^{-m})$, $m \geqslant 1$ *when* $\omega \to \omega_* = c_o k_*$. *As a result, the multi-frequency solution density* g_z^ϵ *featured in (13) and (14) does not belong to* $L^2(\Omega) \times L^2(F_\omega)$ *when the relevant interior problem over* D *is characterized by eigenvalues* k_*^2 *such that* $\omega_* = c_o k_* \in F_\omega$. *In light of this result it is noted that "serial" indicator function (13), in contrast to its "parallel" companion (14), is not applicable to such configurations – a finding that is illustrated in the sequel.*

5 Results

In what follows, an attempt at multi-frequency obstacle reconstruction via the linear sampling method is made assuming far-field scattering by a square obstacle in $\mathbb{R}^2$ [9], which exposes the performance of the method in a generic computational setting. Here it is noted that both the claim and the structure of the proof of Theorem 1, see [1], is independent of the dimensionality of the problem, and can be extended to scattering in $\mathbb{R}^2$ by invoking the two-dimensional counterparts of Lemma 1, see [10]. With such result in place, consider the inverse scattering of plane waves by a unit square, $D = \{x \in \mathbb{R}^2 : x \in [-0.5, 0.5] \times [-0.5, 0.5]\}$, assuming penetrable obstacle with $n = 4$ and $\beta = 1/4$. To this end, a discrete set of directions of plane-wave incidence and observation is assumed as

$$\Omega^h := \{\hat{x} = (\cos(2\pi mh), \sin(2\pi mh)), \quad h = M^{-1} m \in \{0, 1, \ldots, M-1\}\}, \quad M = 61. \tag{16}$$

In this setting, a "fine" discretization of the example wavenumber band $k \in [3, 8]$ is taken as

$$F_k^h := \{k : k = 3 + m\,h,\ h = 5 \cdot 10^{-2},\ m \in \{0, 1, \ldots, 10^2\}\}.$$

Here it is noted that the featured interval $k^2 \in [9, 64] \subset \mathbb{R}$ contains, at least numerically, several transmission eigenvalues associated with the assumed scattering configuration in terms of D (see [9] for details). Fig. 1 plots the normalized distribution of indicator functions (13) and (14) on the scale $[0, 1]$. Consistent with the theoretical developments, the two-dimensional reconstruction of a square scatterer via the "serial" indicator $\check{\Pi}_F^{(1)}$ is inferior to that obtained using its "parallel" companion $\check{\Pi}_F^{(2)}$, not only in terms of the contrast of an image, but also in terms of the reconstructed shape.

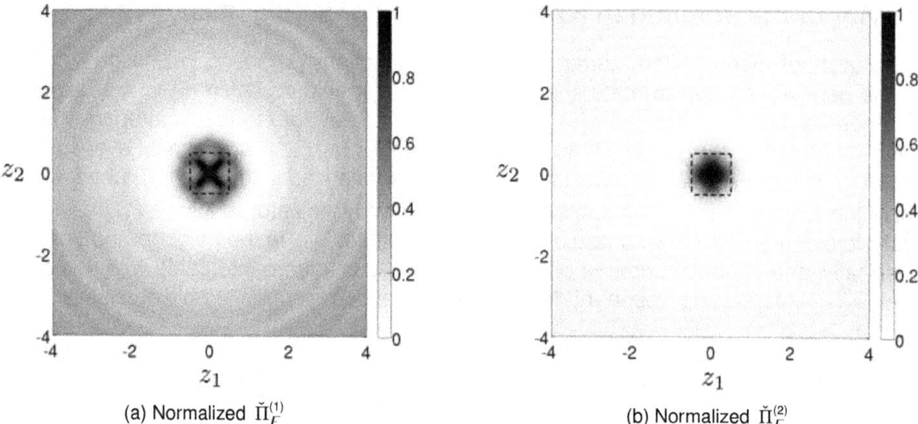

(a) Normalized $\breve{\Pi}_F^{(1)}$ (b) Normalized $\breve{\Pi}_F^{(2)}$

Figure 1: Reconstruction of a square obstacle from the far-field data taken over a "fine" wavenumber set F_k^h.

6 Conclusions

In this study, multi-frequency reconstruction of penetrable obstacles is examined in the context of the linear sampling method entailing far-field measurements and penetrable scatterers. On establishing a suitable approximate solution to the linear sampling equation under the premise of continuous frequency sweep, two possible choices for a cumulative multi-frequency indicator function of the scatterer's support are proposed. The first alternative, termed the "serial" indicator, is taken as a natural extension of its customary monochromatic counterpart in the sense that its computation entails *space-frequency* (as opposed to space) L^2-norm of a solution to the linear sampling equation. As a second possibility, the so-called "parallel" indicator is proposed as an L^2-norm, in the frequency domain, of the monochromatic indicator function. On the basis of the perturbation analysis, it is found that the "serial" indicator is unable to distinguish the interior from the exterior of a scatterer in situations when the prescribed frequency band traverses at least one such eigenvalue. In contrast the "parallel" indicator is, due to its particular structure, shown to be insensitive to the presence of pertinent interior eigenvalues. A set of numerical results is included to illustrate the theoretical developments.

References

[1] B. B. Guzina, F. Cakoni, and C. Bellis, "On the multi-frequency obstacle reconstruction via the linear sampling method," *Inverse Problems*, vol. 26, pp. 125005, 2010.

[2] D. Colton and A. Kirsch, "A simple method for solving inverse scattering problems in the resonance region," *Inverse Problems*, vol. 12, pp. 383–393, 1996.

[3] L. Paivarinta and J. Sylvester, "Transmission eigenvalues," *SIAM Journal on Mathematical Analysis*, vol. 40, pp. 738–753, 2008.

[4] A. Kirsch, "On the existence of transmission eigenvalues," *Inverse Problems and Imaging*, vol. 3, pp. 155–172, 2009.

[5] D. Colton and R. Kress, *Inverse acoustic and electromagnetic scattering theory*, Springer, Berlin, 1992.

[6] A. Kirsch and N. Grinberg, *The Factorization Method for Inverse Problems*, Oxford University Press, New York, 2008.

[7] F. Cakoni, D. Gintides, and H. Haddar, "The existence of an infinite discrete set of transmission eigenvalues," *SIAM J. Math. Anal.*, vol. 42, pp. 237–255, 2010.

[8] F. Cakoni, D. Colton, and H. Haddar, "On the determination of Dirichlet and transmission eigenvalues from far field data," *C. R. Acad. Sci. Paris, Ser. I*, vol. 348, pp. 379–383, 2010.

[9] F. Cakoni, D. Colton, P. Monk, and J. Sun, "The inverse electromagnetic scattering problem for anisotropic media," *Inverse Problems*, vol. 26, pp. 074004, 2010.

[10] F. Cakoni and D. Colton, *Qualitative Methods in Inverse Scattering Theory*, Springer,-Verlag Berlin, 2006.

A boundary integral equation method for a Cauchy problem for the heat equation in semi-infinite domains

R. Chapko[1], B. T. Johansson[2] and V. Vavrychuk[1]

[1] Faculty of Applied Mathematics and Informatics, Ivan Franko National University of Lviv, 79000 Lviv, Ukraine
e-mail: chapko@is.lviv.ua

[2] School of Mathematics, University of Birmingham, Edgbaston, Birmingham B15 2TT, UK
e-mail: b.t.johansson@bham.ac.uk

We consider a Cauchy problem for the time-dependent heat equation in semi-infinite domains (for example a half-plane, quadrant, strip or half-strip) containing an inclusion. For the stable reconstruction of the solution, we employ an iterative method where, in each step of the iterations, one has to solve mixed initial boundary value problems in the semi-infinite domain, where data is specified on the boundary of the semi-infinite domain as well as on the boundary of the inclusion.

For some solution domains, such as a strip, there is no computationally useful non-stationary Green function for the heat equation. In order to address this issue, we propose and investigate a method for these mixed problems where we first perform semi-discretization in time to reduce the parabolic equation to a sequence of elliptic equations. Then, similar to [2], a sequence of Green functions can be defined for the obtained sequence of problems. We show that one can construct these Green functions for the corresponding sequence of elliptic equations for various kinds of canonical semi-infinite regions.

Finally, after applying results from potential theory and using integral equations, we obtain a fully discrete sequence of systems of linear algebraic equations with the same matrix and a recurrent data vector ("right-hand side"). It is shown that the proposed numerical solution approach for the mixed problems has error of first or second order in time and exponential convergence with respect to the spatial variable, provided that the given input data are sufficiently smooth. Numerical examples are also presented.

1 Problem statement

We assume that $D_0 \subset \mathbf{R}^2$ is a semi-infinite region with boundary Γ_0 and $D_1 \subset D_0$ is a simply connected bounded domain with sufficiently smooth boundary Γ_1. We define $D = D_0 \setminus \bar{D}_1$, see Fig. 1, and let $T > 0$ and $c > 0$ be given constants, and the unit normal vector ν is directed outwards relative to the boundary. We consider a Cauchy problem for the heat equation, which consists of the reconstruction of

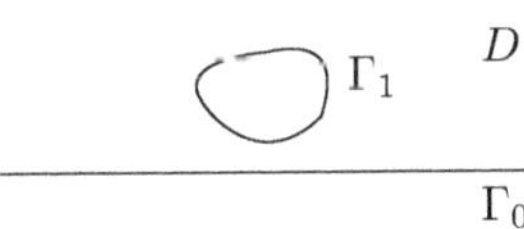

Figure 1: Semi-infinite domain with inclusion

the temperature field u from Cauchy data given on the boundary part Γ_0:

$$\begin{cases} \dfrac{1}{c}\dfrac{\partial u}{\partial t} = \Delta u & \text{in } D \times (0,T), \\ u = f_1 & \text{on } \Gamma_0 \times (0,T), \\ \dfrac{\partial u}{\partial \nu} = f_2 & \text{on } \Gamma_0 \times (0,T), \\ u(x,0) = 0 & \text{for } x \in D, \end{cases} \tag{1}$$

where f_1 and f_2 are known and sufficiently smooth functions. These functions are compatible such that there exists a solution. Moreover, here and later it is assumed that the solution to heat equation should satisfy the following regularity condition at infinity $u(x,t) \to 0$, $|x| \to \infty$ uniformly in all directions and for all $t \in (0,T)$.

The problem (1) is ill-posed in the sense of Hadamard since in general the solution (if it exists) does not depend continuously on the input data, thus traditional (numerical) methods are not suitable for its solution. Furthermore, using standard Tikhonov regularization leads to a change of the governing heat operator. The iterative regularizing method presented in [1], based on the Landweber procedure, does not have this disadvantage and we shall therefore apply that strategy. In this work, we extend the method in [1] to the case of semi-infinite planar domains with a bounded inclusion. In each iterative step some direct initial boundary value problems have to be solved. We shall employ boundary integral equations to reduce the dimension of these problems, which also reduces the computational time.

2 An iterative regularizing method

The aim is to construct the element $u|_{\Gamma_1\times(0,T)}$ in (1). To find an equation for this element, we need the following operators. The operator $K : L^2(\Gamma_1 \times (0,T)) \to L^2(\Gamma_0 \times (0,T))$ is defined as $K\eta := u|_{\Gamma_0\times(0,T)}$, where u is a solution to the following mixed initial boundary value problem

$$\begin{cases} \dfrac{1}{c}\dfrac{\partial u}{\partial t} = \Delta u & \text{in } D \times (0,T), \\ u = \eta & \text{on } \Gamma_1 \times (0,T), \\ \dfrac{\partial u}{\partial \nu} = 0 & \text{on } \Gamma_0 \times (0,T), \\ u(x,0) = 0 & \text{for } x \in D. \end{cases}$$

The operator $G : L^2(\Gamma_0 \times (0,T)) \to L^2(\Gamma_0 \times (0,T))$ is defined as $G\psi := u|_{\Gamma_0\times(0,T)}$, where u is the solution to the following mixed initial boundary value problem

$$\begin{cases} \dfrac{1}{c}\dfrac{\partial u}{\partial t} = \Delta u & \text{in } D \times (0,T), \\ u = 0 & \text{on } \Gamma_1 \times (0,T), \\ \dfrac{\partial u}{\partial \nu} = \psi & \text{on } \Gamma_0 \times (0,T), \\ u(x,0) = 0 & \text{for } x \in D. \end{cases}$$

Using the second Green formula one can show that adjoint operator $K^* : L^2(\Gamma_0\times(0,T)) \to L^2(\Gamma_1\times(0,T))$ is given by $K^*\zeta = -\frac{\partial}{\partial\nu}v|_{\Gamma_1\times(0,T)}$, where v is determined from the mixed initial boundary value problem

$$\begin{cases} \dfrac{1}{c}\dfrac{\partial v}{\partial t} = -\Delta v & \text{in } D \times (0,T), \\ v = 0 & \text{on } \Gamma_1 \times (0,T), \\ \dfrac{\partial v}{\partial \nu} = \zeta & \text{on } \Gamma_0 \times (0,T), \\ v(x,T) = 0 & \text{for } x \in D. \end{cases} \tag{2}$$

The Cauchy problem (1) of finding the element $u|_{\Gamma_1\times(0,T)}$ is equivalent to solving the operator equation $K\eta = f_1 - Gf_2$. Using the representation of the operator K via the Green function for the corresponding Dirichlet problem, one can show that K is a compact integral operator with continuous kernel. Thus, to solve for η, we have to apply a regularization strategy, for example, the Landweber method [1, 6, 7]

$$\eta_k = \eta_{k-1} - \gamma_L K^*(K\eta_{k-1} + Gf_2 - f_1), \tag{3}$$

where $\gamma_L > 0$ is a relaxation parameter, and η_0 is an arbitrary initial approximation.

According to [6], the following convergence result holds in the case of exact input data.

Theorem 1 *Assume that f_1, $f_2 \in L^2(\Gamma_0 \times (0,T))$ and let u be a solution to the Cauchy problem* (1). *If $0 < \gamma_L < 1/\|K\|^2$ then for the k-th approximation u_k from iterative procedure* (3) *the estimate*

$$\lim_{k\to\infty} \|u - u_k\|_{L^2(\Omega\times(0,T))} = 0,$$

holds for an arbitrary initial element $\eta_0 \in L^2(\Gamma_1 \times (0,T))$.

In the case of noisy input data, the discrepancy principle can be used as a stopping criteria. Furthermore, in each iteration two direct mixed Dirichlet-Neumann initial boundary value problems in a strip have to be solved, and solving these in an efficient way will be discussed in the next section. Note that for the adjoint problem we use the substitution $v(x,t) = u(x, T-t)$.

3 Numerical solution of the direct problems

We first consider the mixed Dirichlet-Neumann initial boundary value problem

$$\begin{cases} \dfrac{1}{c}\dfrac{\partial u}{\partial t} = \Delta u & \text{in } D \times (0,T), \\ u = g_1 & \text{on } \Gamma_1 \times (0,T), \\ \dfrac{\partial u}{\partial \nu} = g_0 & \text{on } \Gamma_0 \times (0,T), \\ u(x,0) = 0 & \text{for } x \in D. \end{cases}$$

To numerically solve this problem we use a combination of Rothe's method and boundary integral equations [2, 4] as explained below.

On the interval $[0,T)$ we introduce a uniform mesh $t_n = (n+1)h$, $h = T/(N+1)$, and put $u_n(x) \approx u(x,t_n)$, $g_{\ell n}(x) = g_\ell(x,t_n)$ and $u_{-1} = g_{\ell,-1} = 0$, for $n = 0,\dots,N-1$, $\ell = 0,1$. After applying a finite-difference approximation of the time derivative, we obtain a sequence of N stationary mixed problems

$$\Delta u_n - \gamma^2 u_n = \sum_{m=0}^{n-1} \beta_{n-m} u_m \qquad \text{in } D, \tag{4}$$

$$u_n = g_{1n}, \text{ on } \Gamma_1, \qquad \frac{\partial u_n}{\partial \nu} = g_{0n} \text{ on } \Gamma_0, \tag{5}$$

where $\gamma > 0$ and β_i are known coefficients. To guarantee that the regularity condition at infinity is satisfied, we impose that $u_n(x) \to 0$ when $|x| \to \infty$ uniformly in all directions.

Definition 2 *The sequence of functions* Φ_n, $n = 0,\dots,N-1$, *is denoted a fundamental solution to the system of equations* (4), *provided that*

$$\Delta_x \Phi_n(x,y) - \sum_{m=0}^{n} \beta_{n-m}\Phi_m(x,y) = \delta(x-y).$$

Theorem 3 *The sequence of functions*

$$\Phi_n(x,y) = \frac{1}{2\pi}\{K_0(\gamma|x-y|)v_n(|x-y|) + K_1(\gamma|x-y|)w_n(|x-y|)\},$$

for $n = 0,\dots,N-1$, *is a fundamental solution of* (4) *in the sense of Definition* 2. *Here,* K_0 *and* K_1 *are modified Bessel functions of the second kind, and* v_n, w_n *are known polynomials, with coefficients* a_{nm} *that can be recursively evaluated via* γ *and* β_n *(see* [2]*).*

We note that in [5] another representation of Φ_n can be found. To handle a Neumann condition on the boundary of the semi-infinite region we need a sequence of Green functions.

Definition 4 *The functions* N_n *for* $n = 0,\dots,N-1$, *are called a Green functions sequence for the Neumann problem for the system of elliptic equations* (4) *in the domain* D_0, *if*

- *For every* $x, y \in D_0$, $N_n(x,y)$ *is fundamental solution according to the Definition* 2.
- *For every* $x \in D_0$ *and for every* $y \in \Gamma_0$ *holds* $\frac{\partial N_n}{\partial \nu(y)}(x,y) = 0$.

Theorem 5 *The solution u_n of the Neumann problem for the sequence of elliptic equations in D_0 can be expressed as*

$$u_n(x) = \sum_{m=0}^{n} \int_{\Gamma_0} \left[\frac{\partial u_m}{\partial \nu}(y) - \frac{\partial u_{m-1}}{\partial \nu}(y)\right] N_{n-m}(x,y)ds(y).$$

Proof. The idea of the proof consists of applying the analogue of the second Green formula to $u_n(y)$ and $N_n(\cdot, y)$, and using properties of N_n.

Theorem 6 *The Green function for the Neumann problem of the sequence of elliptic equations can be written as*

$$N_n(x,y) = \Phi_n(x,y) + \phi_n(x,y),$$

where Φ_n is fundamental solution, and ϕ_n is determined from the boundary value problem

$$\Delta_y \phi_n(x,y) - \sum_{m=0}^{n} \beta_{n-m}\phi_m(x,y) = 0 \quad \text{in } D_0,$$
$$\frac{\partial \phi_n(x,y)}{\partial \nu(y)} = -\frac{\partial \Phi_n(x,y)}{\partial \nu(y)} \quad \text{on } \Gamma_0.$$

Let us write the solution u_n to the mixed problem (4)-(5) in the form

$$u_n(x) = \sum_{m=0}^{n} \int_{\Gamma_1} \varphi_m(y) N_{n-m}(x,y)ds(y) + \omega_n(x), \tag{6}$$

where φ_n are unknown densities and

$$\omega_n(x) = \sum_{m=0}^{n} \int_{\Gamma_0} \left[g_{0m}(y) - g_{0,m-1}(y)\right] N_{n-m}(x,y)ds(y).$$

Using Theorem 5 in combination with the behaviour of the single-layer potential in $\mathbf{R}^2$ on the boundary, we obtain sequence of integral equations,

$$\int_{\Gamma_1} \varphi_n(y) N_0(x,y)ds(y) = g_{1n}(x) - \omega_n(x) - \sum_{m=0}^{n-1} \int_{\Gamma_1} \varphi_m(y) N_{n-m}(x,y)ds(y), \; x \in \Gamma_1. \tag{7}$$

The integral equations (7) are well-posed in the corresponding Hölder spaces or Sobolev spaces [2, 4].

To further explain the numerical method, we assume that Γ_1 have the parametric representation $\Gamma_1 = \{x(s) = (x_1(s), x_2(s)), s \in [0, 2\pi]\}$. Applying this parametrisation and additive extraction of the logarithmic singularity in the kernels of the integral operators in (7) lead to a sequence of equations

$$\frac{1}{2\pi} \int_0^{2\pi} \mu_n(\sigma) \left[H_{00}(s,\sigma) \ln \frac{4}{e} \sin^2 \frac{s-\sigma}{2} + H_{01}(s,\sigma)\right] d\sigma = G_n(s), \tag{8}$$

where $\mu_n(\sigma) = \varphi_n(x(\sigma))|x'(\sigma)|$. The right-hand side has the representation

$$G_n(s) = g_{1n}(s) - \omega_n(s) -$$
$$\frac{1}{2\pi} \sum_{m=0}^{n-1} \int_0^{2\pi} \mu_m(\sigma) \left[H_{n-m,0}(s,\sigma) \ln \frac{4}{e} \sin^2 \frac{s-\sigma}{2} + H_{n-m,1}(s,\sigma)\right] d\sigma,$$

where $g_{1n}(s) = g_{1n}(x(s))$, $\omega_n(s) = \omega_n(x(s))$ and the integral operator kernels are

$$H_{n0}(s,\sigma) = -\frac{1}{2} I_0(\gamma|r(s,\sigma)|) v_n(|r(s,\sigma)|) + \frac{1}{2} I_1(\gamma|r(s,\sigma)|) w_n(|r(s,\sigma)|),$$
$$H_{n1}(s,\sigma) = H_n(s,\sigma) - H_{n0}(s,\sigma) \ln \frac{4}{e} \sin^2 \frac{s-\sigma}{2},$$

where $H_n(s,\sigma) = 2\pi N_n(x(s), x(\sigma))$, $r(s,\sigma) = x(s) - x(\sigma)$ and I_0, I_1 are the modified Bessel functions of the first kind. The function $H_{n1}(s,\sigma)$ can be continuously extended at $s = \sigma$ as

$$H_{n1}(s,s) = -\frac{1}{2}\ln\frac{e\gamma^2|x'(s)|^2}{4} - C_E + \frac{a_{n1}}{\gamma} + \phi_n(x(s), x(s)),$$

where C_E is the Euler constant, and a_{n1} is one of the coefficients of the polynomial w_n.

For the further discretization we introduce a uniform mesh $s_j = \sigma_j = \frac{j\pi}{M}$, $M \in \mathbb{N}$. The following trigonometric quadratures are used for evaluating the integrals in the equations (8)

$$\begin{aligned} \frac{1}{2\pi}\int_0^{2\pi} f(\sigma)d\sigma &\approx \frac{1}{2M}\sum_{j=0}^{2M-1} f(\sigma_j), \\ \frac{1}{2\pi}\int_0^{2\pi} f(\sigma)\ln\frac{4}{e}\sin^2\frac{s-\sigma}{2}d\sigma &\approx \frac{1}{2M}\sum_{j=0}^{2M-1} R_j(s)f(\sigma_j), \end{aligned} \tag{9}$$

where R_j are known weight functions [8]. The integrals of ω_n are evaluated using quadratures based on *sinc*-approximation and conformal mappings.

Finally, using the discrete collocation method, we obtain a system of linear algebraic equations

$$\sum_{j=0}^{2M-1} \tilde{\mu}_{nj}\left\{R_{|j-k|}H_{00}(s_k, s_j) + \frac{1}{2M}H_{01}(s_k, s_j)\right\} = G^1_{nk},$$

where $\tilde{\mu}_{nj} \approx \mu_n(s_j)$, $R_j = R_j(0)$, and G^1_{nk} is a known (computable) right-hand side.

Convergence and error analysis of this type of numerical scheme for solving the integral equations are presented in [3].

The proposed iterative regularizing method requires evaluation of the trace of the solution and its normal derivative on the boundary for the direct problem. Taking into account the behaviour of the single-layer potential on the boundary, the trace of the solution u_n to the boundary Γ_0 can be calculated via (6). To calculate the normal derivative of the solution on the boundary Γ_1 we use the jump relations for the normal derivative of the single-layer potential

$$\frac{\partial u_n}{\partial \nu}(x) = -\frac{1}{2}\sum_{m=0}^{n}\varphi_m(x) + \sum_{m=0}^{n}\int_{\Gamma_1}\varphi_m(y)\frac{\partial \Phi_{n-m}(x,y)}{\partial \nu(x)}ds(y) + \frac{\partial \omega_n(x)}{\partial \nu(x)}, \quad x \in \Gamma_1.$$

After parametrisation and additive extraction of the logarithmic singularity, we obtain

$$\begin{aligned} \frac{\partial u_n}{\partial \nu}(x(s)) = {} & \frac{1}{2\pi}\sum_{m=0}^{n}\int_0^{2\pi}\mu_m(\sigma)\left[\tilde{H}^{11}_{n-m,0}(s,\sigma)\ln\frac{4}{e}\sin^2\frac{s-\sigma}{2} + \tilde{H}^{11}_{n-m,1}(s,\sigma)\right]d\sigma \\ & -\frac{1}{2}\sum_{m=0}^{n}\frac{\mu_m(s)}{|x'(s)|} + \frac{\partial \omega_n(x(s))}{\partial \nu(x(s))}, \qquad s \in [0, 2\pi]. \end{aligned}$$

Here, the functions $\tilde{H}^{11}_{n0}$ and $\tilde{H}^{11}_{n1}$ are continuous, therefore in the numerical implementation we can use the trigonometric quadratures (9).

4 Numerical experiments

We consider a Cauchy problem in a strip D_0 with boundary $\Gamma_0 = \Gamma_0^1 \cup \Gamma_0^2$, where $\Gamma_0^1 = \{(s,0), s \in \mathbb{R}\}$, $\Gamma_0^2 = \{(s,\pi), s \in \mathbb{R}\}$, and the diamond-shaped inclusion with rounded corners D_1 is bounded by

$$\Gamma_1 = \left\{x(s) = \left(r(s)\cos\left(s - \frac{\pi}{4}\right), r(s)\sin\left(s - \frac{\pi}{4}\right) + \frac{\pi}{2}\right), s \in [0, 2\pi]\right\},$$

where radial function is $r(s) = \left\{(2\cos s)^4 + (2\sin s)^4\right\}^{-\frac{1}{4}}$. We choose the parameters $T = c = 1$, and define $u(x(s),t)|_{\Gamma_1\times(0,T)} = \sin(\pi t)$; this is the function to be numerically reconstructed. Then, given $u|_{\Gamma_1\times(0,T)}$ we choose $f_2 = 0$ and generate f_1 as the solution of the corresponding direct Dirichlet-Neumann problem. To avoid the inverse crime, random 5% noise is introduced in value of f_1. Discretization parameters for solving direct problem are choosen as $N = 9$, $M = 32$, $M_\infty = 100$, and the relaxation parameter is $\gamma_L = 2.0$. Results for the reconstruction using the proposed iterative procedure are presented in Fig. 2. Absolute error observed there is $\|u - u_{2567}\|_{L^2(\Gamma_1\times(0,T))} \approx 0.1919$.

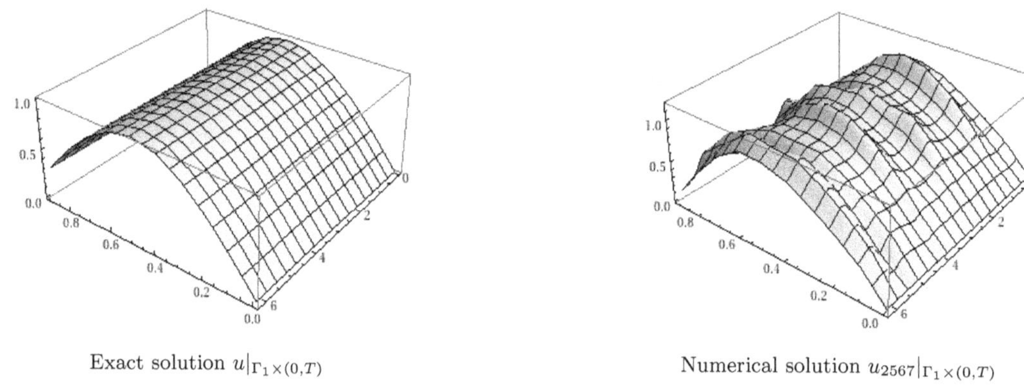

Exact solution $u|_{\Gamma_1\times(0,T)}$ Numerical solution $u_{2567}|_{\Gamma_1\times(0,T)}$

Figure 2: Numerical results in the case of a strip

As we can see, given the introduced level of noise in the input data, the procedure finds a reasonable accurate and stable reconstruction of the boundary data on Γ_1. Due to the effective implementation of solving the direct problems, it is possible to perform large number of iterations in the Landweber procedure with small computational effort.

References

[1] Bastay, G., Kozlov, V. A. and Turesson, B. O., "Iterative methods for an inverse heat conduction problem," J. Inverse Ill-posed Probl.- 9, 2001.- P. 375-388.

[2] Chapko, R. and Kress, R., "Rothe's Method for the Heat Equation and Boundary Integral Equations," J. Integral Equations Appl.- 9, 1997.- P. 47-69.

[3] Chapko, R. and Kress, R., "On a quadrature method for a logarithmic integral equation of the first kind," Agarwal, ed., World Scientific Series in Applicable Analysis. Contributions in Numerical Mathematics, Vol. 2 (World Scientific, Singapore, 1993) 127–140.

[4] Chapko, R. S. and Vavrychuk, V. G., "On the numerical solution of a mixed initial boundary value problem for the heat equation in a double-connected planar domain," J. Numer. Appl. Math.- 97, 2009.- P. 26-38.

[5] Gavrilyuk I.P., Makarov V.L. "An explicit boundary integral representation of the solution of the two-dimensional heat equation and its discretization," J. Integral Equations Appl. 2000. Vol. 12. P.63–83.

[6] Engl, H. W., Hanke, M., and Neubauer, A., Regularization of Inverse Problems, Kluwer, Dordrecht, 1996.

[7] Johansson, B. T., "An iterative method for a Cauchy problem for the heat equation," IMA J. Appl. Math.- 71, 2006.- P. 262–286.

[8] Kress R., Linear Integral Equations, 2nd. ed., Springer-Verlag, Heidelberg 1999.

Image Reconstruction with $p(x)$-Parabolic Equations

George Baravdish and Olof Svensson

Department of Science and Technology, Linköpings University,
SE 601 74 Norrköping, SWEDEN
Email: george.baravdish@liu.se, olof.svensson@liu.se

Abstract

In this paper we study the $p(x)$-parabolic equation to enhance images. We propose to consider the inverse problem for the $p(x)$-parabolic equation to reconstruct the original image from a given noisy image. The purpose is twofold. Solving the inverse problem by a sequence of forward problem we obtain a smooth image which is denoised and by choosing the initial data properly we try to avoid to blur the image. Numerical results shows the feasibility of our approach applied to medical images.

1. Introduction

Image reconstruction is a fundamental task in image processing, i.e., given a degraded image, usually blurred or noisy, how to reconstruct an enhanced image.

Let $u_0 : \Omega \subset \mathbb{R}^2 \to \mathbb{R}$ be an observed image of an original image u describing a real scene. We assume that the degraded image u_0 is the sum $Au(x) + n(x)$ of the original image blurred by a smooth operator A and noise $n(x)$. We also assume that the noise $n(x)$ has a zero mean and an estimated variance. To solve the original image $u = A^{-1}(u_0 - n)$ is an ill-posed problem, since A^{-1} need not to exist or might be impossible to compute. In the case of denoising, i.e. A is the identity, one tries to minimize the following energy functional

$$J(u) = \int_\Omega \left(\frac{1}{2}(u - u_0)^2 + \lambda \Phi(|\nabla u|) \right) dx, \qquad \lambda > 0, \tag{1}$$

where $\Phi(\cdot)$ is a strict convex function. In the energy functional J, the first term measures the fidelity to the noisy image and the second term imposes a smotthness condition on the clean image u.The Euler-Lagrange equation associated with the energy functional J is the partial differential equation (PDE) $\frac{\partial J}{\partial u} = 0$ with Neumann boundary condition zero on the boundary of Ω.

By introducing a "time" variable, we instead solve a time dependent PDE $u_t = -\frac{\partial J}{\partial u}$ with the noisy image u_o as initial data and with Neumann boundary condition zero on the boundary of Ω. Assume $\Phi(|\nabla u|) = |\nabla u|^{p(x)}$. In the case of constant p, for $p = 2$ we get the heat equation and for $p = 1$ we get the total variation approach of ROF. Running the time dependent PDEs forward in time gives a cleaner image than the initial noisy image u_0. Thus, we need a stopping time T to stop the process.

In this paper, we assume that the given degraded image u_0 is given at a later time $t = T$ and we seek for the original image u at time $t = 0$. This is an inverse problem and is backward in time. We solve it by considering a sequence of well-posed problems. Each problem is the forward version where the initial data is chosen properly to reconstruct the original image u. The aim of this approach is twofold. By solving the PDE forward in time we get a smoother image which is cleaner and the choice of the initial data is such that the image is also get deblurred.

2. Variational formulation

Let $u : \Omega \subset \mathbb{R}^2 \to \mathbb{R}$ be an original image describing a real scene and let u_0 be an observed image of the same scene. Furhermore, we assume that u_0 is a degradation of u where $u_0 - u$ is a perturbation (noise) causing oscillations in the image. Our goal is to remove these oscillations while preserving essential details such as edges.

We search for u that best fits the data u_0 and such that the oscillations are removed (i.e. the magnitude of the gradient is small).

Consider the variational problem

(2)

$$\inf_u J(u)$$

where

$$J(u) = \int_\Omega \left(\frac{1}{2}(u-u_0)^2 + \lambda\Phi(|\nabla u|)\right)dx, \qquad \lambda > 0, \tag{3}$$

where Φ is a strictly convex nondecreasing function. A unique solution to Eq. (2) is given by Euler-Lagrange equation:

$$u - u_0 - \lambda\,\mathrm{div}\left(\frac{\Phi'(|\nabla u|)}{|\nabla \mathrm{u}|}\nabla u\right) = 0 \tag{4}$$

with the Neumann condition $\partial_n u = 0$ on $\partial\Omega$. We solve Eq. (4) by iterating

$$\frac{u_{k+1} - u_k}{\lambda} - \mathrm{div}\left(\frac{\Phi'(|\nabla u|)}{|\nabla \mathrm{u}|}\nabla u\right) = 0 \tag{5}$$

Let $\lambda \to 0, \lambda k \to t$, then $u_k \to u(x,t)$ solving the partial differential equation

$$\begin{cases} \partial_t u - \mathrm{div}\left(\frac{\Phi'(|\nabla u|)}{|\nabla \mathrm{u}|}\nabla u\right) = 0, & \Omega\times(0,T) \\ \partial_n u(x,t) = 0, & \partial\Omega\times(0,T) \\ u(x,0) = \psi(x), & x\epsilon\Omega \end{cases} \tag{6}$$

where $\psi = u_0$ is the degraded image.

3. A PDE framework

As we saw in the last section, images can be denoised by solving a time dependent PDE forwards starting with a noisy image. Let us put the coefficient $a(|\nabla u|) = \frac{\Phi'(|\nabla u|)}{|\nabla \mathrm{u}|}$ and consider the parabolic equation

$$\begin{cases} \partial_t u - \mathrm{div}(a(|\nabla u|)\nabla u) = 0, & \Omega\times(0,T) \\ \partial_n u(x,t) = 0, & \partial\Omega\times(0,T) \\ u(x,0) = \psi(x), & x\epsilon\Omega \end{cases} \tag{7}$$

The solution to problem Eq. (7) is assumed to be in the weak sense. Furthermore, we assume that the nonlinear coefficient $a(\cdot) \geq c > 0$. For more details, see DiBenedetto [5]. Due to the smoothness property of Eq. (7) the image $u(x,t)$ is blurred and the process should be terminated after a final time T. Thus, we need to determine a final time T. In the next chapter, we propose an approach from the inverse problem theory, which aim to denoise but not blur.

4. The p-parabolic equation

A special case which has attract attention is when $\Phi(|\nabla \mathrm{u}|) = \frac{1}{p}|\nabla u|^p, 1 \leq p < \infty$, in Eq. (3) then the equation in Eq. (7) is the p-parabolic equation defined by

$$\begin{cases} \partial_t u - \Delta_p u = 0, & \Omega\times(0,T) \\ \partial_n u(x,t) = 0, & \partial\Omega\times(0,T) \\ u(x,0) = \psi(x), & x\epsilon\Omega \end{cases} \tag{8}$$

where $\Delta_p u = \operatorname{div}(|\nabla u|^{p-2}\nabla u)$ is the p-Laplacian.

For the case $p = 2$ the equation in Eq. (8) is reduced to the linear heat equation and in the case $p = 1$ the equation is of total variation (TV), see Rudin , Osher and Fatemi [3]. These two cases are by now well studied.

5. A backward parabolic equation

Instead of solving Eq. (7), we are thinking that the noisy image is given at a later time T and we want to reconstruct the original image at earlier time, $T = 0$. We consider the inverse problem

$$\begin{cases} \partial_t u - \operatorname{div}(a(|\nabla u|)\nabla u) = 0, & \Omega\times(0,T) \\ \partial_n u(x,t) = 0, & \partial\Omega\times(0,T) \\ u(x,0) = \varphi(x), & x\epsilon\Omega \end{cases} \tag{9}$$

where u and φ are unknown and an additional data

$$u(x,T) = \psi \tag{10}$$

is given. Observe that φ is the original image u and ψ is the given noisy image. There are by now several methods of solving the inverse problem Eq. (9). One of them is the nonlinear Landweber method. The PDE problem in Eq. (9) can be reduced to a nonlinear operator equation. In other word, there exists a nonlinear operator A such that at time T, the equation $u(x,T) = \psi$ can be written as

$$A(\varphi) = \psi. \tag{11}$$

The nonlinear Landweber method for solving the operator equation Eq. (11) is

$$\varphi_{k+1} = \varphi_k - A'^*(A\varphi_k - \psi), \tag{12}$$

where φ s arbitrary and A'^* is the adjoint Fréchet derivative of A. For the convergence rates and results on this method, see Engl, Hanke and Neubauer [2].

6. The backward p-parabolic equation

In this paper we will consider the case where $a(|\nabla u|) = |\nabla u|^{p-2}$. Since in this case $a(0) = 0$, we define the modified p-Laplacian $\Delta_{\varepsilon,p}$ by

$$\Delta_{\varepsilon,p} u = \operatorname{div}((|\nabla u|^{p-2} + \varepsilon)\nabla u), \tag{13}$$

where $\varepsilon > 0$ is a small fix number. Hence, we consider the following inverse problem

$$\begin{cases} \partial_t u - \Delta_{\varepsilon,p} u = 0, & \Omega\times(0,T) \\ \partial_n u(x,t) = 0, & \partial\Omega\times(0,T) \\ u(x,T) = \psi(x), & x\epsilon\Omega. \end{cases} \tag{14}$$

In the next section we propose an iterative method to solve the backward p-parabolic equation Eq. (14).

7. The iterative method

To solve the problem Eq. (14), we propose the following iterative method. Let $\varphi_0 \in L^2(\Omega)$ be arbitrary. Assume that u_k has been constructed. Then we proceed to solve the linear adjoint problem

(15)

$$\begin{cases} \partial_t v_k + \operatorname{div}(L_\varepsilon|\nabla u_k|)\nabla v_k) = 0, & \Omega\times(0,T) \\ \partial_n v_k(x,t) = 0, & \partial\Omega\times(0,T) \\ v_k(x,T) = u_k(x,T) - \psi(x), & x\epsilon\Omega \end{cases}$$

where

$$L_\varepsilon(|\nabla u|) = \varepsilon I + |\nabla u|^{p-2} I + (p-2)|\nabla u|^{p-4}\nabla u(\nabla u)^t, \tag{16}$$

and for $k+1$ solve

$$\begin{cases} \partial_t u_{k+1} - \Delta_{\varepsilon,p} u_{k+1} = 0, & \Omega\times(0,T) \\ \partial_n u_{k+1}(x,t) = 0, & \partial\Omega\times(0,T) \\ u_{k+1}(x,0) = u_k(x,0) - v_k(x,0), & x\epsilon\Omega \end{cases} \tag{17}$$

Notice that for obvious numerical reasons, the $|\nabla u|$ is considered as a regularized version $(|\nabla u|^2 + \delta^2)^{1/2}$, where $\delta > 0$ is some fixed small number. This scheme is in analogue with the one given in Eq. (12).

8. Unknown p and T

As in the case of the forward approach in Section 3, the final time T is unknown and has to be determined. Moreover, for the approach in Section 6, besides the unknown T, the exponent p is also unknown. One way to determine the parameters p and T is to minimize the error obtained after the first iteration in the iterative method in Section 7. This error is measured by the function

$$E(p,T) = \int_\Omega \left(\left(u_1(x,T) - u_0(x)\right)^2 + \frac{\mu}{p}|\nabla u_1(x,T)|^p \right) dx, \qquad \mu > 0, \qquad 1 < p < \infty \tag{18}$$

where $u_1(x,t)$ is the first approximate solution to Eq. (17) and $\mu > 0$ is some fixed number. Table 1 shows the values of the function E for some values of T and p, where T varies from 0.5 to 2, and p varies from 1.7 to 2:

0	0,5	1	1,5	2
1,7	157,90476	140,68862	134,62435	134,5067
1,775	150,79804	135,60408	133,39202	137,18838
1,85	144,00108	132,86536	135,61896	143,83651
1,925	138,25723	133,13139	141,62277	154,44921
2	134,28629	136,79145	151,39242	168,75129

Table 1

We see that we obtained the minima of E at p = 1.85 and T =1. We use these values to improve the degraded image u_0 shown in Fig. 1 in Section 10. The result after 5 iteration is shown in Fig. 2.

9. Variable exponent $p(x)$

Here, we allow the exponent to vary and let $p(x) = p(|\nabla u|)$. Then, the functional to minimize is given by

$$J(u) = \int_\Omega \left(\frac{1}{2}(u - u_0)^2 + |\nabla u|^{p(|\nabla u|)} \right) dx, \qquad \lambda > 0. \tag{18}$$

The associated inverse problem is then

$$
\begin{cases}
\partial_t u - \operatorname{div}\left(\frac{\Phi'(|\nabla u|)}{|\nabla u|}\nabla u\right) = 0, & \Omega\times(0,T) \\
\partial_n u(x,t) = 0, & \partial\Omega\times(0,T) \\
u(x,0) = \psi(x), & x\epsilon\Omega
\end{cases}
\tag{19}
$$

where $\psi = u_0$ and

$$
\Phi'(|\nabla u|) = p(|\nabla u|)|\nabla u|^{p(|\nabla u|)-1} + \log(|\nabla u|)p'(|\nabla u|)|\nabla u|^{p(|\nabla u|)}. \tag{20}
$$

For simplicity of notation let $s = |\nabla u|$. The matrix coefficient in the adjoint linear problem to Eq. (19) is given by

$$
\begin{aligned}
L_\varepsilon(s) = \varepsilon I + \tfrac{\Phi'(s)}{s} I + s^{p-4}\{(p - 1 + sp' \log s)(p + sp' \log s + 2p' + p' \log s + sp'' \log s) - \\
-(p + sp' \log s)\}\nabla u(\nabla u)^t
\end{aligned}
\tag{21}
$$

where I is the identity matrix. Some choices of p which satisfies the interesting case, $1 \le p(s) \le 2$ such that $p(s) \to 1$ as $s \to \infty$ and $p(s) \to 2$ as $s \to 0$, are $p(s) = 1 + \frac{1}{1+s}$ and $p(s) = 1 + e^{-s}$. In Fig. 4 the result after one iteration is shown for the case $p(s) = 1 + e^{-s}$ and $T = 1$.

10. Numerical results

The resulting images in Fig. 2 and 3 are notably sharper than the damaged image in Fig. 1. We can also compare the forward approach in Section 4 and the inverse approach in section 5. It is obvious that the forward approach in Fig. 4 gives a more blurred image then the images from the inverse approach in Fig. 2 and 3. This is due to the construction in the method for solving the inverse problem, since we are solving the inverse problem by a sequence of forward problems we obtain smooth images that are denoised and by choosing properly the initial data we get less blurred images as result. We believe this is an idea to investigate further.

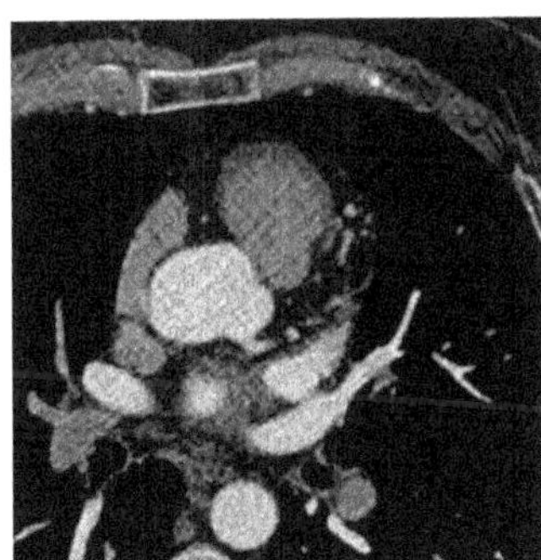

Figure 1: Degraded image

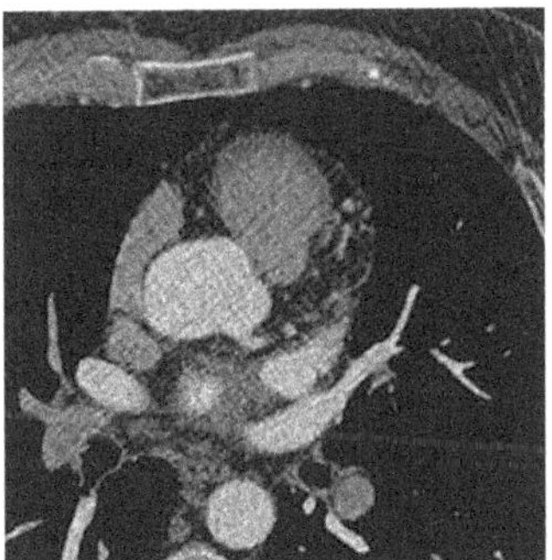

Figure 2: Image improved with inverse p-parabolic

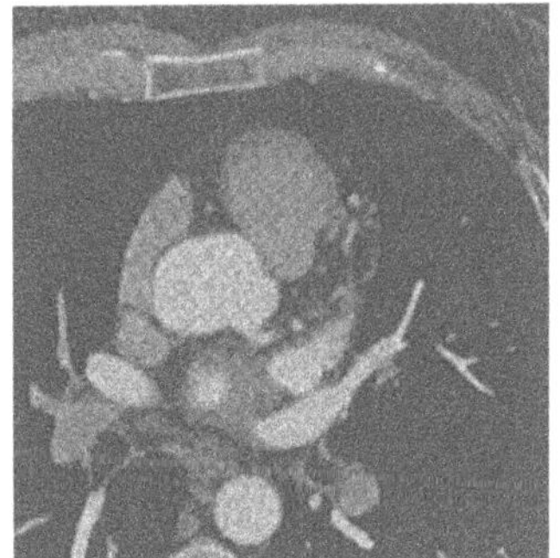

Figure 3: Image improved with inverse p(x)-parabolic

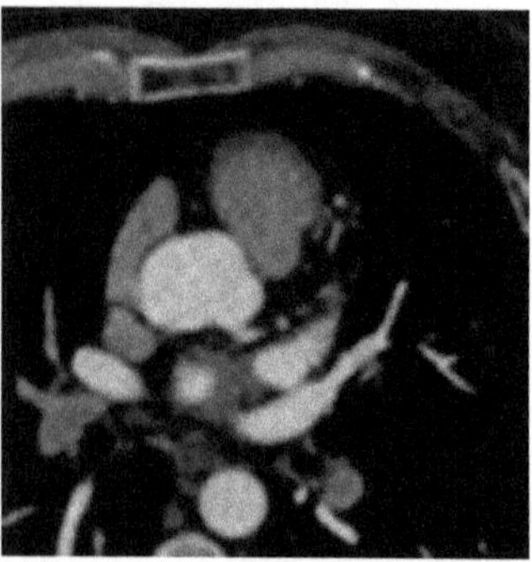

Figure 4: Image improved with forward p-parabolic

The authors would like to thank Professor Anders Ynnerman head of the division Media and Information Technology at Linköping University for providing them with medical images.

References

[1] L. Alvarez, P.-L. Lions, and J.-M. Morel, "Image selective smoothing and edge detection by nonlinear diffusion II", SIAM J. Numer. Anal., 29, pp. 845-866, 1992.

[2] H. W. Engl, M. Hanke and A. Neubauer, "Regularization of inverse problems", Kluwer Academic Publishers, Netherlands, 2000.

[3] L. Rudin , S. Osher and E. Fatemi, "Nonlinear Total Variation Based Noise Removal Algorithms", Physics D, 60: pp. 259-268, 1992.

[4] Y. Chen, S. Levine, and M. Rao, "Variable exponent, linear growth functionals in image restoration", SIAM J. Appl. Math., 66, pp. 1383-1406, 2006.

[5] E. DiBenedetto, " Degenerate parabolic equations", Springer-Verlag, New York, 1993.

[6] P. Harjulehto and P. Hästö, "An overview of variable exponent Lebesgue and Sobolev spaces", in Future trends in geometric function theory, vol. 92 of Rep. Univ. Jyväskylä Dep. Math. Stat., Univ. Jyväskylä, pp. 85-93, 2003.

[7] O. Kováčik and J. Rákosník, "On spaces $L^{p(x)}$ and $W^{k,p(x)}$", Czechoslovak Math. J., 41(116) , pp. 592-618, 1991.

[8] S. Antontsev and S. Shmarev, "Parabolic equations with anisotropic nonstandard growth conditions", in Internat. Ser. Numer. Math., vol. 154, Birkhäuser, Verlag Basel/Switzerland, pp. 33-44, 2006.

[9] S. Antontsev and S. Shmarev, "Existence and uniqueness of solutions of degenerate parabolic equations with variable exponents of nonlinearity", Fundamental and Applied Mathematics, 12 (4) pp. 3-19, 2006.

Using of metaheuristics methods for solving inverse problems in liquid chromatography

Reynier Hernández Torres[1], Mirtha Irizar Mesa[1], Leoncio Diogenes T. Camara[2], Antonio Jose Silva Neto[2], Orestes Llanes Santiago[1]

[1]Departamento de Automática y Computación. Facultad de Eléctrica, ISPJAE, Cuba. (reynier, mirtha, orestes@electrica.cujae.edu.cu)

[2]Departamento de Ingeniería Mecánica y Energía. Instituto Politécnico de la Universidad del Estado de Río de Janeiro, IPRJ-UERJ. Nova Friburgo, Brasil. (ajsneto, dcamara@iprj.uerj.br)

Abstract

In chromatography, complex inverse problems related to the parameters estimation and process optimization are presented.

Metaheuristics methods are known as general purpose approximated algorithms which seek and hopefully find good solutions at a reasonable computational cost. These methods are iterative process to perform a robust search of a solution space.

In this work, the effectiveness of two metaheuristics methods is investigated to satisfy the requirements of the parameter estimation in liquid chromatography. A comparison based on different criteria is made in order to select the most adequate method under determined economical and design conditions.

Key Words: Chromatography, Metaheuristics methods, Inverse problems, Parameters estimation

1 Introduction

Liquid chromatography (LC) is a common separation method very important in chemistry, pharmaceutical and biotechnological industries. It is used to isolate one or more compounds in a mixture. In LC, a sample of molecules are injected into a column of adsorbing porous material (stationary phase), frequently silica, and a liquid (mobile phase) is pumped through the column, and the different kinds of molecules are distributed differently between the two phases. [1-3]

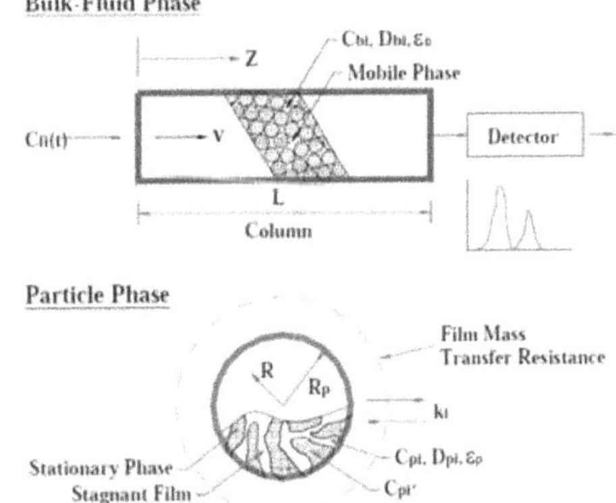

Fig.1 Anatomy of a chromatographic column.[4]

In LC, numerical values of several parameters have to be well estimated. Since geometrical and physical characteristics of the column and of the absorbent particles are usually known, other parameters like adsorption equilibrium, mass transport and axial mixing parameters have to be estimated, usually through experimental measurements.

Parameters like film mass transfer coefficient, axial dispersion coefficient and effective diffusivity are often not available from literature, or not easily measured by experiments. Nevertheless, they can be estimated with certain accuracy. Opportunely, rate models are not very sensitive to mass transfer parameters. Errors up to a certain degree do not affect the outcome largely.[4,5]

2 Mathematical Modeling of LC

Consider a fixed-bed adsorption column piqued with uniform porous, spherical and solid adsorbents. The process is supposed isothermal and there is no concentration gradient in the radial direction of the column. Another assumption is that there exist local equilibrium for each component between the pore surface and the liquid phase in the macropores inside particles.

Under these suppositions, two differential equations are used to modeling the process of chromatography in this work.[4,6]

2.1 Continuity Equation in the Flowing Mobile Phase

The flowing mobile phase equation is:

$$-D_{bi}\frac{\partial^2 C_{bi}}{\partial Z^2} + v\frac{\partial C_{bi}}{\partial Z} + \frac{\partial C_{bi}}{\partial t} + \frac{(1-\varepsilon_b)}{\varepsilon_b}\frac{3k_i}{R_p}\left(C_{bi} - C_{pi,R=R_p}\right) = 0 \quad (1)$$

where C_{bi} is the bulk-fluid phase concentration of component i, D_{bi} is the axial dispersion coefficient of component i, Z is the axial coordinate, v is the interstitial velocity, ε_b is the bed void volume fraction, k_i is the film mass transfer coefficient of component i and R_p is the particle radius.

2.2 Continuity Equation inside the Macropores

The particle phase equation in spherical coordinates is:

$$\left(1-\varepsilon_p\right)\frac{\partial C_{pi}^*}{\partial t} + \varepsilon_p\frac{\partial C_{pi}}{\partial t} + \varepsilon_p D_{pi}\left[\frac{1}{R^2}\frac{\partial}{\partial R}\left(R^2\frac{\partial C_{pi}}{\partial R}\right)\right] = 0 \quad (2)$$

whereC_{pi} is the concentration of component i in the stagnant fluid phase inside the particle macropores, C_{pi}^* is the concentration of component i in the solid phase of particle (based on unit volume of particle skeleton), ε_p is the particle porosity, D_{pi} is the effective diffusivity of component i, porosity not included and R is the radial coordinate for particle.

The column is initially equilibrated and the initial state is the following:

$$t = 0,\ C_{bi} = C_{bi}(0,Z), C_{pi} = C_{pi}(0,R,Z) \quad (3,4)$$

$$Z = 0,\ \frac{\partial C_{bi}}{\partial Z} = \frac{v}{D_{bi}}\left(C_{bi} - C_{fi}(t)\right) \quad (5)$$

$$Z = L,\ \frac{\partial C_{bi}}{\partial Z} = 0 \quad (6)$$

$$R = 0,\ \frac{\partial C_{pi}}{\partial R} = 0 \quad (7)$$

$$R = R_p,\ \frac{\partial C_{pi}}{\partial R} = \frac{k_i}{\varepsilon_p D_{pi}}\left(C_{bi} - C_{pi,R=R_p}\right) \quad (8)$$

2.3 Model Solution

A robust and efficient numerical procedure has been developed by [6](and well used in [4]) to solution the system. First, the model is converted to a dimensionless model. Later, the spatial axes, z and r, are discretized.

The bulk-fluid phase equation is discretized using the finite elements method (FE) and the particle equation using orthogonal collocation method (OC).

2.4 Solution to the ODE system

Finally, if N_z quadratic nodes are used for z-axis in the bulk fluid phase equation and N_r interior OC points are used for the r-axis in the particle phase equation, the discretization procedure gives a total of $N_s N_z (N_r+1)$ equations that are solved simultaneously by any of the stiff ordinary differential equations solver (*ode15s* is used in the simulation).

3 Parameter Estimation

The proposed technique assumes a first principle model structure where some of the parameters are unknown.

The identification objective consists of finding a set of parameters ($\hat{\zeta}$) which minimizes differences between the real response vector of the process ($y(t)$) and the model output vector ($\hat{y}(t)$). A cost function for a minimization in a time interval$[0, t_e)$, known as Sum of Squared Errors, is:

$$J(\zeta) = \sum_{j=1}^{N} \left(y_i(j) - \hat{y}_i(j) \right)^2 \tag{9}$$

where N is the number of samples, $y_i(j)$the component i of vector $y(j)$ and $\hat{y}_i(j)$ the component i of vector $\hat{y}(j)$.

Once the cost function is established, it is necessary to choose an optimization technique. For complicated models (nonlinearities, saturation, high order, etc.) the optimization problem can be very complicated, and then a powerful optimization technique is required.

In this paper, genetic algorithms and particle swarm optimization are used as possible techniques for the role of optimizers in the parameter estimation.

3.1 Parameter Estimation in Adsorption Models

In chromatography models, the objective is to estimate the parameters related with mass transfer as k_i,D_{bi} and D_{pi} and particle porosity (ε_p). Component concentration in the liquid phase (c_b) is the variable to be simulated.

In this case, the codification of the parameters is $\langle k_i D_{bi} D_{pi} \varepsilon_p \rangle$.

4 Metaheuristics methods

4.1 Genetic Algorithms

Genetic algorithms method (GA) is an optimization technique based on the principles of genetics and natural selection. It was pioneered by J. Holland and his collaborators in the 1960s and 1970s [7]. This technique is based on applying natural selection laws onto a population to achieve individuals that are better adjusted to their environment.

A population is a set of points in the search space. Each individual of the population represents a point in that space by means of his chromosomes. The adaptation degree of the individuals is given by the objective function.

The evolution mechanism of individuals is achieved by genetic operators. The usual operators are:

- Selection: its main goal consists of selecting the chromosomes to integrate the next population (these would depend on the cost function for each individual).

- Crossover: new individuals are generated and integrated by combining the chromosomes of two individuals.
- Mutation: randomly varying of some part of the chromosome of an individual in the population generates new individuals.

GA implementation have some inherent parameters such as size population, number of generations, crossover and mutation probability. There are others parameters that have to be determined before the operation, but definitive general approaches do not exist. These algorithms should work in a wide interval of their parameters, but with differences in the efficiency. Another aspect to consider in genetic algorithms is the fitness function, which offers information about the quality from the possible solutions to a problem. Execution parameters and fitness function define the genetic algorithms completely.

4.2 Particle Swarm Optimization

In Particle Swarm Optimization (PSO), simple agents, called particles, move in the search space of an optimization problem. PSO was developed by Kennedy and Eberhart in 1995.[8]

The position of a particle represents a candidate solution to the optimization problem. Each particle searches for better positions in the search space by changing its velocity according to rules originally inspired by behavioral models of bird flocking or fish schooling.

Unlike other population-based methods, PSO does not resample populations to produce new ones; there is not any kind of selection.

Each particle consists of two parts: the location (x) in space and the velocity (v), which is the speed and direction of the particle, each time step. Initially, each particle starts at a random location and with a random velocity. Also, it is necessary keep track of the fittest known location that a particle discovered so far (x^*) and the fittest known location that has been discovered by anyone so far $(x^!)$.

Each time step, these operations are performed: assess the fitness of each particle and update the best discovered locations; determine how to change each particle. Its velocity is updated by adding in, to some degree, vectors pointing towards x^* and $x^!$, respectively; and change the position of each particle by moving it along its velocity vector.

The update equations are $v_{t+1} = \alpha v_t + c(x^* - x_t) + s(x^! - x_t)$ and $x_{t+1} = x_t + v_{t+1}$, where v_{t+1} is the updated velocity, v_t is the current velocity, x_{t+1} is the updated position, x_t is the current position, α is the inertia weight, c is the cognitive acceleration coefficient and s is the social acceleration coefficient.

5 Results and Discussion

All calculations were performed on a PC with a 1.67 GHz Intel Core 2 Duo processor and 1 GB of internal memory. The solver and estimation routines were implemented in MATLAB 7.6 (R2008a), over Windows 7.

In the GA implemented, a tournament selection scheme by two individuals was used. A uniform crossover operator is applied with a probability of 0.95, and a uniform mutation operator where each individual has a probability 0.05 of being mutated. Population size is 20.

In the PSO algorithm implemented, swarm size (number of particles) is 10, the inertia weight is 1.0, the cognitive acceleration coefficient is 2.8 and the social acceleration coefficient is 1.3.

The maximum number of iterations was 100. Stop conditions were set to 20 iterations without improvements and a tolerance below or equal to 10^{-5}.

For simulation, synthetic data were generated running the direct solution of the model. Parameters values were estimated in ten runs of algorithms with equal operation characteristics. The results are presented in Table 1 and Table 2. In figures 2 to 4 is shown breakthrough curves resulting for average estimated values.

Table 1.Best, worst and average estimated values with noiseless data.

	k_i	D_{bi}	D_{pi}	ε_p	**Fitness function**	**Stall iterations**
Exact	0.01670	0.000200	0.009600	0.400	0	
			GA			
Best parameter set	0.01699	0.000228	0.008819	0.411	$4.339 \cdot 10^{-4}$	18 of 21
Worst parameter set	0.01903	0.000306	0.009656	0.381	$4.884 \cdot 10^{-3}$	17 of 21
Average	0.01798	0.000218	0.009203	0.401	$3.170 \cdot 10^{-4}$	
			PSO			
Best parameter set	0.01601	0.000186	0.009646	0.400	$3.976 \cdot 10^{-6}$	3 of 41
Worst parameter set	0.01679	0.000168	0.009165	0.402	$3.178 \cdot 10^{-5}$	5 of 36
Average	0.01740	0.000192	0.009347	0.401	$5.437 \cdot 10^{-5}$	

Table 2.Best, worst and average estimated values with 2% noisy data.

	k_i	D_{bi}	D_{pi}	ε_p	**Fitness function**	**Stall iterations**
2% Noisydata	0.01670	0.000200	0.009600	0.400	$6.286 \cdot 10^{-2}$	
			GA			
Best parameter set	0,01971	0,00006909	0,00799160	0,39306	$1,881 \cdot 10^{-3}$	17 of 21
Worst parameter set	0,01316	0,00004412	0,00677520	0,54625	$1,408 \cdot 10^{-1}$	16 of 21
Average	0.01577	0.000132	0.008652	0.424	$3.576 \cdot 10^{-3}$	
			PSO			
Best parameter set	0.01503	0.000101	0.009113	0.395	$5.319 \cdot 10^{-4}$	4 of 40
Worst parameter set	0.01643	0.000137	0.009158	0.390	$1.036 \cdot 10^{-3}$	10 of 56
Average	0.01609	0.000160	0.009496	0.392	$6.757 \cdot 10^{-4}$	

6 Conclusions

In this work, the parameter estimation for liquid chromatography process was made. Two metaheuristics methods (GA and PSO) were used for this purpose. The PSO method showed better results, since the fitness function value obtained was lower than with the GA. Furthermore, in PSO the convergence was achieved in less time, and the number of stall iterations was lower, getting better results in the optimization process.

In further studies, other metaheuristics techniques like Ant Colony Optimization and Cuckoo Search will be used in order to search for more effective and robust techniques for parameter estimation in liquid chromatography.

7 Bibliography

[1] J.G. Dorsey, W.T. Cooper, B.A. Siles, C. William, J.P. Foley, and H.G. Barth, "Liquid Chromatography : Theory and Methodology," Analytical Chemistry, vol. 70, 1998, pp. 591-644.

[2] V. Grosfils, "Modelling and parametric estimation of simulated moving bed chromatographic processes (SMB)," PhD thesis. Universite Libre de Bruxelles. 2009. Available at: http://theses.ulb.ac.be/ETD-db/collection/available/ULBetd-06172009-155602/unrestricted/phd_thesis_VGrosfils.pdf.

[3] G. Guiochon, "Preparative liquid chromatography: Review," Journal of Chromatography A, vol. 965, 2002, pp. 129-161.

[4] C. Lazo, "Simulation of Liquid Chromatography and Simulated Moving Bed (SMB) Systems,"Studienarbeit,TechnischeUniversitat Hamburg-Harburg.1999
Available at: http://www.itap.uni-kiel.de/theo-physik/heinze/cesar/SMB.pdf.

[5] K.-U. Klatt, F. Hanisch, G. Dünnebier, and S. Engell, "Model-based Optimization and control of Chromatographic Processes," Computers & Chemical Engineering, vol. 24, 2000, pp. 1119-1126.

[6] T. Gu, G.-J. Tsai, and G.T. Tsao, "New Approach to a General Nonlinear Multicomponent Chromatography Model," AIChE Journal, vol. 36, 1990, pp. 784-788.

[7] J. Holland, Adaptation in Natural and Artificial Systems, University of Michigan Press, Ann Arbor., 1975.

[8] Kennedy, J. and Eberhart, R. C.."Particle swarm optimization", Proceedings of IEEE International Conference on Neural Networks.1995, pp. 1942–1948.Piscataway, NJ

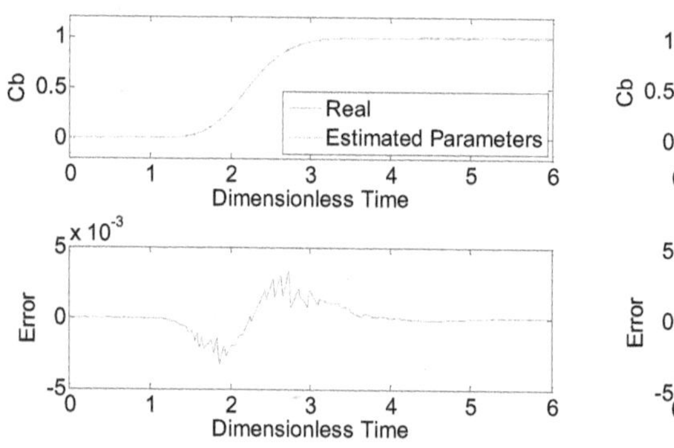

Fig.2. Breakthrough curves (up) and error (down) for real and estimated parameters with GA.

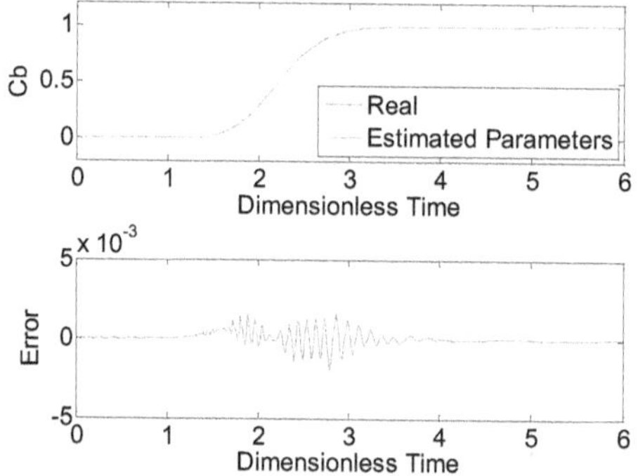

Fig.3. Breakthrough curves (up) and error (down) for real and estimated parameters with PSO.

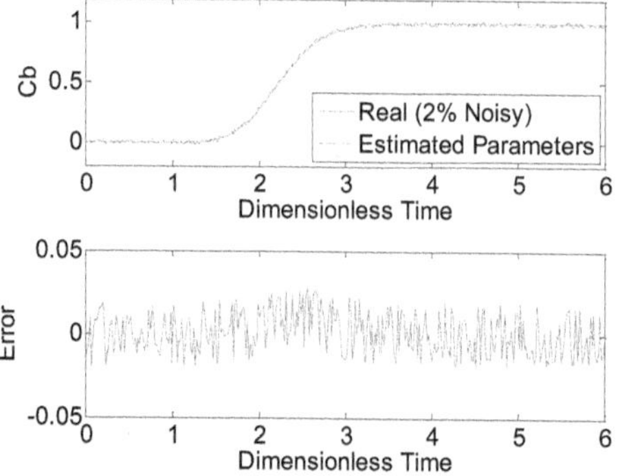

Fig.4. Breakthrough curves (up) and error (down) for real and estimated parameters with GA. (2% noisy data)

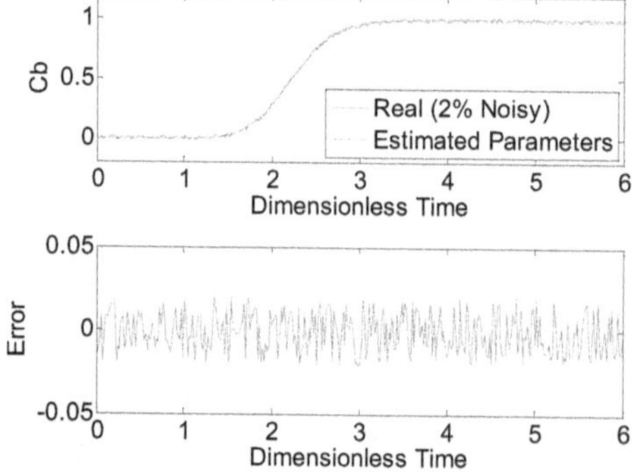

Fig.5. Breakthrough curves (up) and error (down) for real and estimated parameters with PSO.(2% noisy data)

Parameter estimation in column chromatography with RBF neural networks for biotechnological applications.

Mirtha Irizar Mesa[1], Leôncio Diogenes T. Câmara[2], Orestes Llanes Santiago[2] and Antônio J. Silva Neto[2]

[1]*Departamento de Automática y Computación*
Facultad de Eléctrica, ISPJAE, Cuba.
Email: mirtha, orestes@electrica.cujae.edu.cu)

[2]*Departamento de Ingeniería Mecánica y Energía*
Instituto Politécnico de la Universidad del Estado de Río de Janeiro, IPRJ-UERJ. Nova Friburgo, Brasil.
Email:ajsneto, dcamara@iprj.uerj.br

Abstract

Parameter estimation in chromatography models is a complex inverse problem and novel methods should be applied to obtain valid results with low computational costs. In the general mass transfer model, which describes the relevant phenomena in chromatographic columns, the coefficients of effective diffusivity and film mass transfer are difficult to estimate by means of chemical experiments.

Radial basis function neural networks (RBFNN) require a short design time and have been applied to several practical problems. In this work they are applied in the estimation of the aforementioned parameters from synthetic data of protein chromatography, leading to a better understanding of the mass transfer phenomena involved and development of new techniques to be applied at industrial level.

Key Words: mass transfer, chromatography, biotechnological processes, RBF neural networks

1. Introduction

Chromatography is a separation method which nowadays allows the production of pure chemicals in large amounts for a variety of purposes, mostly in the chemical, pharmaceutical and biotechnological industries [1]. Modeling of Chromatography allows a better understanding and development of new techniques to be applied at industrial level. Some parameters are difficult to estimate by means of chemical experiments therefore computational methods should be applied.

In previous works those methods have been developed for parameter estimation in chromatography models starting from experimental data (Fig. 1). It's a complex inverse problem [2] for this type of generally nonlinear processes and can be formulated as a problem of optimization. In those papers, local search [3, 4, 5, 6, 7, 8] or computational intelligence based methods with encouraging results are considered nevertheless more precise values are required for some parameters.

In the general mass transfer model, which describes the relevant phenomena in chromatographic columns, it's important to improve the estimation of the coefficients of effective diffusivity and film mass transfer. Artificial neural networks emulate the biological neuron and have been used successfully in many problems of diverse areas so in this work they are used with that purpose.

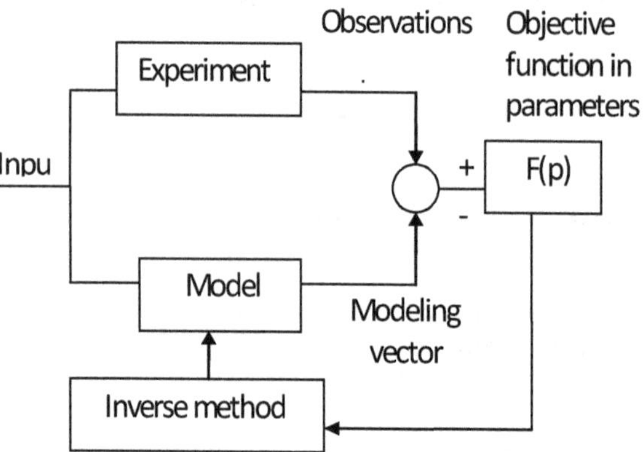

Figure 1: Parameter estimation method scheme.

In section 2 the characteristics of the general rate model for column chromatography are presented. In Section 3 the inverse problem and its solution method by means of radial basis function networks (RBFN) is reformulated and the developed experiments for parameter estimation are explained. In section 4 the discussion of results is shown and future works are addressed.

2. Modeling of column chromatography

The theory of non-linear, non-ideal chromatography has given rise to numerous models whose advantages, disadvantages and ranges of application are now well understood [8]. The general rate model attempts simultaneously to account for all the possible contributions to the mass transfer kinetics arising in chromatography. It does this by including their contributions in the system of partial differential equations which states mass conservation and transport. Because this model considers separately the stagnant mobile phase, inside the particle, and the percolating mobile phase, outside the particles, two mass balance equations are written for the solute. The level of mathematical difficulty depends much on the nature of isotherm. The equations of the model are partial differential equations which are combined with proper initial and boundary conditions.

One of these equations represents the mass balance of component i in the mobile phase, written for a fluid percolating through a bed of spherical particles of radius R_p:

$$\varepsilon_e \frac{\partial \mathbf{C_i}}{\partial \mathbf{t}} + \mathbf{u}\frac{\partial \mathbf{C_i}}{\partial \mathbf{Z}} = \varepsilon_e \mathbf{D_L}\frac{\partial^2 \mathbf{C_i}}{\partial \mathbf{Z}^2} - (\mathbf{1}-\varepsilon_e)\boldsymbol{k_{exp,i}} a_p \times \left[\boldsymbol{C_i} - \boldsymbol{C_{pi}}(\boldsymbol{r} = \boldsymbol{R_p})\right] \quad (1)$$

where C_i and $C_{p,i}$ are the concentration of component i in the mobile phase and its concentration in in the particle pores, respectively, z and t are the abscissa and time, respectively, ε_e is the external or interstitial porosity, u is the mobile phase velocity, D_L is the coefficient of axial dispersion, k_{exp} the external mass transfer coefficient, and a_p the external surface area of the adsorbent particles.

The mass balance inside the particles is written:

$$\varepsilon_p \frac{\partial C_{pi}}{\partial t} + (1-\varepsilon_p)\frac{\partial q_i}{\partial t} = D_{eff}\frac{1}{r^2}\frac{\partial}{\partial r}\left(r^2 \frac{\partial C_{pi}}{\partial r}\right) \quad (2)$$

where ε_p is the internal porosity of the particles, *q* is the concentration of the studied component in the adsorbed phase, and D_{eff} is the effective diffusion coefficient, with the following initial conditions:

$C_i(0,z) = C_i^0$, $C_{pi}(0,r,z) = C_{p,i}^0(r,z)$;

$q_i(0,r,z) = q_i^0(r,z)$; $for\ 0 < z < L;\ 0 < z < R_p$

The boundary conditions for Eq. (1) are:

for t>0 and z=0,

$$u_f C_{fi}^{'} - u(0)C(0) = -\varepsilon_e D_L \frac{\partial C_i}{\partial z}$$

$$C_{fi}^{'} = C_{fi}\ for\ 0 < t < t_p$$

$$C_{fi}^{'} = 0\ \ for\ t_p < t$$

for t>0 and z=L,

$$\frac{\partial C_i}{\partial z} = 0$$

The boundary conditions for Eq. (2) are:

for t>0 and r=R_p

$$D_{eff}\frac{\partial C_{p,i}(t,r)}{\partial r} = k_{ext,i}\left[C_i - C_{p,i}(t,r)\right]$$

and for t>0 and r=0

$$\frac{\partial C_{p,i}(t,r)}{\partial r} = 0$$

In this case the Langmuir isotherm relates the concentration q_i in Eq. (2), its mobile phase concentration and the other parameters of the system.

A robust and efficient numerical procedure for the solution of differential equations system has been developed in [10]. The bulk-fluid phase equation is discretized using the finite elements method and the particle equation using orthogonal collocation method. This procedure is applied in the current work as the direct solution that lets the synthetic data generation for the experiments, which is explained in section 4.

3. Inverse problem formulation and solution

In most of the scientific disciplines and particularly in engineering there are problems characterized by differential equations with associated initial and boundary conditions. When these problems are solved in a direct way, the result is generally a functional relationship or a system of equations, which can be used to calculate values of the dependent variable for given values of the independent variable.

The interest in the solution of problems involving the inverse solution of systems of partial differential equations has grown in recent years, with relevant applications in many different areas. This constitutes a complex problem, for which there are no universally accepted methods.

Given an applicable direct solution to a system of partial differential equations, it is possible to propose an inverse problem as a problem of optimization. An algorithm to achieve this is [11]:

- Suppose a solution to the inverse problem. This can include the supposition of an initial or boundary condition, or a typical parameter for a given problem.
- Feed the supposed condition to the direct solution of the partial differential equation system, calculating in this way values of the dependent variable *y*. Here, the output of the direct solution is a vector of values corresponding to the times in which the values of *y* are measured. This vector of solutions will be denoted as calculated and it will be represented as $\hat{y}$.
- Compare the calculated values $\hat{y}$ with the values of the dependent variable *y* measured in consistent times with those for which $\hat{y}$ was calculated.

The success of this approach is the mechanism for which the supposed condition is improved in the subsequent invocations of the first step. Optimization is the procedure to upgrade the suppositions of the conditions. The most applied function in the measure of prediction error is the sum of the square error (SSE).

$$SSE(\hat{y},\hat{\theta}) = \sum_{t_i=1}^{N_T} (y(t_i) - \hat{y}(t_i,\hat{\theta}))^2$$

where θ represents the parameters to be estimated and N_T is the total number of experimental data.

A radial basis function neural network (RBFNN) is a feedforward neural network where hidden units do not implement an activation function, but represents a radial basis function. It approximates a desired function by superposition of nonorthogonal, radially symmetric functions [12]. It has been shown that RBFNNs are universal approximators [13, 14].

RBFNN require a short design time and have been applied to several practical problems with satisfactory results so in this work their possibilities for parameter estimation in the model aforementioned are explored.

In Fig. 2 the RBFNN structure for parameter estimation is showed. The network consists of an input layer, a hidden layer, and an output layer. The input layer receives a data set $x = [x_1,\ldots, x_d]^T$ and submits it to the hidden layer. The nodes in the ouput layer correspond to the number of parameters.

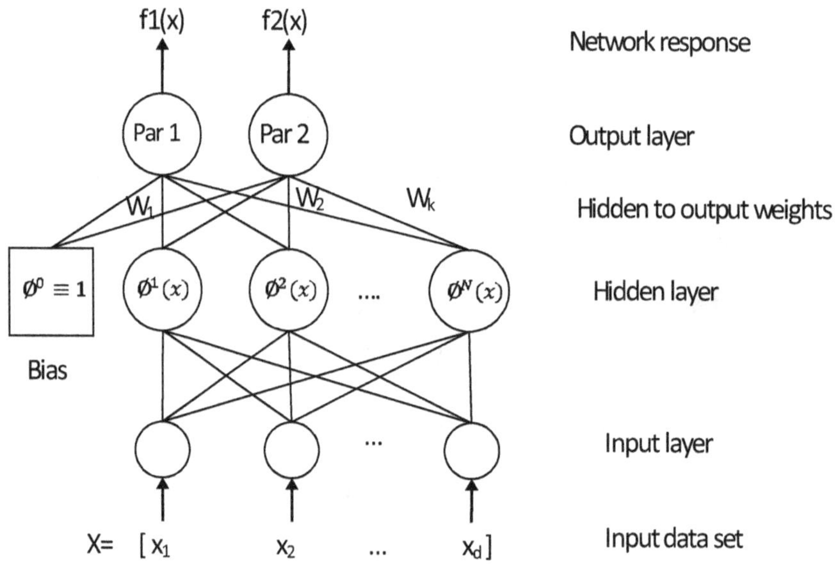

Figure 2: RBFNN structure for parameter estimation.

Nodes at the hidden layer are activated according to:

$\Phi^n(x) = \Phi(\|x - c_n\|, \sigma_n), n= 1,\ldots,N$

where $\|.\|$ denotes Euclidean distance and $\Phi^n(.)$ is a radial basis function localized around c_n with a spread constant σ_n that determines the width of an area in the input space to which each neuron responds. The radial basis function most commonly used in the neural network community is the Gaussian, with $\Phi(z, \sigma) = exp(-z^2/2\sigma^2)$.

The activations of all hidden nodes are weighted and sent to the output layer. Each output node represents a parameter and has its own hidden-to-output weights. The weighted activations of the hidden nodes are summed at each output node to produce the output for the associated parameter. Denoting $w_k = [w_{0k}, w_{1k},\ldots, w_{Nk}]^T$ as the weights connecting hidden nodes to the k-th output node, the output for the k-th parameter, in response to input x, takes the form:

$f_k(x) = w_k^T\Phi(x)$, where $\Phi(x) = [\Phi^0(x), \Phi^1(x),\ldots, \Phi^N(x)]^T$ is a column containing $N+1$ basis functions with $\Phi^0(x) \equiv 1$ a dummy basis accounting for the bias.

RBFNN training considers methods to find the best values for parameters which influence its accuracy, such as the number of basis functions used and their location, defined by the center vector, and the spread σ_n. In this case the following steps are repeated until the network's mean squared error falls below a prefixed goal:

- The network is simulated.
- The input vector with the greatest error is found.
- A neuron with a Gaussian transfer function is added with weights equal to that vector.
- The output layer weights are redesigned to minimize error.

As it was stated in section 2, to estimate the coefficients of effective diffusivity and film mass transfer, whose intervals are shown in Table1, a direct solution for the column chromatography model is used. Protein concentration in the liquid phase was selected as input to the network and a set of synthetic data was generated for its training and validation (Fig. 3). The number of chosen inputs is 35, in correspondence to the samples quantity of concentration in the mobile phase and there are 2 outputs, one for each parameter to estimate.

Table 1: Intervals for coefficients to estimate.

Parameter	Lower limit	Upper limit
D_{eff}	0	2.0 x 10^-12
k	0	2.0 x 10^-6

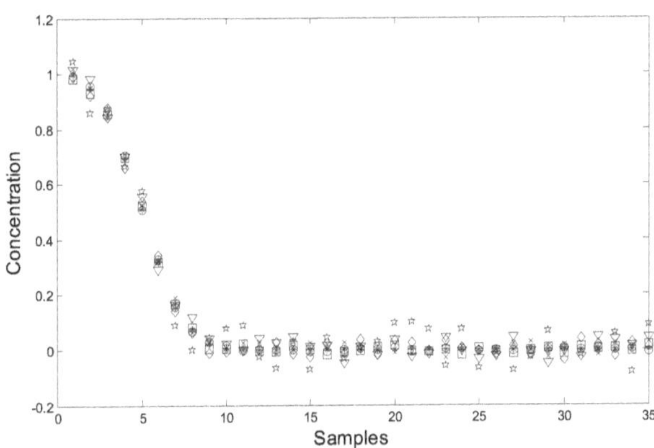

Figure 3: Training and validation data for the RBFNN.

Many experiments were carried out to solve this inverse problem, whose results are summarized in Table 2. The RBFNN was trained The first test showed was developed with noiseless synthetic data for validation and the following by adding simulated noise from 1 to 5 and 10 percent respectively. Boldface numbers correspond to D_{eff} and k value between the limits of their intervals, used for the RBFNN design.

Table 2: Tests results with different data sets.

Tests	D_{eff} estim (x 10^-12)				k estim(x 10^-6)				t(s)			
	0.50	**1.00**	**1.50**	**2.00**	**2.00**	**1.50**	**1.00**	**0.50**	**D_{eff} values**			
									0.50	**1.00**	**1.50**	**2.00**
0	0.50	1.00	1.50	2.00	2.00	1.50	1.00	0.50	0.0158	0.0158	0.0156	0.0157
1	0.49	0.99	1.49	1.99	1.99	1.49	0.99	0.49	0.0158	0.0188	0.0154	0.0155
2	0.49	0.99	1.49	1.99	1.99	1.49	0.99	0.49	0.0157	0.0155	0.0158	0.0155
3	0.49	0.99	1.48	1.98	1.98	1.48	0.99	0.49	0.0157	0.0154	0.0156	0.0156
4	0.49	0.98	1.48	1.97	1.97	1.48	0.98	0.49	0.0162	0.0155	0.0157	0.0155
5	0.49	0.97	1.46	1.95	1.96	1.46	0.97	0.48	0.0157	0.0154	0.0156	0.0156
10	0.45	0.91	1.36	1.82	1.83	1.36	0.91	0.45	0.0156	0.0155	0.0156	0.0156

Comparisons between the estimated and initial parameter values demonstrate that in all the cases the RBFNN is able to approximate both coefficients with precission in a short time, in correspondence with the target of this research.

4. Conclusions

The good results obtained in the application of radial basis neural network to estimate the coefficients of effective diffusivity and film mass transfer in this type of chromatography can contribute to the development of models and studies which let to optimize the experimental and production conditions for this type of process. New experiments with practical experimental data should be done in the near future to validate and extend this results.

5. References

[1] Guiochon, G., "Preparative liquid chromatography: Review", Journal of Chromatography A, Vol. 965, pp. 129-161, 2002.
[2] Tarantola, A., Inverse Problem Theory and Model Parameter Estimation, SIAM, 2005.
[3] Horstmann, B. J.; Chase, H. A., Modelling the affinity adsorption of immunoglobulin g to protein a immobilized to agarose matrices, Chem. Eng.Res. Des. 67, 1989.
[4] Gu, T., Mathematical Modelling and Scale-Up of Liquid Chromatography, Springer Verlag New York, 1995.
[5] Altenhöner, U., Meurer, M., Strube, J., Schmidt-Traub, H., "Parameter estimation for the simulation of liquid chromatography", Journal of Chromatography 769, 59-69, 1997.
[6] Persson, P.; Nilsson, B., "Parameter estimation of protein chromatographic processes based on breakthrough curves", *in* D.Dochain and M.Perrier, eds, *Proceedings of the 8th International Conference on Computer Applications in Biotechnology,* 2001.
[7] De Vasconcellos, J.F.V.; Silva Neto, A.J.; Santana, C.C.; Soeiro, F.J.C.P. "Parameter estimation in adsorption columns with stochastic global optimization methods", *4th International Conference on Inverse Problems in Engineering,* 2002, *Rio de Janeiro, Brazil.*
[8] De Vasconcellos, J.F.V.; Silva Neto, A.J.; and Santana, C.C. "An inverse mass transfer problem in solid-liquid adsorption systems", *Inverse Problems in Engineering* 11(5), 391-408, 2003.
[9] Guiochon, G., Shirazi, D., Felinger, A., Katti, A., Fundamentals of preparative and nonlinear chromatography, 2nd edn. Academic Press, 2006.
[10] Lazo, C., "Simulation of Liquid Chromatography and Simulated Moving Bed (SMB) Systems" Sudienarbeit Technische Universitat Hamburg-Harburg, 1999.
[11] Karr, C., Yakushin, I., Nicolosi, K., "Solving inverse initial-value, boundary-value problems via genetic algorithm", Engineering Applications of Artificial Intelligence 13, pp. 625-633, 2000.
[12] Engelbrecht, A., Computational Intelligence. An Introduction, 2nd edn. Wiley, 2007.
[13] P. Baldi. "Computing with Arrays of Bell-Shaped and Sigmoid Functions". In R.P. Lippmann, J.E. Moody, and D.S. Touretzky, editors, *Neural Information Processing Systems,* volume 3, pages 735–742, San Mateo, C.A., 1991. Morgan Kaufmann.
[14] E.F. Hartman, J.D. Keeler, and J.M. Kowalski. "Layered Neural Networks with Gaussian Hidden Units as Universal Approximators". *Neural Computation,*2(2):210–215, 1990.

Thermal characterization of thin sub-micrometric layers in liquid state

A. Kusiak [1], J.-L. Battaglia [1], A. Cappella [1,2], V. Schick [1] C. Wiemer [3], M. Longo [3], L. Lamagna[3], B. Hay [2]

[1] Laboratoire TREFLE, UMR 8508, University of Bordeaux, 33405 Talence, France
2 Scientific and Industrial Metrology Direction, Laboratoire National d'Essais, Optical Division, Bureau Nationale de Metrologie – LNE, Trappes, France
[3] Laboratorio MDM, IMM-CNR, via C. Olivetti 2, 20041 Agrate Brianza, (MB), Italy

Abstract

A method for the estimation of thermal properties of thin sub-micrometric layers in liquid state is proposed. The measurement is based on Photo Thermal Radiometry (PTR), largely used for studies of thin solid films on a substrate.

The application concerns the framework of Phase Change Memories (PCM) based on the $Ge_2Sb_2Te_5$ (GST) chalcogenide semi conducting alloy. GST is characterized by two stable solid phases: amorphous at room temperature and hexagonal close-packed (*hcp*) above 350°C; finally, it exhibits one metastable - face centered cubic (*fcc*) phase between 130°C and 350°C. The corresponding different electrical properties of each phase are used for data storage. The amorphous-to-crystalline phase change is induced by heating the GST above the crystallization temperature but below the melting temperature and the crystalline-to-amorphous state is obtained by fast quenching after GST is heated above its melting temperature, around 600°C.

An experimental configuration have been developed to address the thermal characterization of the GST in liquid phase. It rest on sub-micrometric clusters of material to be characterized, deposited on a silicon substrate using Metal Organic Chemical Vapour Deposition. The clusters are encapsulated with Al_2O_3 layer, in order to prevent their evaporation in liquid state. In the first approximation these configurations are considered as a three layers system.

The data from modulated PTR are used together with a heat transfer model in the configuration of the experiment to identify the thermal resistance of the fused material included the thermal contact resistance at interfaces.

Key Words: identification, thermal resistance, photo thermal radiometry, phase change memory, liquid state

1. Introduction

Chalcogenide phase change semi conducting alloys are widely studied due to their promising applications in the framework of Phase Change Memory (PCM) [1,2]. One of the most popular candidates is $Ge_2Sb_2Te_5$, commonly denoted GST [3]. It can be reversibly brought from the amorphous to the crystalline state, so that the corresponding different electrical properties can be used for data storage. It is stable at room temperature in the amorphous and hexagonal crystalline phases (*hcp*), and metastable in the face centered cubic phase (*fcc*) [4,5]. The phase transition temperature is ~130°C for the *fcc*-crystalline phase and ~350°C for *hcp*-crystalline phase, whereas the melting temperature is approximately 600°C. The transformation between crystalline and amorphous phases is reversible: heating the amorphous GST to a temperature above the glass-transition temperature leads to the crystalline phase; subsequent heating to a temperature above the melting temperature with fast quenching leads to retrieving the amorphous phase.

The thermal characterization of the GST at functioning temperatures is a key point for design and optimization of the final PCM. The thermal conductivity and the thermal barrier resistance of the GST solid thin films were well studied and described in the literature [4, 5]. Nevertheless, there is still lack of data concerning these parameters at liquid state, necessary for retrieving the amorphous phase.

In this work, we are interested in developing an experimental configuration that will permit the thermal characterization of chalcogenide alloys in liquid state by modulated Photo Thermal

Radiometry (PTR). In our previous work, the PTR measurements were performed on solid GST thin films (several hundreds of nanometers) deposited on silicon substrate samples [5]. Due to high temperatures, the measurements had to be carried out under high purity argon atmosphere in order to avoid any oxidation effects and evaporation of the material. Thus, it is obvious that at this scale, for PTR measurements above melting point, the studied alloy must be retained in a special sample configuration.

The main idea retained in this work, is to develop small portions (several hundreds of cubic nanometer) of material to be melted on a silicon substrate. Then, a protective layer is applied in order to prevent their evaporation. This protective layer must resist to high temperature (above the melting temperature of the studied alloy). Due to the difficulty of deposition of small portions of the GST using conventional techniques, in this work we replaced it by Ge doped tellurium. The developed samples were measured by PTR technique in order to obtain the thermal resistance of the melted tellurium.

2. Samples

The developed samples are therefore constituted of a random lattice of lightly Ge-doped tellurium hemispherical structures deposited by Metal Organic Chemical Vapour Deposition process (MOCVD) on a silicone substrate. Tellurium melts at a temperature around T_m~450°C, lower than GST, nevertheless the thermal properties of the melted tellurium at the nanoliter scale, might be useful for understanding the thermal behavior of the melted sample. These hemispheres were encapsulated by a protective layer in order to prevent their evaporation during the melting (see Fig. 1). Two types of protective layers: SiO_2 and Al_2O_3 were tested in order to find the most adapted one, ensuring the best integrity and stability during long heating. The Al_2O_3 encapsulation deposited by Atomic Layer Deposition technique appeared more stable in temperature than SiO_2. The posteriori SEM (Scanning Electron Microscopy) observations revealed that Al_2O_3 presents better integrity and resists better to thermal stress.

Finally, a 30 nm platinum capping layer was deposited by e-beam evaporation technique. This Pt layer acts as a transducer for the incident laser beam in PTR measurement. Indeed, the heat flux is absorbed by the Pt layer due to high extinction coefficient for the wavelength of the laser what is not a case for Al_2O_3. On the other hand the Pt is assumed to be isothermal for all the frequency range swept during the experiment.

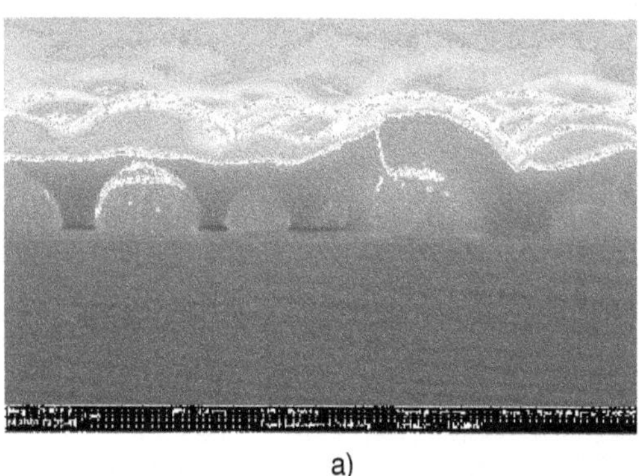

a)

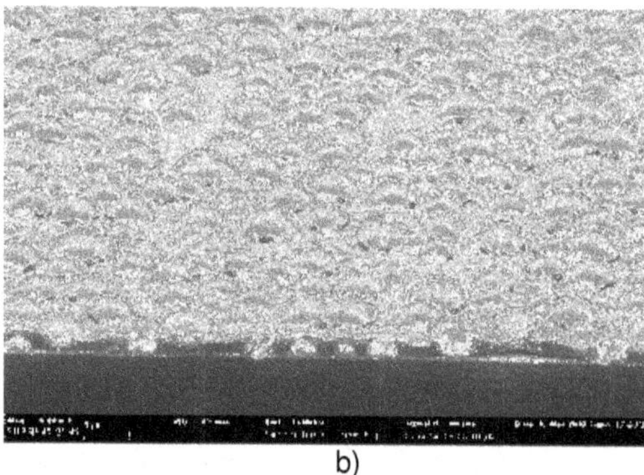

b)

Figure 1: SEM cross section of the Te- Al_2O_3 sample. The hemispherical tellurium structures deposited on silicon substrate are covered with Al_2O_3 layer. The 30 nm platinum capping layer is present on the top.

As presented in Fig 1 a), the geometrical configuration of the sample is relatively complex to describe. In the first approximation, the system can be considered as three layer system. The hemispherical bulbs of tellurium can be represented as homogenous layer with an equivalent thickness. In this way, considering the spatial frequency of Te bulbs on the sample surface and the mean value of radius of the bulbs, the volume of deposited Te can be evaluated. The rapport of this volume to the sample surface leads to the equivalent thickness of the Te layer. This evaluation is of course based on SEM images of the sample as presented in Fig 1 b). According to this approach, the sample can be assimilated as bi-layer deposit on a silicon substrate.

The thickness of the protective layer on bulbs is considered homogenous and was evaluated equal 310 nm. The thickness of the Te layer was evaluated equal 290 nm. Finally, the thickness of the silicon substrate is 0.6 mm.
It is important to notice, that the thickness of the Te layer is given here as indicative parameter. In fact, considering the thicknesses of films, the PTR measurement leads to the total thermal resistance R of the bi-layer, defined as:

$$R = R_t + R_{\mathrm{Al_2O_3}} = \underbrace{R_{\mathrm{Te}} + R_i}_{R_t} + \frac{e_{\mathrm{Al_2O_3}}}{k_{\mathrm{Al_2O_3}}} \qquad (1)$$

In this relation R_i denotes the thermal resistance at the Te interfaces, R_{Te} is the thermal resistance of the Te layer, $R_{\mathrm{Al_2O_3}}$ denotes the thermal resistance of the Al_2O_3 layer. Knowing the thermal conductivity $k_{\mathrm{Al_2O_3}}$ and the thickness $e_{\mathrm{Al_2O_3}}$ of the Al_2O_3 layer, one can obtain the thermal resistance R_t.

3. PhotoThermal Radiometry

Photo thermal radiometry (PTR) experiments were implemented to measure the total thermal resistance R of the sample, as a function of temperature. The basic principle is to measure the phase lag and the amplitude of the periodic temperature response produced on the sample surface by a modulated laser beam [6]. A thermal model describing the heat transfer in the sample during the experiment enables the calculation of the theoretical phase lag and the amplitude as a function of frequency. The identification the bi-layer thermal resistance is performed by minimization of the gap between the theoretical and the experimental data.
The schematic view of the PTR experimental setup is presented in Fig 2.

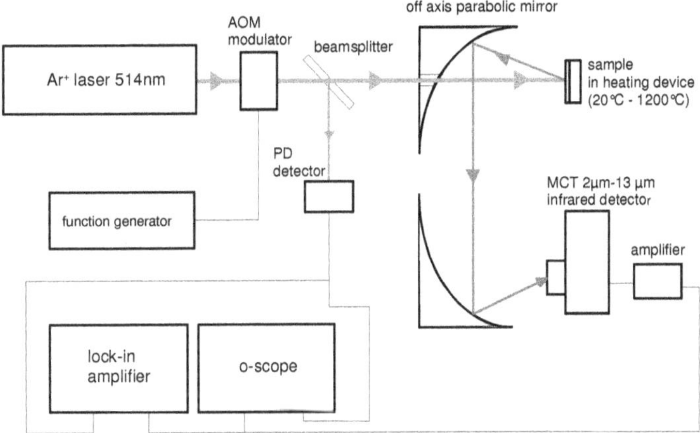

Figure 2: PhotoThermal Radiometry experimental setup. The sample is put inside a heating device that permits reaching 1200°C under vacuum or under inert gas.

In order to perform experiments at different temperatures, the temperature of the sample was controlled by a commercially available heating device, working under Argon as inert gas. The sample was heated at a rate of 20 °C/min and annealed for 5 min at the required temperature before starting the measurement. Optical access to the sample located inside the crucible of the heating device was ensured by a BaF_2 window that is transparent in the visible and infrared radiation range. The thermal excitation was generated on the sample front face by an Ar+ laser of 514 nm wavelength and 1.7 W maximum power. The laser was modulated by an acousto-optic modulator using a square signal issued from a function generator and was reflected to the sample surface by a set of mirrors. The laser beam had a Gaussian profile of power repartition on the spot of 1 mm in diameter at $1/e^2$. A very fast photodiode was used to measure the reference signal, in order to avoid the phase lag due to the acousto-optic modulator driver. The thermal response was measured by an infrared HgCdTe detector.

The wavelength measurement range of this IR detector was comprised between 2 and 13 µm. Parabolic mirrors coated with high reflective rhodium (reflectivity of 98% in the infrared detector wavelength band) were used to collect the emitted infrared radiation and to focus it on the infrared detector. The detector wavelength operating range was higher than that of the laser, therefore the measurement was not disturbed by the photonic source; moreover an optical filter was used in order to reject all the visible radiation arriving on the IR detector. The zone viewed by the detector was the image of the infrared sensitive element on the sample corresponding to a circle of 1 mm in diameter. A lock-in amplifier was used to measure the amplitude and the phase lag between the reference and the detector output, according to the frequency. The frequency range [1-100] kHz, where the phase and the amplitude are sensitive to the thermal resistance of the stack was swept during the experiment. Each measurement for the amplitude and phase was affected by a 5% standard deviation. The periodic temperature variation ΔT at the sample surface was small enough to assume that the measured radiative emission by the IR detector was linearly proportional to ΔT.

4. Model of heat transfer in the sample

Considering the large diameter of the laser beam with respect to the small bi-layer film thickness, the heat transfer in the sample in this experimental configuration is described by the following relations in 1D.

$$\frac{1}{\alpha_i}\frac{\partial T}{\partial t}=\frac{\partial^2 T}{\partial z^2},\quad 0\le z\le e_i,\ t>0,\ i=\mathrm{Al_2O_3},\ \mathrm{Te},\ \mathrm{Si} \tag{2}$$

with related boundary conditions:

$$-k_{\mathrm{GST}}\frac{\partial T}{\partial z}=\varphi_0\cos(\omega t),\ z=0,\ t>0 \tag{3}$$

$$T=0,\ z=e_T=e_{\mathrm{Al_2O_3}}+e_{\mathrm{Te}}+e_{\mathrm{Si}},\ t>0 \tag{4}$$

$$T_{\mathrm{Te}}-T_{\mathrm{Al_2O_3}}=R_i\,\varphi,\ z=e_{\mathrm{Te}},\ t>0 \tag{5}$$

and the initial condition:

$$T=0,\quad 0\le z\le e_T,\quad t=0 \tag{6}$$

where α_i is the thermal diffusivity ($k_i/\rho_i\ C_i$) of the respective layer i, T is the temperature, t is the time and φ the heat flux. Condition (3) corresponds to the periodic heat flux with angular frequency $\omega=2\pi f$. Eq. (4) gives the condition of prescribed temperature (after a change of variable) and initial condition (6). The boundary condition (5) introduces the thermal resistance at the interfaces of Te.

The system of partial differential equations was solved using the Laplace integral transform on time [6]. The unknown thermal resistance $R=R_{\mathrm{Te}}+R_i+R_{\mathrm{Al_2O_3}}$ is identified using a minimization procedure based on the large-scale algorithm. This algorithm is a subspace trust region method and is based on the interior-reflective Newton method described in ref. [7]. It must be noticed that each iteration involves the approximate solution of a large linear system using the method of preconditioned conjugate gradients.

5. Results

In order to retrieve the thermal resistance R_t of the Te layer the properties of the silicon substrate and the Al_2O_3 layer must be perfectly known. The temperature dependent thermal conductivity of silicone wafer equal $k_{\mathrm{Si}}=982.98\,T^{-0.4737}\ \mathrm{W\,m^{-1}K^{-1}}$ and specific heat per unit volume $\rho C_{p\mathrm{Si}}=2300\left(705+0.428\,T\right)\ \mathrm{J\,m^{-3}K^{-1}}$ of the 0.6 mm thick Si substrate were measured using Hot Disk and Differential Scanning Calorimetry respectively. The thermal conductivity of the Al_2O_3 layer were measured by PTR technique in a precedent work. It was found that it does not vary in temperature and is equal $k_{Al_2O_3} = 1.40\,\mathrm{W\,m^{-1}\,K^{-1}}$. The estimation of the total thermal resistance R was performed using the experimental data. The resistance R_t of the Te layer was obtained by subtraction of $R_{Al_2O_3} = e_{Al_2O_3} / k_{Al_2O_3}$ from R. The measurements were carried out on the temperature range included between 50°C and 500°C with 50°C step. The sample was heated at a rate 20 °C/min and annealed for 5 min at the required temperature in argon atmosphere. The fitting of the experimental measurement were performed for each temperature. It appeared that the model does not fit the measurements at temperatures exceeding 350°C. Pure tellurium melts at 450°C, but the estimation at higher than 350°C temperatures was not reliable. This problem may be due to the destruction of the sample integrity at higher temperatures. The analyses of samples with SiO_2 protective layer have shown that up to 300°C the tellurium hemispheres withstand the thermal stress [8], above this temperature, the tellurium migrates through the SiO_2 layer joining the Pt capping layer, finally forming some Pt-Te chemical phases. This point must be verified in our Al_2O_3 case by SEM observations. The estimated thermal resistance R_t of the Te layer is represented in Fig. 3.

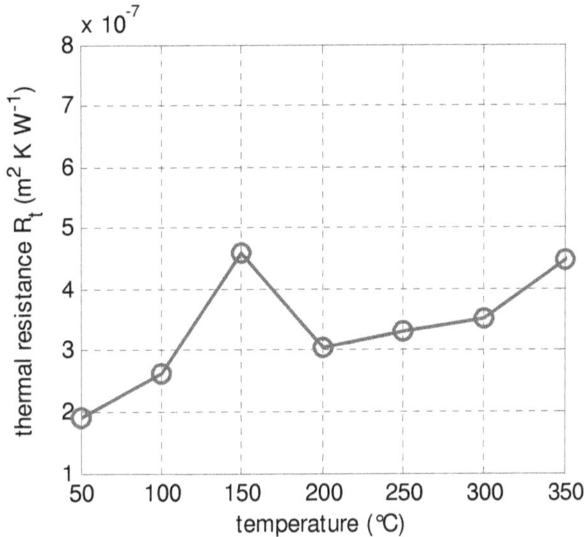

Figure 3: The thermal resistance R_t of the Te layer versus temperature.

The evolution plotted in Fig. 3 show the R_t values increasing from $2 \times 10^{-7}\ \mathrm{m^2 K\,W^{-1}}$ to $4 \times 10^{-7}\ \mathrm{m^2 K\,W^{-1}}$. One singular point appears at 150°C, where the R_t increases to maximum value on this temperature range. The variation of R_t is not very drastic from the thermal point of view. The increase can be attributed to the evolution of R_i during annealing. In fact, the tellurium grain size grows for temperature increasing from room temperature, leading to worse interface with silicon substrate and protective layer [8].

6. Conclusion

An experimental arrangement for thermal characterization of materials in liquid state at micro scale was tested. The measurement based on photo thermal radiometry was used in order to retrieve the thermal resistance of liquid layer. The main difficulty remains in sample development. It must keep its integrity at high temperature, exceeding the melting point of the studied material.

The proposed configuration was based on hemispherical structures of tellurium encapsulated with Al_2O_3 protective layer.

The measurements permitted the estimation of the thermal resistance of the tellurium from 50°C up to 350°C. The estimation at liquid state could not be achieved due to probably migration of the Te at temperatures exceeding 350°C.

Further analysis by SEM and X-ray diffraction must be carried out for better comprehension of studied sample under high temperature. These analyses should be helpful for development of more reliable sample structure.

7. References

[1] Wuttig M, *Nature Mater.* **4**, 265, 2005.

[2] Lankhorst M, Ketelaars B and Wolters R, *Nature Mater.* **4**, 347, 2005.

[3] Kolobov A, Fons P, Frenkel A, Ankudinov A, Tominaga J and Uruga T, *Nature Mater.* **3**, 703, 2004.

[4] Lyeo H, Cahill D, Lee B, Abelson J, Know M, Kim K, Bishop S, and Cheong B, *Appl. Phys. Lett.* **89**, 151904, 2006.

[5] Battaglia J-L, Kusiak A, Schick V, Cappella A, Wiemer C, Longo M, Varesi E, *J. Appl. Phys.* **107**, 2010. doi. 10.1063/1.3284084

[6] Battaglia J-L., Kusiak A., International Journal of Thermophysics 28, 1563, 2007.

[7] Coleman, T.F. and Y. Li, SIAM Journal on Optimization 6, 418, 1996.

[8] Cappella A., Battaglia J.-L., Schick V., Kusiak A., Wiemer C., Longo M., Hay B., *Photothermal Radiometry applied in nanoliter melted tellurium alloys*, Work Materials Research Institues Forum, Workshop for Young Materials Scientists, August 31 - September 3, 2010, 2nd International Workshop, Berlin,Germany.

Conductivity Reconstructions Using Real Data from a New Planar Electrical Impedance Device

Cristiana Sebu

School of Technology, Oxford Brookes University, Wheatley Campus, Oxford, OX33 1HX.
Email: csebu@brookes.ac.uk

ABSTRACT

We present conductivity reconstructions obtained using real data from a new planar electrical impedance tomography device developed at the Institut für Physik, Johannes Gutenberg Universität, Mainz, Germany. This prototype was designed mainly for breast cancer detection. The inverse problem to solve is different from the classical inverse conductivity problem. We reconstruct the electrical conductivity of a circular two-dimensional domain from boundary measurements of currents and interior measurements of potential. The reconstruction algorithm used is an integral equation approach for smooth conductivity distributions.

Key Words: electrical impedance tomography, integral equation methods, breast cancer detection, nonlinear inverse problems, ill-posed problems.

1. Introduction

Electrical Impedance Tomography (EIT) is a non-invasive, low cost technology developed to image the electrical conductivity distribution of a conductive medium. The technique works by performing simultaneous measurements of electric currents and voltages on the boundary of an object. These are the data used by an image reconstruction algorithm to determine the electrical conductivity distribution within the object which can be used to infer the internal structure of the object under consideration. Thus, EIT can be used as a method of industrial, geophysical and medical imaging [1]. One of the possible medical applications is the breast cancer detection [2-5].

The inverse conductivity problem is mathematically challenging being both nonlinear and extremely ill-posed in the Hadamard sense. Substantial progress has been made in determining the class of conductivity distributions that can be recovered from boundary data [6, 7], as well as in designing reconstruction algorithms from noisy measurement data [8].

In vitro studies have discovered a difference of three times or more in electrical conductivity between healthy and cancerous tissue [9]. Therefore, in mammography, the interest is to find regions in the interior where the conductivity changes rapidly in comparison to the background conductivity. In this way, the ill-posedness is circumvented to a certain extent and some of the theoretical difficulties are overcome.

The success of EIT in medical applications has been rather disappointing, except for two-dimensional situations or detection of objects situated at shallow depths (e.g. breast tumours). The issues are both practical: errors in electrode positions or boundary shape, high and uncontrollable contact impedance of the skin, and theoretical: poor spatial resolution [10].

In this paper, we use an integral equation approach designed for a planar electrical impedance tomography device developed at the Institut für Physik, Johannes Gutenberg Universität, Mainz, Germany, to obtain reconstructions from real data.

The latest EIT device developed in Mainz is similar to the one described in [11, 12]. It consists of a planar sensing head of circular geometry, and it was designed mainly for breast cancer detection. There are 12 large outer electrodes arranged on a ring of radius $R \simeq 4.8$ cm where the external currents are injected, and a set of 54 point-like high-impedance inner electrodes where the induced potentials are measured. To avoid any problems due to the unknown contact impedance, are not measured at the outer electrodes. At the inner electrodes very high impedance voltage measurements are taken and the problem of contact impedance does not arise. It is important to note that the device has a fixed geometry, the positions of the electrodes are exactly known, and there are no problems related to the contact impedance.

Since most of breast tumours are situated at shallow depths, a map of the conductivity at the surface provides enough information on the existence and location of tumours. Hence, our aim is to determine the electrical conductivity of a two dimensional circular domain from boundary measurements of currents and interior measurements of the potential. This inverse problem is different from the classical inverse conductivity problem and the numerical simulations showed that it has a much better spatial resolution [12].

2. Description of the device

The Mainz tomograph consists of three parts: a sensing head, an electronic device to apply and measure the electric potential and the current patterns, and a computer for image reconstruction. More technical details of the Mainz EIT system can be found in [11]. The sensing head (see Fig. 1) has a diameter of 10 cm and consists of 12 large electrodes arranged on the outer ring of a disk of radius $R \simeq 4.8$ cm and a set of 54 small high-impedance electrodes placed in a hexagonal pattern in the interior which are used to measure the voltage.

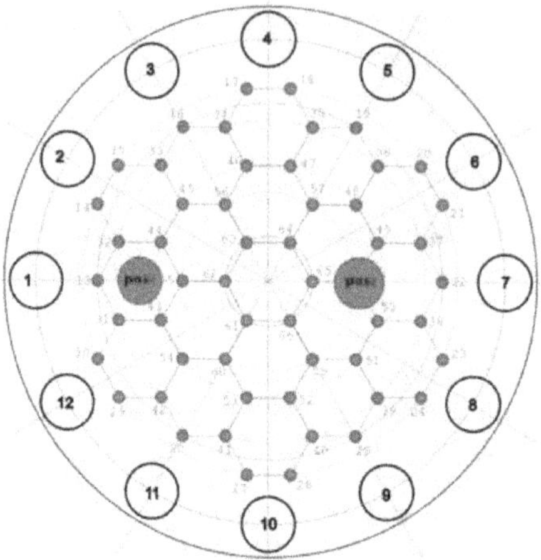

Figure 1: Layout of the electrode array of the Mainz EIT device. The small circles labeled 1 to 12 show the outer electrodes for current injection. The position of the inner electrodes for potential measurements are drawn as thick points and labeled 13 to 66.

The experimental data have been obtained by placing the sensing head into the bottom of a cylindrical tank of the same diameter as the sensing head which was filled with a conducting liquid (salt water). Cylindrical objects of various diameters made out of different materials have been immersed in the liquid at different distances z from the sensing head. In this paper we will present the conductivity reconstructions obtained using the real data that have been obtained in this way.

3. An integral equation reconstruction method

Let $\Omega=\left\{\mathbf{x}=(r\cos\theta, r\sin\theta)\in\mathbb{R}^2:\ \|\mathbf{x}\|<R\right\}$, and σ be an isotropic conductivity distribution in Ω. For an applied current density $j\in H^{1/2}(\partial\Omega)$, the electric potential $\Phi\in H^1(\Omega)$ satisfies

$$\nabla\cdot(\sigma\nabla\Phi)=0 \text{ in } \Omega, \tag{1}$$

$$j=\sigma\frac{\partial\Phi}{\partial n} \text{ on } \Omega\text{, such that } \int_{\partial\Omega} j(\mathbf{x})\, d\mathbf{x}=0. \tag{2}$$

The boundary value problem described by Eqs. (1) and (2) has a unique solution Φ up to an additive constant which we could fix by choosing the ground as

$$\int_{\partial\Omega}\Phi(\mathbf{x})\, d\mathbf{x}=0. \tag{3}$$

The inverse problem under consideration can be formulated as follows: find the conductivity σ in Ω satisfying Eqs. (1), (2) and (3) from the measurement operator

$$\begin{aligned}\mathcal{M}:\mathcal{L}^2_\diamond(\partial\Omega)&\to\mathcal{L}^2(\partial\Omega)\\ j\ &\mapsto\ \Phi,\end{aligned} \tag{4}$$

where $\mathcal{L}^2_\diamond(\partial\Omega)=\left\{f\in\mathcal{L}^2(\Omega):\ \int_{\partial\Omega} f(\mathbf{x})\, d\mathbf{x}=0\right\}$.

Let us consider that $\sigma_0=1$ is our reference or background conductivity. The associated potential Φ_0 is the solution of the boundary value problem given by Eqs. (1), (2) and (3) for $\sigma=\sigma_0$

$$\Phi_0(\mathbf{x})=\int_{\partial\Omega} G_N(\mathbf{x},\mathbf{y}) j(\mathbf{y})\, d\mathbf{y}. \tag{5}$$

G_N is the Neumann Green's function for the Laplace equation in the disk Ω [13].

If σ is smooth and non-zero everywhere, Eq. (1) can be rewritten as

$$\Delta\Phi=-\nabla\ln\sigma\cdot\nabla\Phi, \tag{6}$$

and if we assume that $\sigma=\sigma_0$ near $\partial\Omega$, the solution of the problem given by Eqs. (1), (2) and (3) is

$$\Phi(\mathbf{x})=\Phi_0(\mathbf{x})+\int_{\Omega} G_N(\mathbf{x},\mathbf{y})\nabla\ln\sigma(\mathbf{y})\cdot\nabla\Phi(\mathbf{y})\, d\mathbf{y}. \tag{7}$$

Furthermore, it has been showed in [12] that if $\delta\Phi=\Phi-\Phi_0$ is small then $\nabla\Phi\approx\nabla\Phi_0$, and a partial integration in Eq. (8) leads to the following integral equation for $\ln(\sigma)$

$$\delta\Phi(\mathbf{x})\approx-\int_{\Omega}\nabla_{\mathbf{y}} G_N(\mathbf{x},\mathbf{y})\cdot\nabla\Phi_0(\mathbf{y})\ln\sigma(\mathbf{y})\, d\mathbf{y}. \tag{8}$$

In practice, we only have partial knowledge of the measurement operator $\mathcal{M}$ given in Eq. (4). Only ten standard varying-frequency patterns of the input current are applied on $\partial\Omega$:

$$j^{(k)}(\theta)=\begin{cases} j_0^{(k)}\cos\left(k(\pi-\theta)\right), & k=1,\ldots,5,\\ j_0^{(k)}\sin\left((k-5)(\pi-\theta)\right), & k=6,\ldots,10.\end{cases} \tag{9}$$

The corresponding potentials for $\sigma=\sigma_0$ are explicitly known

$$\Phi_0^{(k)}(r,\theta)=R\int_0^{2\pi} G_N(r,\theta;\rho,\vartheta) j^{(k)}(\vartheta)\, d\vartheta=\frac{1}{n}R\left(\frac{r}{R}\right)^n j^{(k)}(\theta),\ k=1,\ldots,10, \tag{10}$$

and Eq. (9) reduces to a system of integral equations for $\ln\sigma$

$$\delta\Phi^{(k)}(\mathbf{x}) \approx -\int_{\Omega} \nabla_{\mathbf{y}} G_N(\mathbf{x},\mathbf{y}) \cdot \nabla\Phi_0^{(k)}(\mathbf{y}) \ln\sigma(\mathbf{y})\, d\mathbf{y}, \quad k=1,\ldots,10. \tag{11}$$

Moreover, since there are only 54 interior points of potential measurement, $\mathbf{x}_i \in \Omega$, $i=1,\ldots,54$, the above system of equations becomes

$$\delta\Phi^{(k)}(\mathbf{x}_i) \approx \int_{\Omega} K^{(k)}(\mathbf{x}_i,\mathbf{y}) \ln\sigma(\mathbf{y})\, d\mathbf{y}, \quad k=1,\ldots,10, \quad i=1,\ldots,54, \tag{12}$$

where $K^{(k)}(\mathbf{x},\mathbf{y}) = -\nabla_{\mathbf{y}} G_N(\mathbf{x},\mathbf{y}) \cdot \nabla\Phi_0^{(k)}(\mathbf{y})$ are the integration kernels.

In order to find σ, we discretise the system in Eq. (13) as follows

$$\delta\Phi^{(k)}(r_i,\theta_i) \approx \sum_{l=1}^{172} w_l K^{(k)}(r_i,\theta_i;\rho_l,\vartheta_l) \ln\sigma(\rho_l,\vartheta_l), \quad k=1,\ldots,10, \quad i=1,\ldots,54, \tag{13}$$

where (r_i,θ_i) are the polar coordinates of the interior points of measurement, and $\{w_l;\rho_l,\vartheta_l\}$ is a set of quadrature weights and points for the disk of radius R given by Engels [14]. The above system of equations is over-determined (540 equations for 172 unknowns) and it is solved by means of the generalized inverse and truncated value decomposition is used as regularization scheme. In this way very small singular values (i.e. smaller than ε) will be cut out and will not enter the reconstruction. The regularization parameter λ is given by the L-curve criterion [15]. More details can be found in [12].

4. Numerical results

In this paper we will only present the numerical results for two cylindrical metallic objects, M15 and M20, which have 15 mm and 20 mm in diameter, respectively. The objects have been immersed in the liquid at different distances z from the sensing head and placed either in between the inner electrodes, 'pos1' in Fig. 1, or on the top of an inner electrode, 'pos2' in Fig. 1.

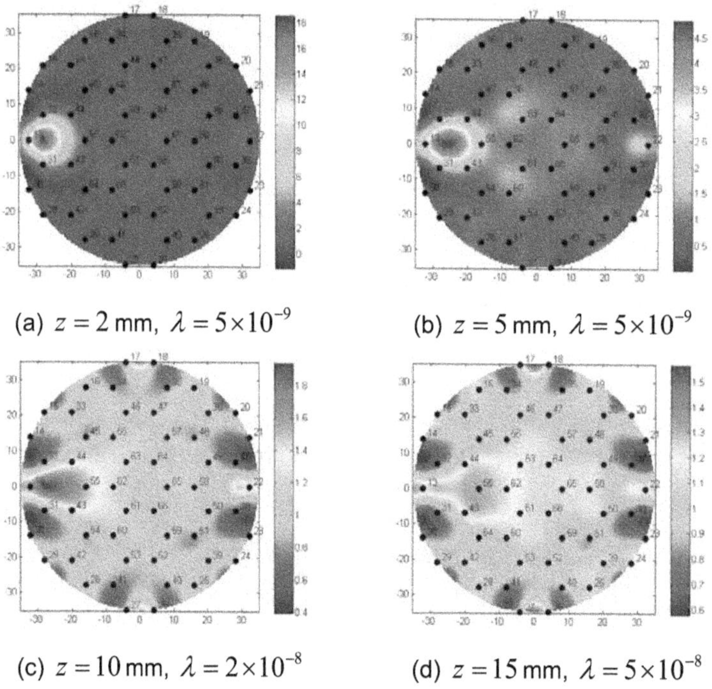

(a) $z=2$ mm, $\lambda=5\times10^{-9}$ (b) $z=5$ mm, $\lambda=5\times10^{-9}$

(c) $z=10$ mm, $\lambda=2\times10^{-8}$ (d) $z=15$ mm, $\lambda=5\times10^{-8}$

Figure 2: Conductivity reconstructions for M20 placed at 'pos1'.

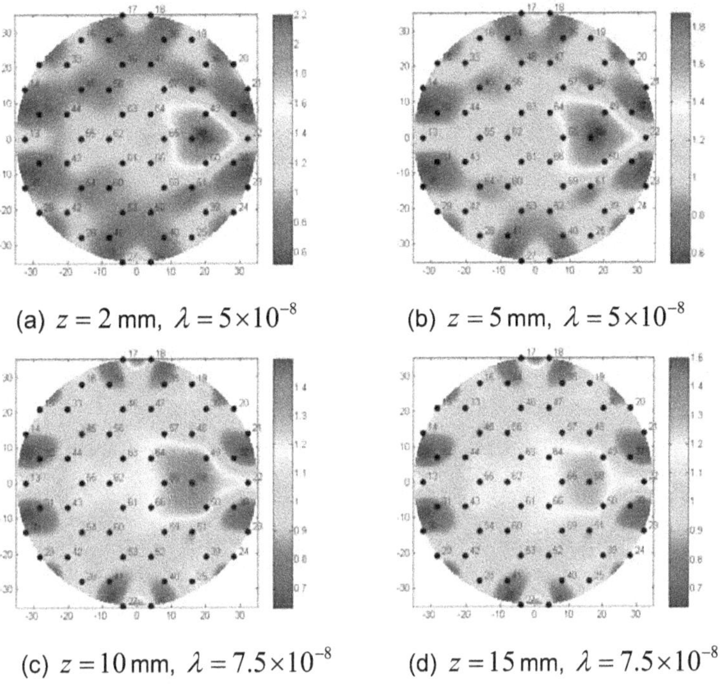

(a) $z = 2$ mm, $\lambda = 5\times10^{-8}$ (b) $z = 5$ mm, $\lambda = 5\times10^{-8}$

(c) $z = 10$ mm, $\lambda = 7.5\times10^{-8}$ (d) $z = 15$ mm, $\lambda = 7.5\times10^{-8}$

Figure 3: Conductivity reconstructions for M20 placed at 'pos2'.

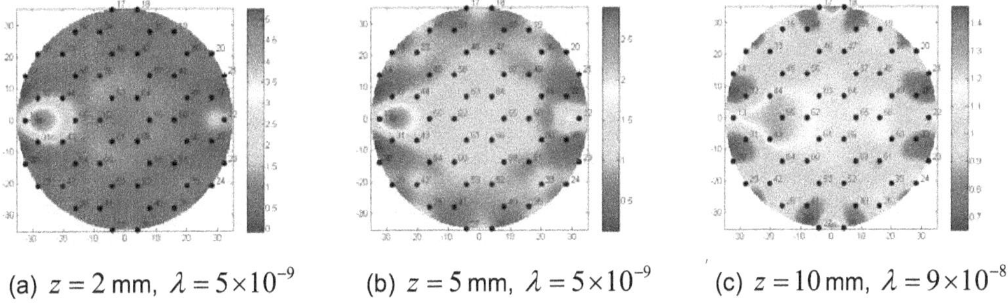

(a) $z = 2$ mm, $\lambda = 5\times10^{-9}$ (b) $z = 5$ mm, $\lambda = 5\times10^{-9}$ (c) $z = 10$ mm, $\lambda = 9\times10^{-8}$

Figure 4: Conductivity reconstructions for M15 placed at 'pos1'

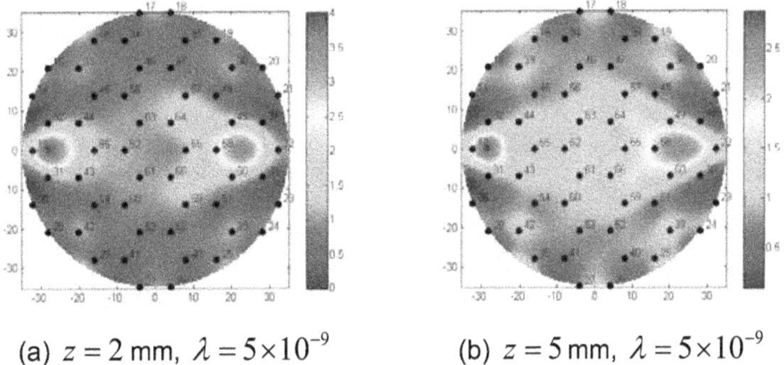

(a) $z = 2$ mm, $\lambda = 5\times10^{-9}$ (b) $z = 5$ mm, $\lambda = 5\times10^{-9}$

Figure 5: Conductivity reconstructions for M15 placed at 'pos1' and M20 placed at 'pos2'.

In all the numerical reconstructions considered, singular values smaller than $\varepsilon = 0.0005$ were cut off. Note that the metallic objects could still be detected up to a distance z approximately equal to their diameters (see Figs. 2 and 4), and that it does not matter whether the object is placed in between the inner electrodes, 'pos1', or on the top of an inner electrode, 'pos2' (see Figs. 2 and 3). In the future, we plan to present more results obtained using real data and to extend our technique to a fully three-dimensional model of the problem in order to obtain additional information about the depths.

5. REFERENCES

[1] Borcea, L., "Electrical impedance tomography", Inverse Problems, Vol. 18, pp. R99-R136, 2002.

[2] Cherepenin, V., Karpov, A., Korjenevsky, A., Kornienko, V., Mazaletskaya, A., Mazourov, D., and Meister, D., "A 3D electrical impedance tomography (EIT) system for breast cancer detection", Phys. Meas., Vol. 22, No. 1, pp. 9-18, 2001.

[3] Cherepenin, V., Karpov, A., Korjenevsky, A., Kornienko, V., Kultiasov, Y., Ochapkin, M., Tochanova, O., and Meister, D., "Three dimensional EIT imaging of breast tissues: system design and clinical testing", IEEE Trans. Medical Imaging, Vol. 21, No. 6, pp. 662-667, 2002.

[4] Kerner, T.E., Paulsen, K.D., Hartov, A., Soho, S.K., and Poplack, S.P., "Electrical impedance spectroscopy of the breast: clinical imaging results in 26 subjects", IEEE Trans. Medical Imaging, Vol. 21, No. 6, pp.638-645, 2002.

[5] Poplack, S.P., Tosteson, T.D., Wells, W.A., Pogue, B.W., Meaney, P.M., Hartov, A., Kogel, C.A., Soho, S.K., Gibson, J.J., and Paulsen, K.D., "Electromagnetic breast imaging: results of a pilot study in women with abnormal mammograms", Radiology, Vol. 243, pp. 350-359, 2007.

[6] Calderón, A.P., "Seminar on Numerical Analysis and its Applications to Continuum Physics", Soc. Brasileira de Matemàtica, Rio de Janeiro, pp. 65-73, 1980.

[7] Astala, K., and Païvärinta, L., "Calderon's inverse conductivity problem in plane", Ann. Math., Vol. 163, pp. 265-299, 2006.

[8] Holder, D.S., Electrical Impedance Tomography, Institute of Physics, Bristol, 2005.

[9] Rigaud, B., Morucci, J.P., and Chauveau, N., "Bioelectrical impedance techniques in medicine. Part I: Bioimpedance measurement. Second section: impedance spectrometry", Crit. Rev. Biomed. Eng., Vol. 24, No. 4-6, pp. 257-351, 1996.

[10] Sabatier, P.C., and Sebu, C., "On the resolving power of Electrical Impedance Tomography", Inverse Problems, Vol. 23, pp. 1895-1913, 2007.

[11] Azzouz, M., Hanke, M., Oesterlein, C., and Schilcher, K., "The factorization method for electrical impedance tomography data from a new planar device", International J. Biomedical Imaging 2007, Article ID 83016, 2007.

[12] Hähnlein, C., Schilcher, K., Spiesberger, H., and Sebu, C., "Conductivity imaging using interior potential measurements", accepted for publication in Inverse Probl. Sci. Eng.

[13] Kervokian, J., Partial Differential Equations. Analytical Solutions Techniques, Texts in Applied Mathematics 25, 2nd ed., Springer Verlag, New York, 2000.

[14] Engels, H., Numerical Quadratures and Cubatures, Academic Press, London, 1980.

[15] Hansen, P.C., "Regularization Tools: A Matlab Package for Analysis and Solution of Discrete Ill-Posed Problems", Numer. Algorithms, Vol. 6, No. 1, pp.1-35, 1994.

Electromagnetic Properties of Metals Based on Transmission Measurements at 64 MHz

Sen-Yong Chen[1], Othmane Benafan[2], Rajan Vaidyanathan[2] and Aravinda Kar[1]

[1]Laser-Advanced Materials Processing Laboratory
Center for Research and Education in Optics and Lasers (CREOL)
College of Optics and Photonics
University of Central Florida, Orlando, FL 32816
[2]Advanced Materials Processing and Analysis Center (AMPAC)
Mechanical, Materials, and Aerospace Engineering
University of Central Florida, Orlando, Florida 32816

Abstract

A mathematical model is developed to determine the refractive and absorption indices of metals based on transmission measurements. Thin Ti sheets of different thicknesses are placed in a magnetic field of frequency 64 MHz, and the incident and transmitted magnetic field strengths are measured. Selected Ti sheets were treated with a laser beam to diffuse Pt into them. The transmission data were used to determine material properties based on the approach of inverse problems. The electrical conductivity of the laser Pt-diffused samples increased by 9.1% when compared to the as-received, untreated Ti sheets. The reflectivity of the treated Ti samples also increased by 0.2% compared to the untreated samples. These data indicate that lasers enable modifying the electromagnetic properties of materials.

Key words: inverse, electromagnetic properties, metal

1. Introduction

Several instruments, such as spectrum analyzers and network analyzers, are available for measuring the electrical and electromagnetic properties of materials at high radiofrequencies. For lower frequencies (e.g., 30 MHz, 100 MHz), however, there is a need for reliable instruments for measuring the electromagnetic response of materials to magnetic fields of high flux densities (e.g., 10-100 μT). There are numerous applications for such measurements including non-destructive evaluation of materials and new materials development for microelectronics, nanoelectronics and biomedical applications. This paper presents a model to determine the electromagnetic properties of Ti based on experimental data pertaining to the incident and transmitted magnetic field strength.

2. Mathematical model

A mathematical model is developed for the transmission of the magnetic field through the Ti sheets by considering a three-medium system as illustrated in Fig. 1. The magnetic field is assumed to propagate in the z direction and the three media are considered non-magnetic and homogeneous. The magnetic field generator, which is indicated by a current loop in Fig. 1, is placed at a distance d_1 from the front surface of the Ti sheet and its back surface is at a distance d_2 from the current loop. The incident magnetic field is partially reflected at both the air-Ti interfaces and absorbed inside the sheet. There are forward and backward propagating fields in medium 1 (air in front of the Ti sheet) and medium 2 (Ti sheet) due to reflection, and there is only a forward propagating field in medium 3 (air behind the sheet). H_{in} and H_r are the incident and reflected magnetic field strengths in medium 1, respectively, while H_{2+} and H_{2-} denote the forward and backward moving magnetic field strengths in the sheet, respectively. H_{3+} is the forward moving magnetic field strength in medium 3.

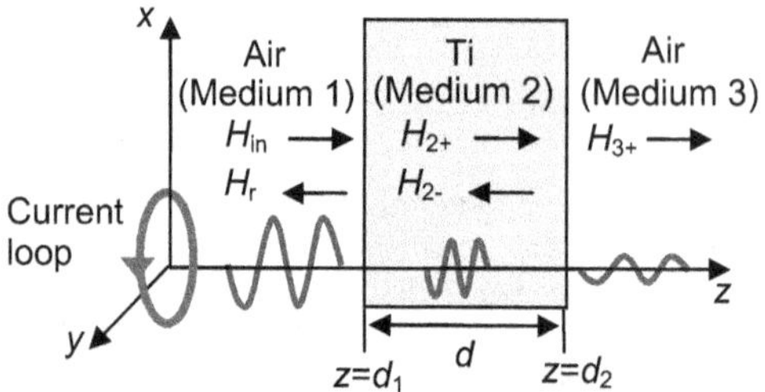

Figure1: Schematic of air-Ti sheet-air system.

Representing t_{12} and t_{23} as the transmission coefficients for the first interface between media 1 and 2 and the second interface between media 2 and 3, respectively, and denoting r_{12} and r_{23} as the corresponding reflection coefficients, the overall transmission coefficient of the Ti sheet, $t(d)$, can be written as [1]:

$$t(d)=\frac{H_{3+}(d)}{H_{in}(d_1)}=\frac{t_{12}t_{23}\exp(i\beta_2 d)}{1-r_{12}r_{23}\exp(i2\beta_2 d)} \tag{1}$$

based on the solutions of Maxwell's electromagnetic equations, where β_2, which is the propagation constant in the medium 2, is given by the following expression:

$$\beta_j=\frac{\omega}{c}\hat{n}_j \text{ for } j=1,2,3 \tag{2}$$

Here, $\hat{n}_j$ is the complex refractive index of medium j, j=1,2,3, which can be expressed in terms of the angular frequency of the magnetic field ω, permittivity of air ε_0, relative permittivity ε_{rj}, electrical conductivity σ_j, refractive index n_j, and absorption index k_j as:

$$\hat{n}_j=\sqrt{\varepsilon_{rj}+i\frac{\sigma_j}{\omega\varepsilon_0}}=n_j+ik_j \tag{3}$$

Based on the boundary conditions for the magnetic field, the transmission and reflection coefficients for each interface can be expressed as:

$$r_{lm}=\frac{Z_l-Z_m}{Z_l+Z_m} \text{ for } l,m=1,2,3 \tag{4}$$

$$t_{lm}=\frac{2Z_l}{Z_l+Z_m} \text{ for } l,m=1,2,3 \tag{5}$$

where Z_j, which is the impedance of medium j, for j=1,2,3, and is given by [2]:

$$Z_j(z)=-\sqrt{\frac{\mu_0}{\varepsilon_0}}\frac{1}{\hat{n}_j}\frac{\frac{-i}{\beta_j z}+\frac{1}{(\beta_j z)^2}}{\frac{-i}{\beta_j z}+\frac{1}{(\beta_j z)^2}+\frac{i}{(\beta_j z)^3}} \tag{6}$$

where μ_0 is the permeability of air. The transmittance, $T(d)$, can be written as:

$$T(d)=\frac{\mathrm{Re}[Z_3(d_2)]}{\mathrm{Re}[Z_1(d_1)]}t(d)t^*(d) \tag{7}$$

Substituting Eqs. (1) to (6) into Eq. (7), the transmittance can be expressed as:

$$T(d)=\frac{16n_1n_3(u^2+v^2)}{p(d)^2+q(d)^2} \tag{8}$$

where u, v, $p(d)$ and $q(d)$ are defined by the following expressions in terms of $C(d)$, $D(d)$, $F(d)$, $G(d)$, $x_a(z)$, $y_a(z)$, $x_m(z)$ and $y_m(z)$:

$$u=n_2[x_a(d_1)x_m(d_1)-y_a(d_1)y_m(d_1)]-k_2[x_a(d_1)y_m(d_1)+x_m(d_1)y_a(d_1)] \tag{9}$$

$$v=k_2[x_a(d_1)x_m(d_1)-y_a(d_1)y_m(d_1)]+n_2[x_a(d_1)y_m(d_1)+x_m(d_1)y_a(d_1)] \tag{10}$$

$$p(d)=\exp\left(\frac{\omega}{c}k_2 d\right)\left[C\cos\left(\frac{\omega}{c}n_2 d\right)+D\sin\left(\frac{\omega}{c}n_2 d\right)\right]+\exp\left(-\frac{\omega}{c}k_2 d\right)\left[F\cos\left(\frac{\omega}{c}n_2 d\right)-G\sin\left(\frac{\omega}{c}n_2 d\right)\right] \tag{11}$$

$$q(d)=\exp\left(\frac{\omega}{c}k_2 d\right)\left[D\cos\left(\frac{\omega}{c}n_2 d\right)-C\sin\left(\frac{\omega}{c}n_2 d\right)\right]+\exp\left(-\frac{\omega}{c}k_2 d\right)\left[G\cos\left(\frac{\omega}{c}n_2 d\right)+F\sin\left(\frac{\omega}{c}n_2 d\right)\right] \tag{12}$$

$$\begin{aligned}C(d)=&\left[x_a(d_1)n_2-y_a(d_1)k_2+x_m(d_1)n_1\right]\left[x_m(d_2)n_3+x_a(d_2)n_2-y_a(d_2)k_2\right]\\&-\left[x_a(d_1)k_2+y_a(d_1)n_2+y_m(d_1)n_1\right]\left[y_m(d_2)n_3+x_m(d_2)k_2+y_m(d_2)n_2\right]\end{aligned} \tag{13}$$

$$\begin{aligned}D(d)=&\left[x_a(d_1)n_2-y_a(d_1)k_2+x_m(d_1)n_1\right]\left[y_m(d_2)n_3+x_a(d_2)k_2+y_a(d_2)n_2\right]\\&+\left[x_a(d_1)k_2+y_a(d_1)n_2+y_m(d_1)n_1\right]\left[x_m(d_2)n_3+x_m(d_2)n_2-y_m(d_2)k_2\right]\end{aligned} \tag{14}$$

$$\begin{aligned}F(d)=&\left[x_a(d_1)n_2-y_a(d_1)k_2-x_m(d_1)n_1\right]\left[x_m(d_2)n_3-x_a(d_2)n_2+y_a(d_2)k_2\right]\\&-\left[x_a(d_1)k_2+y_a(d_1)n_2-y_m(d_1)n_1\right]\left[y_m(d_2)n_3-x_m(d_2)k_2-y_m(d_2)n_2\right]\end{aligned} \tag{15}$$

$$\begin{aligned}G(d)=&\left[x_a(d_1)n_2-y_a(d_1)k_2-x_m(d_1)n_1\right]\left[y_m(d_2)n_3-x_a(d_2)k_2-y_a(d_2)n_2\right]\\&+\left[x_a(d_1)k_2+y_a(d_1)n_2-y_m(d_1)n_1\right]\left[x_m(d_2)n_3-x_m(d_2)n_2+y_m(d_2)k_2\right]\end{aligned} \tag{16}$$

$$x_a(z)=\frac{\left(\frac{\omega}{c}z\right)^4}{1-\left(\frac{\omega}{c}z\right)^2+\left(\frac{\omega}{c}z\right)} \tag{17}$$

$$y_a(z)=\frac{-\left(\frac{\omega}{c}z\right)}{1-\left(\frac{\omega}{c}z\right)^2+\left(\frac{\omega}{c}z\right)^4} \tag{18}$$

$$x_m(z)=\frac{\frac{\omega}{c}(n_2+k_2)z-\left(\frac{\omega}{c}\right)^2\left(n_2^{\,2}-2n_2k_2-k_2^{\,2}\right)z^2+\left[\frac{\omega}{c}k_2z-\left(\frac{\omega}{c}\right)^2\left(n_2^{\,2}-k_2^{\,2}\right)z^2\right]^2}{\left[\frac{\omega}{c}n_2z+2\left(\frac{\omega}{c}\right)^2n_2k_2z^2\right]^2+\left[1+\frac{\omega}{c}k_2z-\left(\frac{\omega}{c}\right)^2\left(n_2^{\,2}-k_2^{\,2}\right)z^2\right]^2} \tag{19}$$

$$y_m(z)=\frac{-\left[\frac{\omega}{c}n_2z+2\left(\frac{\omega}{c}\right)^2n_2k_2z^2\right]}{\left[\frac{\omega}{c}n_2z+2\left(\frac{\omega}{c}\right)^2n_2k_2z^2\right]^2+\left[1+\frac{\omega}{c}k_2z-\left(\frac{\omega}{c}\right)^2\left(n_2^{\,2}-k_2^{\,2}\right)z^2\right]^2} \tag{20}$$

Eq. (8) shows that the transmittance depends on two variables: refractive index n_2 and absorption index k_2 of Ti. After measuring the incident and transmitted magnetic field strengths for two Ti sheets of different thicknesses, and determining their transmittances, one can calculate n_2 and k_2 by the approach of inverse problems. Other electromagnetic properties of Ti such as the relative permittivity ε_{r2}, electrical conductivity σ_2, absorption coefficient α and reflectivity R can be determined from the following expressions:

$$\varepsilon_{r2}=n_2^2-k_2^2 \tag{21}$$

$$\sigma_2=2\omega\varepsilon_0 n_2k_2 \tag{22}$$

$$\alpha=2\frac{\omega}{c}k \tag{23}$$

$$R=\frac{(x_1n_2-y_1k_2-n_1x_{21})^2+(x_1k_2+y_1n_2-n_1y_{21})^2}{(x_1n_2-y_1k_2+n_1x_{21})^2+(x_1k_2+y_1n_2+n_1y_{21})^2} \tag{24}$$

Two sets of Ti sheets were used for determining the material properties. One set of samples was the as-received Ti sheets, while the second set was prepared by difussing Pt into Ti using a laser diffusion technique.

3. Experiments

High purity titanium sheets of thicknesses 25 and 50 µm were used to carry out laser Pt diffusion experiments. Square sheets of side 20 mm were used. The sheets were placed in a laser diffusion chamber as illustrated in Fig. 2. The chamber was filled with a platinum precursor that was prepared by dissolving $Pt(acac)_2$ [platinum(II) acetylacetonate, $Pt(C_5H_7O_2)_2$] in acetylacetone [$CH_3CHCH_2CHCH_3$] and heated inside a bubbler. A carrier gas, argon, was passed through the bubbler to transport the $Pt(acac)_2$ vapor to the chamber. A

Nd:YAG laser beam was used to heat the Ti sheet, resulting in the thermochemical decomposition of the precursor. This process produces Pt atoms which subsequently diffuse into the Ti sheet.

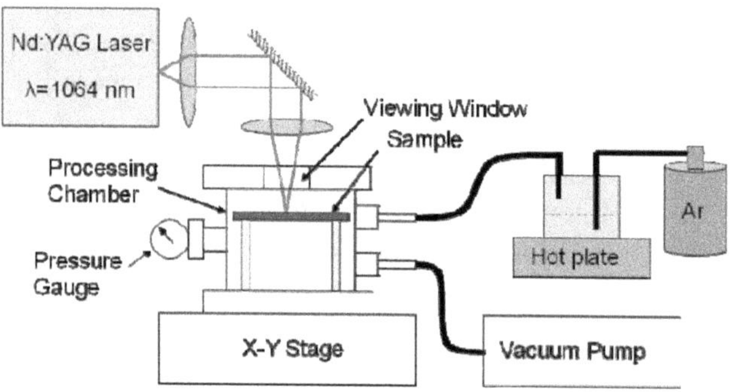

Figure 2: Setup for laser diffusion

Fig. 3 shows the setup for measuring the incident and transmitted magnetic field strengths. The Ti sheet is placed between two copper plates having rectangular windows so that the two surfaces of the sheet are exposed to the air. Two HP 11940A probes, one of which acts as a magnetic field source and the other as the field sensor, are placed on either side of the sheet as shown in Fig. 3 respectively. The magnetic field source was connected to a signal generator to create magnetic fields of strength 13.0 mA/m at 64 MHz. The field sensor was connected to a spectrum analyzer to measure the amplitude of the transmitted magnetic field.

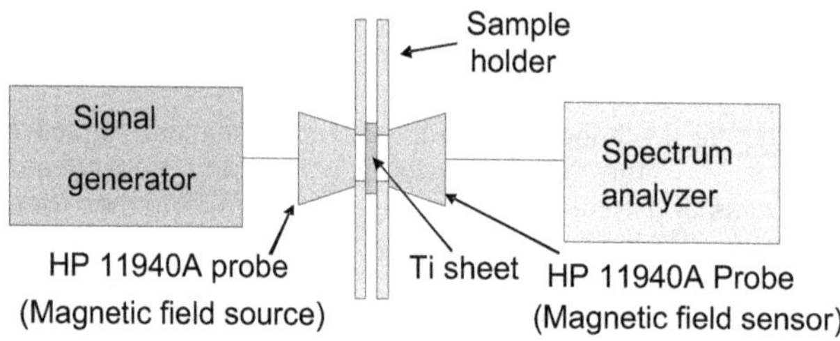

Figure 3: Setup for magnetic field strength measurement

4. Results and discussion

The measured data represent the incident and transmitted magnetic field strengths, H_{in}^* and H_{3+}^*, respectively, in units of dBμV, and the spectrum analyser provides a scaling factor of $10^{(H^*+48.5)/20}$ to convert the field strengths from the measured unit to the unit of μA/m for H_{in} and H_{3+}. The experimental values of transmittance are determined using the expression $T = (H_{3+}/H_{in})^2$. These values are listed in Table I for different samples. Using Eqs. (21) to (24), the properties of Ti can be calculated as listed in Table II. Using the values of n_2 and k_2, the calculated transmittance, T_{cal}, is obtained from the mathematical model. The errors between the calculated and experimental transmittance are determined as:

$$E_r = \left| \frac{T_{cal} - T}{T} \right|, \tag{25}$$

which are also listed in Table II. The laser Pt-diffused sample has higher electrical conductivity, reflectivity and absorption coefficient. The reflectivity, R, of the Pt-diffused sample is 0.9484, while it is 0.9461 for the as-received sample. This increase in the reflectivity of the laser-treated sample is due to the increased conductivity from 2.30×10^6 m$^{-1}\Omega^{-1}$ of the as-received sample to 2.51×10^6 $m^{-1}\Omega^{-1}$ of the treated sample. The conductivity of the treated sample increases by 9.1% when compared to that of the as-received sample.

Table I. Experimental data for magnetic field strengths and transmittance for as-received and laser Pt-diffused Ti samples

Sample	d (μm)	H_{in} (mA/m)	H_{3+} (μA/m)	T (%)
As-received	25	13.0	521.8	0.16
	50	13.0	310.8	0.06
Laser Pt-diffused	25	13.0	498.3	0.15
	50	13.0	263.0	0.04

Based on the conservation of energy, the amount of incident magnetic energy is equal to the sum of the absorbed and transmitted magnetic energies. For the 25 μm thick as-received Ti sheet, the reflectance is 0.9461 and the transmittance is 0.0016, indicating that the absorbance is 0.0523. For the 25 μm thick laser Pt-diffused Ti sheet, on the other hand, the absorbance is 0.0501. Therefore, this laser-treated Ti sheet absorbs less amount of the magnetic energy. The reduction in the absorbance is 4.2% of the absorbance of the as-received sample.

Table II. Calculated properties of Ti

Sample	n_2	k_2	ε_{r2}	σ ($m^{-1}\Omega^{-1}$)	R	α (mm^{-1})	E_r for 25μm sheet	E_r for 50μm sheet
As-received	17986	17987	-3.60×10^4	2.30×10^6	0.9461	24.1	23.8%	23.6%
Laser Pt-diffused	18798	18799	-3.76×10^4	2.51×10^6	0.9484	25.2	13.5%	12.6%

5. Conclusion

A mathematical model is presented to determine the electromagnetic properties of metals based on transmission measurements. Since the properties of interest were just for magnetic fields of low frequencies, such as 30 – 100 MHz, and instruments are not readily available for such frequency ranges, the inverse problem approach proved to be a good technique for determining the materials properties in this study. The properties of Ti sheets of thicknesses 25 μm and 50 μm were modified by diffusing Pt into the sheets using a laser diffusion technique. Using the model, the refractive and absorption indices of Ti were determined for the magnetic field of frequency 64 MHz. Other electromagnetic properties of Ti such as the relative permittivity, conductivity, absorption coefficient and reflectivity are also calculated. The conductivity of the Pt-diffused sample increased by 9.1% when compared to that of the as-received sample and, consequently, the reflectivity of the former was higher than the latter. Furthermore the absorbance of the Pt-diffused sample decreased by 4.2% when compared to that of the as-received sample.

6. References

[1] Heavens O.S., Optical Properties of Thin Solid Films, Dover, New York, pp. 91, 1965.

[2] Kaiser, K.L., Electromagnetic Compatibility Handbook, CRC, New York, pp. 21-48 , 2005

Dynamic Observers Based on Green's Function Applied to Estimate Heat Flux and Temperature Distribution During Drilling Process

Priscila F. B de Sousa, Valério L. Borges, Igor C. Pereira, Marcio B. da Silva, Gilmar Guimarães

Federal University of Uberlândia, School fo Mechanical Engineering, Uberlândia, Brasil
E-mails:priscila@mecanica.ufu.br, vlborges@mecanica.ufu.br, igorcesarp@yahoo.com.br, mbacci@mecanica.ufu.br, gguima@mecanica.ufu.br

Abstract

This study presents an inverse procedure to estimate temperature and heat flux at the tool-piece interface during a dr illing process. The technique uses a m ethod based on G reen's function and dynamic observers. While the most of works found in literature focuses the tool, in this study the work piece is modeled thermally. The thermal model is obtained by a numerical solution of the transient three-dimensional heat diffusion equation that considers the drill as an immaterial heat source moving. Finite volumes are used to solve the heat diffusion equation. Experimental tests were performed using cemented carbide tool drilling a micro alloyed steel piece (HSLA steel- High-Strength Low-Alloy). The heat flux generated in the cutting interface during the drilling process was estimated by using the developed inverse technique. Comparisons between experimental and estimated data present a good accordance.

Key Words: Heat conduction, inverse problems, dynamic observers, drilling.

1. Introduction

Cutting temperatures have strongly influenced both the tool life and the metallurgical state of machined surfaces. Heat generation is a common problem during any type of cutting, but particularly during a drilling due to the difficulty of conducting heat away from the cutting edge and the fact that the chips remain in contact with the tool for a relatively long time in a hole. Since the direct temperature measurements at the tool-piece interface are very complex the inverse solution represents a good approach. The inverse technique based on Green's function and dynamic observers with global heat transfer function is applied to solve the drilling problem represented by a three-dimensional transient model.

The inverse technique can be divided in two simple steps: i) obtaining of the transfer function model G_H; ii) obtaining of the heat transfer functions G_Q and G_N and the building algorithm identification. The inverse technique is applied to estimate the heat flux generated during a micro alloyed steel drilling process. Experimental temperatures were used to reconstruct the unknown boundary. Calculated temperature filed and c omparisons between calculated and ex perimental temperatures show the inverse technique power and robustness dealing with complex geometry and drilling problems.

2. Fundamentals

This section presents the 3D thermal model developed to the inverse problem solution, as well as, the inverse technique based on Green's function and dynamic observers. A complete description of this technique can be found in the work of Blum and Marquardt [1] and Sousa [2].

2.1. Thermal Model

The 3D thermal model represents the work piece being drilled. In order to model the thermal process a circular heat source moving with a diameter of the drill simulates the cutting process. Figure 1 shows the idealization of thermal process with a rectangular work piece being drilling at specific time of cutting, t_a.

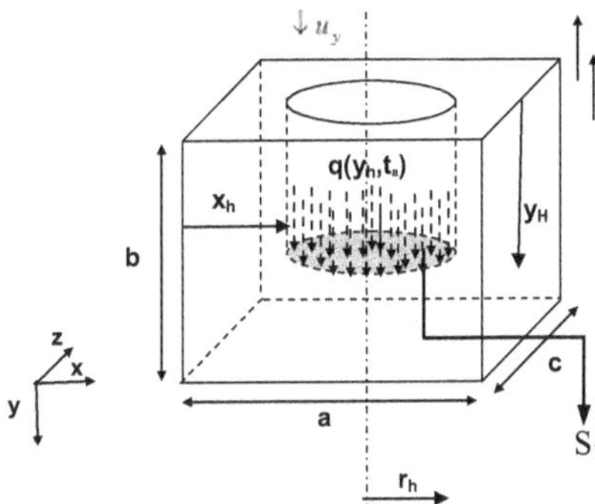

Figure 1: Thermal model scheme.

It can be observed in the literature [3-5] that the maximum temperature during the drilling process is generated on the primary edges of the drill and there is no significant variation on the drill's body. Due to this fact the heat source is just imposed in the cylinder base. All external surfaces are considered exposed to a convection medium.

The thermal model given by Fig.1 can, then, be represented by a sample initially in thermal equilibrium at T_0. The sample is then submitted to a unidirectional and uniform heat flux *q(t)* at surface S_1. The heat flux varies in vertical position according to the time and penetration speed, while the external surfaces of the cube are exposed to a convection medium.

This thermal problem can be described by the diffusion equation as

$$\frac{\partial^2 T}{\partial x^2}+\frac{\partial^2 T}{\partial y^2}+\frac{\partial^2 T}{\partial z^2}=\frac{1}{\alpha}\frac{\partial T}{\partial t} \tag{1a}$$

In the region R (0 < x < a, 0 < y < b, 0 < z < c) and t> 0, subjected to the boundary conditions in the cutting interface as

$$-k\left.\frac{\partial T}{\partial y}\right|_{y=y_h=u\times t_a}=q(y_h,t_a)\ \ on\, S_1 \tag{1b}$$

It should be observed that for $t<t_a$ the drill not yet achieve the y_h location and therefore the respective location ($y>y_h$) represents a point of the internal domain. All the remaining external surfaces are subjected to a convection medium, it means,

$$-k\frac{\partial T}{\partial \eta}=h_i(T-T_\infty) \tag{1c}$$

where $\partial/\partial\eta$ denotes differentiation along the outward-drawn, normal to the boundary surface S, h_i is the heat transfer coefficient, T_∞ is the ambient temperature.

Since material is being retired during the drilling process, the geometry of the work piece modifies at each instant of cutting. The hole that appears inside the work piece (Fig. 1) for each instant *t* represents a new boundary of the piece. In this case, due to the fact that there is no lateral contact with the drill that surface, here, this surface is considered to be exposed to a convection medium with a heat transfer coefficient, h_i. This boundary condition can be better described in curvilinear coordinates as

$$k\left.\frac{\partial T(r,y,t)_{y<y_h}}{\partial r}\right|_{r=r_h}=h_i(T(r_h,y,t)_{y<y_h}-T_\infty) \tag{2}$$

where r_h is the drill radius and is defined in Fig.1. The numerical solution of the direct problem, shown in Fig. 1, was developed using finite volume method [6] with half volumes applied at the boundaries. The constructed mesh was structured and non-uniform. A fully implicit numerical integration formulation was used.

2.2. Inverse Technique

The inverse problem solution technique Based on Green's Functions and Dynamic Observers [2] can be divided in two distinct steps: i) obtaining of the transfer function model G_H; Eq. (1); ii) obtaining of the heat transfer functions G_Q and G_N and the building algorithm identification, respectively. The transfer function model, G_H, is obtained from the equivalent dynamic systems theory and using Green's functions. The G_Q and G_N are obtained by following the procedure presented by Blum and Marquardt [1].

Transfer Function Model identification (G_H).

The solution of Eqs. (1) can be given in terms of Green's function as, by Beck et al. [7]

$$T(x,y,z,t) = \int_{\tau=0}^{t} \left[G_h(x,y,z,t/\tau)\ q(\tau) \right] d\tau \tag{3}$$

where

$$G_h(x,y,z,t/\tau) = \frac{\alpha}{k} \int_0^{x_h} \int_0^{z_h} G_h^{+}(x,y,z,t/x',y',z',\tau)\Big|_{y'=y_H} dx'dz'$$

and $G_h^{+}(x,y,z,t-\tau)$ represents the Green's function of the thermal problem given by Eq.(1).

The Green's function is available for the homogeneous version associated with the problem defined by Eqs. (1). Although the analytical Green's function is available and exists [7], it will not be used in this work. Instead, the solution of the problem defined by Eqs. (1) will be performed numerically. Since the convection boundary can be treated as a homogenous condition, the heat flux imposed is the only one active condition. In this case, just the bounds of S_1 area are need to be calculated in Eq. (3).

Equation (3) reveals that an equivalent thermal model can be associated with a dynamic model. It means, a response of the input/output system can be associated to Eq. (3) in the Laplace domain as the convolution product [8].

$$T(x,y,z,t) = G_h(x,y,z,t-\tau) * q\ (\tau) \tag{4}$$

This dynamic system can be represented as shown in Fig. (2). Equation (4) can also be evaluated in the Laplace domain as a single product

$$\overline{T}(x,y,z,s) = \overline{G}_h(x,y,z,s)\ \overline{q}(s) \tag{5}$$

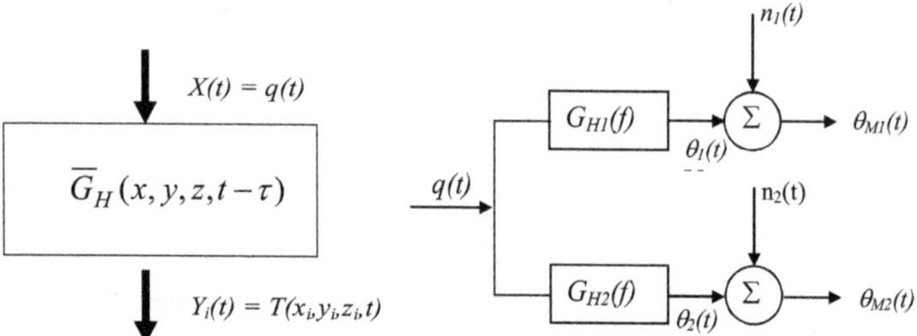

Figure 2: Dynamic thermal model system

The complete G_H identification is obtained using an auxiliary problem. The heat transfer function $\overline{G}_H(x,y,z,s)$ can, then, be obtained through the auxiliary problem which is a homogenous version of the problem defined by Eq.(1) for the same region with a zero initial temperature and unit impulsive source located at the region of the original heating.

Inverse algorithm

A similar representation of Eq.(5) can be used to represent the single-input/two-output system. The only change that must be made is to consider the sum of temperature and the sum of heat transfer function in the system.

It can be observed that if the unknown heat flux ***q(s)*** is applied to the conductor (reference model), $G_{H,}$ and results in a measurement signal θ_M corrupted by noise *N*,

$$\theta_M = \theta + N = G_H \; q + N \tag{6}$$

If the estimate value $\hat{q}$ can be computed from the output data θ_M Thus, the estimator can be represented in a closed-loop transfer function of the feedback loop (Fig. 2b) [1] as

$$\hat{q} = G_Q \, q + G_N \, N \tag{7}$$

where the transfer function G_Q is chosen to have the behavior of type I chebychev filter and G_N is identified by

$$G_N = G_Q G_H^{-1} \tag{8}$$

provided that G_H is obtained.

It can be observed in Eq. (7) if the algorithm estimates the heat flux correctly, G_Q is equal to unity, $G_Q = 1$, and the frequency ***w*** is within the pass band. In this case, the noise transfer function G_N is equal to the inverse transfer function of the heat conductor, G_H^{-1}, (Eq. 8).

The inverse algorithm represented by the closed-loop transfer function of the feedback loop (Fig. 2b) can be resumed in the use of two discrete-time difference equations as

$$q(k) = \sum_{i=0}^{n_n} b_i Y_M(k-i) - \sum_{i=1}^{n_n} a_i q(k-i) \tag{9}$$

and

$$\hat{q}(k) = \sum_{i=0}^{n_n} b_i q(k-i) - \sum_{i=1}^{n_n} a_i \hat{q}(k-i) \tag{10}$$

where the coefficients $\boldsymbol{a_i}$ and $\boldsymbol{b_i}$ that appear in Eqs. (9) and (10) are obtained using Eq.(8). In this case, the inverse procedure is concluded with the $\overline{G}_H(x,y,s)$ identification.

2. Results and Discussion

The inverse technique based on Green's functions and dynamic observers with global transfer function was applied to estimate heat flux in the cutting interface of HSLA micro alloyed steel and the piece temperature field during drilling process. Once this study focuses the heat flux estimation the influence of cutting parameters in the temperature variation was not evaluated. Due to that all the experiments were realized under the same cutting conditions.

3.1. Experimental Setup

Cylindrical plate samples were used in the experimental tests. The drilling tests were carried out, without a coolant fluid on a CNC Di scovery 760 machining using a cemented carbide tool with 10mm diameter. Four k-type thermocouples were attached on the cylindrical plate side. The experiment consists to drill the center of the piece, generating a hole with 36 mm of depth and 10 mm of diameter. Dimensions and thermocouples position are shown in Fig. 3. Experimental tests were performed with cutting speed of 2000 rpm and feed rate of 20 mm/min. Complete experimental set up is shown in Fig. 3.

Heat flux estimative at the cutting interface is then calculated by the inverse technique using two of the thermocouples information, T_1 and T_2.It means that the global heat transfer function is derived by using only the thermocouples at the positions 1 and 2. Figure 4 b shows calculated heat flux for the drilling conditions considered.

Figure 6 shows the complete temperature field calculated using the typical heat flux (Fig. 4 b)

Figure 3: Micro alloyed steel piece and thermocouples position (piece scheme).

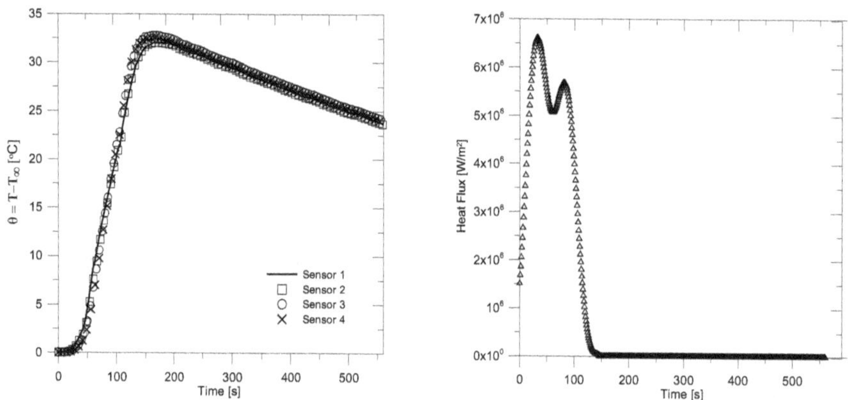

Figure 4: a) experimental temperature for each thermocouple (Fig.3); b) Typical heat flux using 1 and 2.

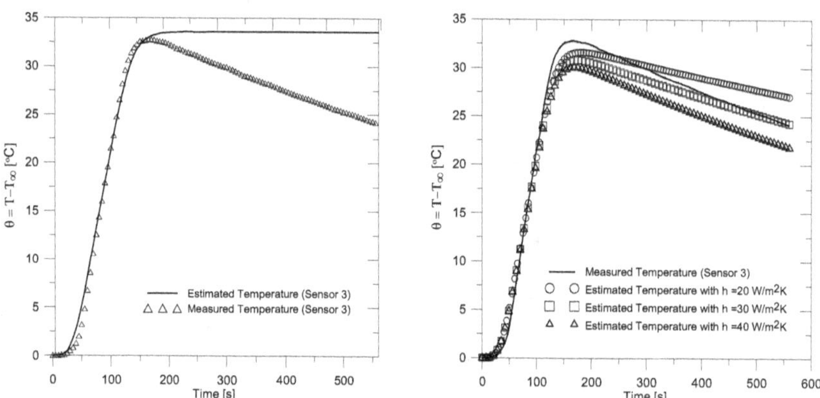

Figure 5: Experimental and calculated temperatures at position 3 considering different values for h_i in Eq. (2) a) h_i=0 and b) h_i=20, 30 and 40 W/m^{2}K

It is observed in Fig. 5 a good ac cordance between the experimental and c alculated temperatures with a maximum deviation of 2 °C during the heating stage. This deviation can be explained due t o the numerical approaches of the direct solution. For times higher than 110s the difference between the temperatures becomes greater due t o the convection effects that were considered null in the direct solution hypotheses (Fig 5 a) .The effect of convection can be better observed in Fig. 5b by using different values of h_i in direct solution, Eq.(1-2). It can be concluded that although these effects are small during the heating time (cutting stage) the heat transfer coefficients developing an important role and must be taken into account in both direct and inverse models.

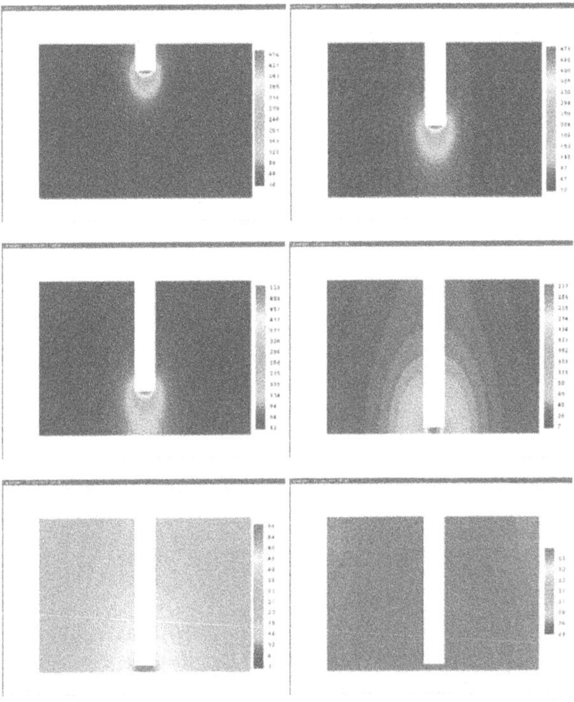

Figure 6: Temperature field θ (T_0 =26.36 °C) for times 10, 50, 70, 110, 150 and 170s.

3. Conclusions

The technique based on Green's function and dynamic observers is applied here to solve with success a drilling inverse problem. This study shows that the temperature and heat flux at the tool-piece interface during a drilling process can be estimated with a suitable experimental apparatus avoiding problems of sensitivity.

4. Acknowledgments

The authors thank to CAPES, Fapemig and CNPq.

5. References

[1] J. Blum, J. and e W. Marquardt, W., "An Optimal Solution to Inverse Heat Conduction Problems Based on Frequency-Domain Interpretation e Observers", Numerical Heat Transfer, parte B, Vol. 32, pp. 453-478, 1997.

[2] Sousa, P. F. B., Development of a Technique Based on Green's Functions and Dynamic Observers to Be Applied in Inverse Problems, Dissertation of Master's degree, Federal University of Uberlândia, MG. Brazil,. 2006

[3] Chen, W-C., "Effect of the Cross-Sectional Shape Design of a Drill Body on Drill Temperature Distributions," Int. Comm. Heat Mass Transfer, Vol 23, No.3. pp. 355-366, 1996.

[4] Shen, Q., Lee, T.C., Lau, W.S., "A Finite-Element Analysis of Temperature Distributions in Spade Drilling," Journal of Materials Processing Technology, Vol. 66, pp.112-122, 1997.

[5] Shatla, M. Altan, T., "Analytical modeling of drilling and ball end milling," Journal of Materials Processing Technology, Vol. 98, pp.125-133, 2000.

[6] Patankar, S.V., Numerical Heat Transfer e Fluid Flow. New York: Hemisphere Publishing Corporation, 1980.

[7] Beck, J. V., and Heat Conduction Using Green´s Function. Washington, DC: Hemisphere publishing, 1992.

[8] Özisik, M. N. Heat Conduction, John Wiley & Sons, New York. 1993.

[9] Sousa., P. F. B, Thermal Studies of Cutting Process. Inverse Problem Application in Drilling, 172p, Doctorate Thesis, Federal University of Uberlândia, MG. Brazil,. 2009

[10] Borges, V. L., Desenvolvimento do Método de Aquecimento Plano Parcial Para a D eterminação Simultânea de Propriedades Térmicas Sem o Uso de Transdutores de Fluxo de Calor. 108f. Doctorate Thesis, Federal University of Uberlândia, MG. Brazil. 2008.

Analyses of the Effects of Cutting Parameters on the Cutting Edge Temperatures Using Inverse Heat Conduction Technique

Marcelo R. Santos[1], Gilmar Guimarães[1], Sandro M. M. Lima e Silva[2] and Solidônio R. Carvalho[1]

[1]*School of Mechanical Engineering*
Federal University of Uberlândia, Uberlândia, MG, Brazil
Email: srcarvalho@mecanica.ufu.br
[2]*Mechanical Engineering Institute*
Federal University of Itajubá, Itajubá, MG, Brazil
Email: metrevel@unifei.edu.br

Abstract

During machining a considerable amount of energy is transformed into heat. This happens due to plastic deformation of the workpiece surface, the friction of the chip on the tool face and the friction between the tool and the workpiece. High temperatures are generated in the region of the tool cutting edge, and these temperatures have a very important influence on the rate of wear of the cutting tool and on the friction between the chip and the tool and mainly on the tool life. In this sense, this work proposes the estimation of the temperature and the heat flux at the chip-tool interface using inverse techniques. Besides an investigation of different factors which influence the temperature distribution at the high speed steel AISI M32 C tool rake face during machining of a ABNT 12L14 steel workpiece is accomplished. The temperature distribution was predicted using finite volume elements with an inverse heat conduction problem technique. A transient 3D numerical code using irregular and non-staggered mesh was developed to solve the non-linear heat diffusion equation. To validate the software some experimental tests are conducted. The inverse problem was solved by using the function specification method. Experimental temperatures were measured by thermocouples on accessible locations of the tool. Heat fluxes at the tool–workpiece interface are estimated using inverse problems and the experimental temperatures. Some tests were performed to study the effect of cutting parameters on the cutting edge temperature. The results were compared to tool-work thermocouple technique and presented good agreement.

Key Words: machining, temperature, inverse problems, heat transfer

1. Introduction

Nowadays, several researches have been proposed the combination of inverse techniques and analytical or numerical heat transfer problems to analyze the thermal fields during machining processes. In this sense, one way to study the heat transfer problem is to consider the heat source known and to determine the temperature fields from the solution of the heat diffusion equation [1, 2, 3, 4, 5 and 6]. In agreement with the literature, this methodology is called direct problem in heat transfer. However, during machining the experimental heat source is unknown and inverse techniques can be used to predict this parameter. The proposal is to estimate the transient heat flux from a numerical model based on heat difusion equation using inverse problems and experimental temperatures measured in accessible regions of the sample.

In Chen et al. [7] the inverse technique used is based on a sequential method with specification function proposed by Beck et al. [8]. The numerical technique used was the finite volume method and the temperatures were obtained by inserting a thermocouple in the tool. In both methods the model just took into account the insert and some discrepancies were observed after calculating and measuring the temperature.

Lazard and Corvisier [9] considered the problem of estimating the transient temperature and the heat flux at the tip-tool interface during a turning process using an inverse approach. The heat transfer model was based on a quadrupole formulation that is commonly used to solve ordinary differential equations in the Laplace domain. The results of temperature obtained with the analytical model used were in good agreement with those obtained with FLUENT. The authors assumed that the temperature measurements were available for the inverse analysis. It means the experimental temperature was calculated with the addition of noise.

In the work of Yvonnet et al. [10] an innovative approach was proposed. This work was based on a simple inverse procedure to identify both the heat flux flowing into the tool through the rake face and the heat transfer coefficient between the tool and the environment during a typical orthogonal cutting process. To determine the heat flux and convection heat transfer coefficient an iterative Newton–Raphson procedure was used to minimize the error defined by the difference between experimental and calculated temperatures. Four different tools were used, three of them were manufactured by electrical discharge machine (EDM) cutting the tight slot with different distances between the tool tip and the slot, namely 0.35, 0.50 and 0.60 mm. The last one was instrumented only with the thermocouple in order to measure the whole average heat flux flowing into the tool through the rake face.

Woodbury et al. [11] demonstrated the solution of a three dimensional inverse heat conduction problem using an Evolutionary Algorithm (EA). The heat flux from a workpiece into a tool during a turning process was determined using evolutionary operations combined with measurements of the surface temperature on the tool. The three-dimensional conduction in the tool and tool holder was simulated using FLUENT.

The aim of the present paper is to estimate the heat flux and the three-dimensional temperature field on the cutting interface of the high speed steel tool. In addition, the influence of the cutting parameters (cutting speed, depth and feed rate) on the tool-chip interface was also investigated. The methodology is validated with a controlled experiment in the laboratory, in which the sample is subjected to a heat flux in one side and the other faces are insulated. The results obtained in this work were compared with the results obtained by using the Tool-workpiece thermocouple technique [12].

2. The direct problem: thermal model

Figure 1 shows the schematic model developed in this work for the high speed steel tool and the dimensions of the sample. The interface contact area, $A_q(x,y)$ is subjected to the heat flux $q''_o(t)$ generated due to contact with the tool and the piece. In addition, as remaining boundary conditions, it was considered a constant convective heat transfer coefficient equals to 20 (W/m²K).

The three-dimensional physical problem was solved in Cartesian coordinates, with a irregular mesh, using finite volumes technique. The objective is to obtain the temperature distribution in the tool using the direct problem and to estimate the heat flux generated in the chip-tool interface based on inverse techniques.

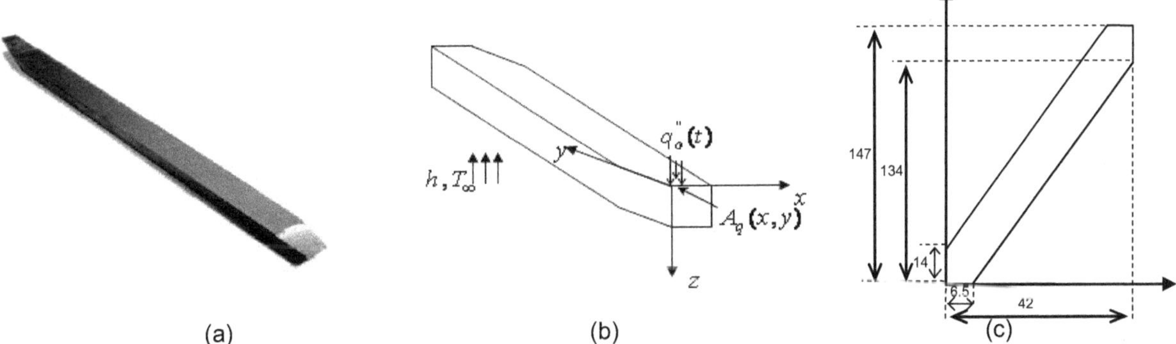

Figure 1: (a) Real high speed steel tool; (b) three-dimensional physical model and (c) dimensions of the tool in millimeters (mm) where the coordinate z is equals to 9.5 (mm)

The choice of the high speed steel tool is due to its great application in industrial area and also because Luiz [12] presents a thermal analysis of this tool using Tool-workpiece thermocouple technique (TWT). It provides to compare the results of this work with other ones obtained through a different technique.

The thermal problem presented in Fig. 1b can be described for the heat diffusion equation as:

$$\frac{\partial}{\partial x}\left(\lambda\frac{\partial T}{\partial x}\right)+\frac{\partial}{\partial y}\left(\lambda\frac{\partial T}{\partial y}\right)+\frac{\partial}{\partial z}\left(\lambda\frac{\partial T}{\partial z}\right)=\rho Cp\frac{\partial T}{\partial t} \quad (1)$$

The boundary conditions imposed to the problem presented in the Eq. (1) can be written by

$$-\lambda\frac{\partial T}{\partial \eta}=h(T-T_{\infty}) \tag{2}$$

on the regions exposed to the environment and

$$-\lambda\frac{\partial T}{\partial \eta}=q_o''(t) \tag{3}$$

at the interface defined as A_q, where η represents the normal outside in the coordinates x, y, and z, T the temperature, T_{∞} the room temperature, λ the thermal conductivity, ρCp the volumetric heat capacity and h the heat transfer coefficient. The initial condition is

$$T(x,y,z,0)=T_o \tag{4}$$

where T_o represents the initial temperature of the tool, shim and tool holder.

In this work the thermal properties of the tool were obtained from [13] varying with temperature as shown in Table 1.

Table 1: Thermal properties of the high speed steel tool (AISI M32 C, with 10 % of cobalt)

Range of temperature (°C)	$0 \geq T \leq 400$	$T > 400$
Thermal conductivity (W/mK)	$0.0105 \cdot T + 23.8$	$-0.005 \cdot T + 30$
Thermal diffusivity (m²/s)	$-5.03 \cdot 10^{-10} \cdot T + 7.02 \cdot 10^{-6}$	$-5.94 \cdot 10^{-9} \cdot T + 9.19 \cdot 10^{-6}$

3. The inverse problem: function specification procedure

A convenient way to apply inverse problems in heat conduction problems is to adopt a functional form for the heat flux variation with time. That is called function specification procedure [8]. In this case the function can be a sequence of constant, linear, parabolic, cubic or exponential segments. This methodology consists of estimating the heat flux sequentially, beginning with initial temperatures and moving forward to successively values according to the time step. In this work, a constant function specification was adopted and future time steps and various sensors/thermocouples were considered during the estimation procedure. Regularization was not implemented, and the sensibilities of the thermal model were obtained previously from the direct problem according to thermocouple's position. The experimental temperatures measured in the tool during turning were also used to estimate the heat flux for each time step.

4. Experimental procedure

The machining test was carried out in a conventional lathe IMOR MAXI–II–520–6CV without coolant. The material used in the experimental tests were cylindrical bars ABNT 12L14 [12] with an external diameter of 50.2 mm. According to Fig. 2 the temperatures were measured on accessible locations of the insert by using thermocouple T type and a data acquisition system HP 75000 Series B with a voltmeter E1326B controlled by a PC.

(a)

(b)

Figure 2: (a) Thermocouples positioned in the tool and (b) bench of tests.

Table 2 presents the location of thermocouple according to the coordinate system shown in Fig. 1. To evaluate the influence of the machining conditions - depth of cut (Ap), feed rate (f) and cutting speed (Vc) - in the temperature in chip-tool interface, the following tests were accomplished, accordingly to Tables 3, 4 and 5.

Table 2: Thermocouple location according to the coordinate system defined in Fig. 1.

Position/Thermocouple	*1*	*2*	*3*	*4*	*5*
x [mm]	0.61	2.70	0.0	3.30	2.00
y [mm]	7.20	8.50	9.0	7.00	3.40
z [mm]	0.0	0.0	5.0	9.50	9.50

Table 3: Depth of cut (constant parameters: Vc= 56 m/min and f = 0.138 mm/rev)

	Unit [mm]			
Ap (measured from the ray of the piece)	0.5	1.0	1.5	2.0

Table 4: Feed rate (constant parameters: Vc= 56 m/min and Ap= 1.0 mm)

	Unit [mm/rev]					
f	0.138	0.162	0.176	0.204	0.242	0.298

Table 5: Cutting speed (constant parameters: f = 0.138 mm/rev and Ap = 1.0 mm)

	Unit (m/min)													
Vc	4.4	7.1	8.8	11.2	14.2	17.7	22.1	28.4	35.3	44.2	56	88.3	112	142

In this work, the chip-tool interface (A_q) was measured with the software GLOBAL LAB Image. After each cutting condition foreseen in the previous tables the tool edge was analyzed with a video camera Hitachi CCD KP-110 connected to a PC. To compare and analyze the temperatures in chip-tool interface, these were also measured using the tool-workpiece thermocouple technique as presented by Luiz [12].

5. Results and discussions

Figure 3 presents the mean temperature in chip-tool interface according to the cutting conditions defined in Tables 3 to 5. These values were calculated using the numerical methodology proposed in this work and measured with the experimental technique TWT [12].

In a general way, as defined in literature [14], the temperature in the cut interface increases with the increase of the cut conditions. In other words, increasing the cut conditions the chip-tool contact area and the rate of deformation also increase, what provides more energy and consequently high temperatures in the cut interface.

Figure 4 shows the difference between the temperature profile measured with TWT and calculated with numerical techniques. Considering the experimental technique TWT as a reference, it can be seen a significant difference between such temperatures in Fig. 4a and Fig. 4b. However, analyzing Fig. 4c it is verified a decrease of this difference when the cutting speed increases. [15] shows that cutting speeds between 100 to 200 m/min provides temperatures in the cutting interface between 600 °C and 800 °C , considering high speed steel tool with low carbon. So, it can be seen a good agreement among the values presented in this work and other ones available in literature

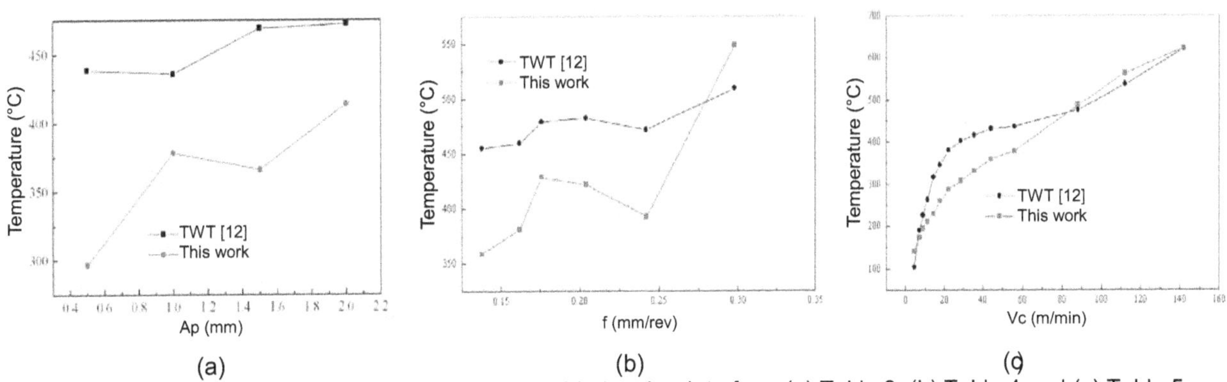

Figure 3: Temperature in chip-tool interface considering the data from (a) Table 3, (b) Table 4 and (c) Table 5

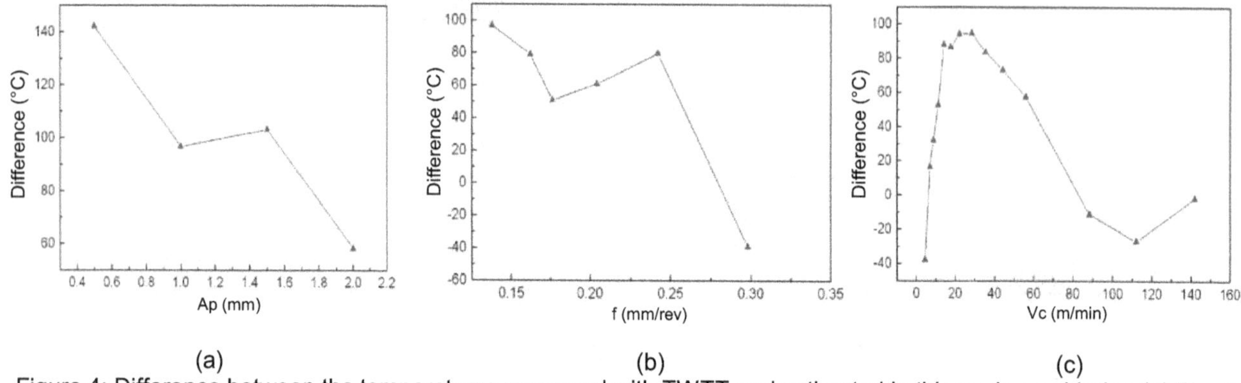

(a) (b) (c)

Figure 4: Difference between the temperatures measured with TWTT and estimated in this work considering (a) Fig. 3a, (b) Fig. 3b and (c) Fig. 3c

Figure 5 presents the temperature profile in the tool for the following cutting conditions Vc=142 m/min, f=0.138 mm/rev and Ap=1.0 mm, where the maximum temperature at the chip-tool interface was indentified: 600 °C.

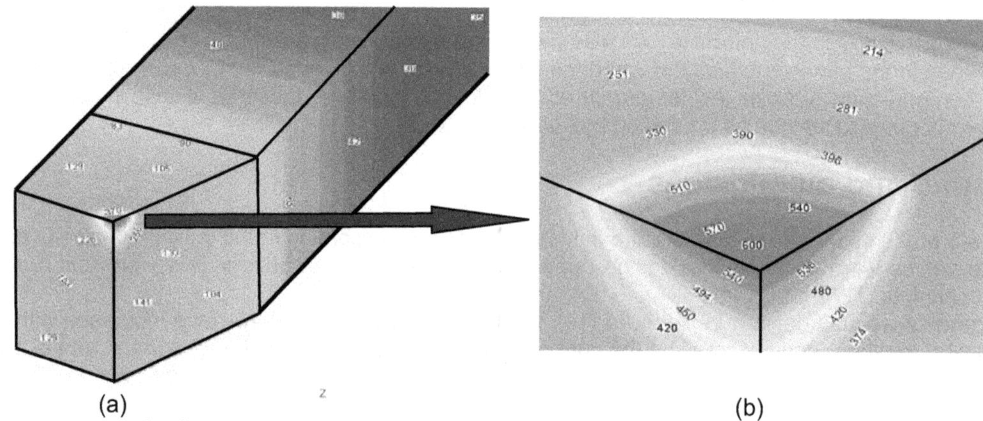

(a) (b)

Figure 5: Temperature profile for the cutting speed of 142 m/min, feed rate of 0.138 mm/rev and depth of cut of 1.0mm: (a) Tool: scale 3:1 and (b) Cutting interface: scale 20:1

It is important to notice that both methodologies have sources of errors that can influence directly in the measured and calculated temperatures and it can justify the differences presented in Fig. 3. Considering the numerical methodology presented in this work, the uncertainties are related to the dimensions and simplifications adopted to simulate the thermal behavior of tool; the correct identification of the thermal properties of the high speed steel and the convective heat transfer coefficient both varying with temperature; the experimental temperature measured in the tool during turning and the correct identification of the chip-tool contact area for each cutting condition.

According to [12] the uncertainties in TWT technique are related to the calibration of the system and the assembly of the experimental components. The TWT basically consists in an electric circuit that involves the tool, the piece and the lathe. The tool and the workpiece as thermocouple has to be calibrated previously as any conventional sensor and the experimental procedure is strongly dependent of the equipments and the ability of the operator. Besides, during turning, a mercury chamber has to be used to connect the components of the system due to the rotation of the piece as presented in Fig. 2a. Also the attrition between the tool and piece provides some noises in measured temperature that have to be minimized using statistical tools to remove excessive noises and to define the mean temperature and the standard deviation.

6. Conclusions

A numerical methodology based on inverse problems is proposed in this work to calculate the temperatures in chip tool-interface during turning of cylindrical bars ABNT 12L14 with high speed steel tool (AISI M32 C, with 10 % of cobalt).

The results were compared to an experimental technique called tool-workpiece thermocouple as presented by [12]. The objective was to obtain and compare the mean temperature in chip-tool interface for several cutting conditions. Once the heat flux generated in the chip-tool interface is unknown, this work only presented a qualitative analysis of the results. In literature is verified that the majority of papers present only the average temperature at the cutting interface. It means, they assume that the machining process happens in steady state. Few works are concerned to analyze the thermal influence of the variation of the cutting conditions and the transient process because, during turning, the thermal analysis at the tool is highly complex and involves a great amount of variables that directly interfere in this process. Based on the presented results and the analyzed works, it can be said that until now none of the existing technique is universally accepted as absolute [15]. In reality, what exist are attempts to understand the basic points of the heat transfer process during the machining by turning.

7. Acknowledgments

The authors thank CNPq, FAPEMIG and CAPES for the financial support, without which this work would not be possible.

8. References

[1] Stephenson, D. A.; Ali, A., "Tool Temperatures in Interrupted Metal Cutting", Journal of Engineering for Industry, Vol. 114, pp. 127-133, 1992.

[2] Radulescu, R.; Kapoor, S. G., "An analytical Model for Prediction of Tool Temperature-fields During Continuos and Interrupted Cutting", Journal of Engineering for Industry, Vol. 116, pp. 135-143, 1994.

[3] Shijun, Z.; Zhanqiang, L., "Analytical and Numerical Solutions of Transient Heat Conduction in Monolayer-coated Tools", Journal of Materials Processing Technology Vol. 209, pp. 2369-2376, 2008.

[4] Grzesik, W.; Bartoszuk, M.; Nielonsy, P., "Finite Difference Analysis of the Thermal Behavior of Coated Tools in Orthogonal Cutting of Steels", International Journal of Machine Tools & Manufacture, Vol. 44, pp. 1451–1462, 2004.

[5] Majumdar, P.; Jayaramachandran, R.; Ganesan, S., "Finite Element Analysis of Temperature Rise in Metal Cutting Processes", Applied Thermal Engineering Vol. 25, pp. 2152–2168, 2005.

[6] Liu, J.; Chou, Y. K. "On Temperatures and Tool Wear in Machining Hypereutectic Al–Si Alloys with Vortex-tube Cooling", International Journal of Machine Tools & Manufacture, Vol. 47, pp. 635–645, 2007.

[7] Chen, W. C.; Tsao, C. C.; Liang, P. W., "Determination of Temperature Distributions on the Rake Face of Cutting Tools Using a Remote Method" International Communications Heat Mass Transfer, Vol. 24, pp. 161-170, 1997.

[8] Beck, J. V.; Blackwell, B.; St. Clair, C., Inverse Heat Conduction: Ill-posed Problems, Wiley-Interscience Publication, New York, 1985.

[9] Lazard, M.; Corvisier, P., , "Inverse Method for Transient Temperature Estimation During Machining", *Proceedings of the 5th International Conference on Inverse Problems in Engineering: Theory and Practice*, 2005, Cambridge, UK.

[10] Yvonnet, J.; Umbrello, D.; Chinestaa, F.; Micari, F., "A Simple Inverse Procedure to Determine Heat Flux on the Tool in Orthogonal Cutting", International Journal of Machine Tools & Manufacture, Vol. 46, pp. 820–827, 2006.

[11] Woodbury, K. A.; Duvvuri, S.; Chou, Y. K; Liu, J., "Use of Evolutionary Algorithms to Determine Tool Heat Fluxes in a Machining Operation" *Proceedings of the Inverse Problems Design and Optimization Symposium, IPDO*, 2007, Miami Beach, Florida, USA.

[12] Luiz, N. E. "Usinabilidade do aço de corte fácil baixo carbono ao chumbo abnt 12l14 com diferentes níveis de elementos químicos residuais (cromo, níquel e cobre)", Doctorate thesis, School of Mechanical Engineering, Federal University of Uberlândia, Brazil, pp. 189, 2007.

[13] Taylor Specials Steels Ltda, http://www.taylorspecialsteels.co.uk/pdfdownload/m35.pdf#search='M35%20thermal%20properties, Accessed in 12th November, 2005.

[14] Trent, E. M., Metal cutting, 2° Ed, Butterworths, London, 1984

[15] Machado, A. R.; Silva, M.B., Usinagem dos Metais, 8ª Edition, Brazil, 2004.

ESTIMATION OF THE EVOLUTION OF THE 2D AXISYMMETRIC SHAPE

Morgan Dal, Philippe Le Masson, Muriel Carin

Laboratory of Material Engineering of Brittany
University of South Brittany, European University of Brittany
Lorient cedex, France.
Email: morgan.dal@univ-ubs.fr

Abstract

The paper deals with heat transfer analysis in a static welding process: a method is developed to estimate the evolution of the 2D axisymmetric shape of a forehead of fusion. The estimation is made from temperature measurements taken in the solid part. The estimation procedure is based on the conjugate gradient method coupled to the iterative regularization algorithm. The gradient is computed using adjoint equations. The goal of this work is to apply an ALE (Arbitrary Lagrangian Eulerian) moving mesh method suitable for solving two-dimensional and axisymmetric moving-boundary problems. In fact, we do not want to model the fluid mechanics in the fused zone. Thus, at each time increment, the shape of the domain is modeled considering the fusion of the material.

Key Words: shape identification, conjugate gradient method, moving mesh, adjoint state, iterative regularization method.

1. Introduction

Welding is an assembly method which is widely used in the industrial context. Moreover, arc welding leads to other methods. The thermal evolution of the workpiece produces mechanical effects, like residual stresses or distortions. To control them, numerical simulation is used. However, the GTAW process is composed of three parts, the cathode, the arc column and the anode (the workpiece), and the ending weld pool shape is conditioned by all of them. The difficulty of the numerical study is the connection between all physical phenomena. In order to avoid it, several simplified models can be distinguished. The first remains multiphysic, but it considers only the anode [1] and has to emit assumptions concerning the boundary heat input. Nevertheless, the fluid mechanics, the electromagnetism, and the free boundary problems are considered. In the second one, only the thermal problem is studied [2] and an equivalent heat source is designed to account for the other physical problems. Finally, only the solid part is considered in the third one. The liquid/solid interface is treated as a free boundary with an imposed temperature [3]. The latter compels one to acquire experimental data to estimate the front motion.

In this study, a gradient based method coupled to a simplified simulation is developed to estimate, in a transient case, the evolution of a liquid/solid interface. The fusion is created by static GTAW welding applied for three seconds. Before the experimental case, the direct problem and the inverse method must be theoretically validated with a known front evolution. In this paper, the direct problem is introduced with the difficulties related to the main assumption. Then, the inverse problem is presented through the sensitivity and adjoint problems. Finally, the identification is applied to more realistic noisy measurements.

2. Direct problem

A simplified axisymmetric finite element model is proposed and compared to a multiphysic simulation (anode only). Both are computed with the commercial software COMSOL Multiphysics®. The validation method begins by the extraction of the solidus temperature shape evolution from the multiphysic problem. Then, this evolution is introduced in the free boundary problem as input data. The validation of this approach is obtained with a comparison of the two thermal fields.

The input data (Fig. 1) are curves representing the coordinates of the free boundary for each step time. Weld pool shapes are imposed by the ALE (Arbitrary Lagrangian Eulerian [4]) moving mesh method. In this geometry, only the energy conservation Eq. (1) is solved. Where θ is a dimensionless temperature ($\theta=(T-T_{fus})/(T_{fus}-T_{\infty})$) to avoid values of thermal gradients that are too high.

$$\rho(\theta)\,cp(\theta)\frac{\partial\theta}{\partial t}=\frac{1}{r}\frac{\partial}{\partial r}\left(r\,\lambda(\theta)\frac{\partial\theta}{\partial r}\right)+\frac{\partial}{\partial z}\left(\lambda(\theta)\frac{\partial\theta}{\partial z}\right) \tag{1}$$

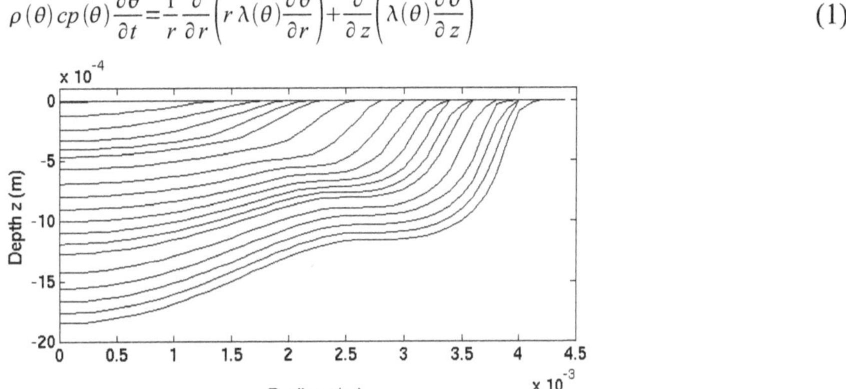

Figure 1: Multiphysic front locations.

An example of the geometry and the mesh size are summarized in Fig. 2. In this simulation, we work on the evolution of the boundary [AC]. Thus, point B is purely theoretical as its location is time dependent. The motion of the free boundary is defined along the z axis. An automatic re-meshing procedure is developed to overcome the mesh quality losses due to the large deformation.

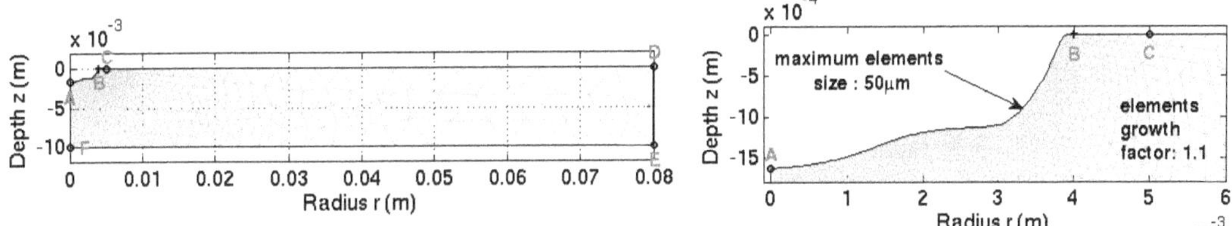

Figure 2: Geometry, mesh, and boundary conditions.

The thermal boundary conditions are expressed as Eq. (2) for convective and radiative heat losses ([BC]∪[CD]∪[DE]∪[EF]), Eq. (3) for the liquid/solid interface ([AB]) and Eq. (4) at the pool center line ([FA]).

$$-\lambda(T)\frac{\partial\theta}{\partial n}=h(\theta-\theta_{\infty})+\varepsilon\sigma(\theta^4-\theta_{\infty}^4)\approx h_{tot}(\theta-\theta_{\infty}) \tag{2}$$

Where h_{tot} is a linearized coefficient to approximate losses.

$$\theta=\theta_{solidus} \tag{3}$$

$$-\lambda(T)\frac{\partial\theta}{\partial n}=0 \tag{4}$$

In practice, the main difficulty is to impose the solidus temperature and heat losses on the same geometric boundary [AC] (Fig. 2). In this work, the Dirichlet condition is approximated by a "convective heat loss" with an infinite coefficient and an "ambient" temperature equal to the solidus one. This numerical tool induces a thermal gradient peak located on the B point. To avoid this singularity, the thermal problem is solved with linear elements [5].

The comparison between the thermal fields given by the multiphysic and the simplified simulations is shown in Fig. 3. Two levels of validation can be distinguished. The first one (Fig. 3 (a)) is a spatial validation at the last time increment (3s), based on the relative discrepancy between the two results. To enhance the readability, the graph is presented only for the section (1cm*1cm) surrounding the fusion zone. It should be noted that the deviations are less than 2%. The second one (Fig. 3 (b)), is a thermal validation for temperature evolutions on several points located at 0.5mm from the free boundary. Here, it should also be noted that the deviations are less than 0.2%.

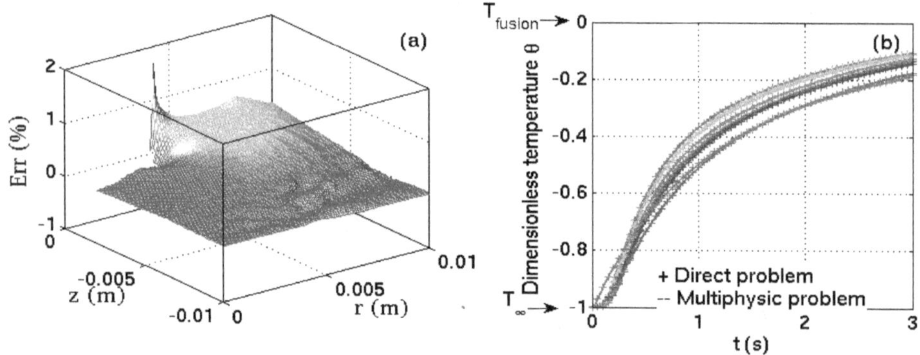

Figure 3: Validation of the direct problem, (a) the relative error (b) the temporal evolutions.

A fairly good agreement is found between the two simulation methods. According to these results, the simplified model is employed for the development of the inverse problem. Nevertheless, it should be noted that the maximal error is localized at B.

3. Estimation and minimization procedure

The present work can be qualified as a shape identification because the aim is to determine the front motion depending on both radius and time $dz=\sigma(r, t)$. Referring to Eq. (5), let the temperature taken by the simulated thermocouples be denoted $Y_m(t)$. As this theoretical study is realized to check a possibly experimental development, the theoretical measurements are located in the solid part at 1.5mm from the interface. Seven measurements are distributed all around the fusion zone to obtain as much information as possible.

In order to reduce the number of unknown variables, the curves representing the interface are parameterized by a piecewise cubic interpolation (7 points + 1 point (at z=0)). This last method is not the most efficient, but it is the most reliable.

The boundary displacement is considered as obtained in such a way that the Eq. (5), the residual functional, is minimized.

$$J[\sigma(r,t)] \equiv J(r,t) = \frac{1}{2}\sum_{m=1}^{M} \frac{r_m}{R} \int_{t_0}^{t_{fin}} [\theta(r_m, z_m, t; \sigma) - Y_m(t)]^2 dt \tag{5}$$

Where M is the total thermal measurements, r_m and z_m are the coordinates of the measurement: m. θ is the dimensionless temperature extracted from the direct problem at this m point.

3.1 The conjugate gradient method for minimization

The following method coupled to the quadratic criterion allows for the global minimization of the residual functional. It is an iterative process as shown by Eq. (6):

$$\sigma(r,t)^{k+1} = \sigma(r,t)^k + \gamma^k d^k \tag{6}$$

where k is the iteration index. γ is the descent step size and d the descent direction defined by Eq (7):

$$d^k = -\nabla J^k + \beta^{k-1} d^{k-1} \tag{7}$$

With ∇J, the gradient of the functional (Eq. (5)) and β, the conjugate coefficient which could be written in different ways [6] (Eq. (8)).

$$(a)\ \beta_{FR}^{k-1} = \frac{\|\nabla J^k\|^2}{\|\nabla J^{k-1}\|^2} \qquad (b)\ \beta_{RP}^{k-1} = \frac{\nabla J^k.(\nabla J^k - \nabla J^{k-1})}{\nabla J^{k-1}.\nabla J^k} \qquad (c)\ \beta_{HS}^{k-1} = \frac{\nabla J^k.(\nabla J^k - \nabla J^{k-1})}{d^{k-1}.(\nabla J^k - \nabla J^{k-1})} \tag{8}$$

At each iteration the optimal step size γ is defined by considering the minimization of $J(\sigma^{k+1})=J(\sigma^k+\gamma^k d^k)$. The literature clearly explains how to find Eq. (9), as in [7].

$$\gamma^k = \sum_{m=1}^{M} r_m \int_{t_0}^{t_{fin}} \left[\theta(r_m, z_m, t;\sigma) - Y_m(t)\right] \delta\theta(r_m, z_m, t;\sigma)\, dt \Big/ \sum_{m=1}^{M} r_m \int_{t_0}^{t_{fin}} \delta\theta(r_m, z_m, t;\sigma)^2\, dt \quad (9)$$

This iterative process is implemented with the software Matlab©, but at each iteration, the search step size γ^k and the gradient ∇J^k, have to be computed. The first one needs the values of $\delta\theta(r_m, z_m, t)$ which is extracted from a sensitivity problem. The second one is expressed after the writing of an adjoint problem. They are both described in the next two subsections.

3.2 *The sensitivity problem*

The sensitivity problem results from the direct problem by applying a small variation $\delta\theta$ on the dimensionless temperature θ. The next expressions (Eq. (10) to Eq. (14)) are found by replacing, in the equation system (Eq. (1) to Eq. (4)), the temperature θ by $\theta+\delta\theta$. The assumption is: a variation $\delta\sigma$ of the boundary leads to a temperature variation $\delta\theta$ and a residual functional variation δJ (Eq. (15)).

$$\frac{\partial(\rho(\theta) cp(\theta) \delta\theta)}{\partial t} = \frac{1}{r}\frac{\partial}{\partial r}\left(r \frac{\partial(\lambda(\theta)\delta\theta)}{\partial r}\right) + \frac{\partial}{\partial z}\left(\frac{\partial(\lambda(\theta)\delta\theta)}{\partial z}\right) \quad (10)$$

$$\delta\theta(r, \sigma, t) = \delta\sigma \frac{\partial\theta}{\partial z}(r, \sigma, t) \quad \text{on [AB] ([8])} \quad (11)$$

$$\frac{-\partial(\lambda(\theta)\delta\theta)}{\partial n} = h_{tot}\,\delta\theta \quad \text{on [BD], [DE] and [EF]} \quad (12)$$

$$\frac{\partial(\lambda(\theta)\delta\theta)}{\partial r} = 0 \quad \text{on [FA]} \quad (13)$$

$$\delta\theta(r, z, t_0) = 0 \quad \text{for } t=0 \quad (14)$$

$$\delta J[\sigma;\delta\sigma] = \sum_{m=1}^{M} \frac{1}{R} \int_{t_0}^{t_{fin}} \left[\theta(r,z,t;\sigma) - Y_m(t)\right] r\, \delta\theta\, \delta(r-r_m)\delta(z-z_m)\, dt \quad (15)$$

In order to account for the shape variation on the boundary [AB] without perturbing the geometry, Taylor's expansion $\theta(\sigma+\delta\sigma)$ is expressed and only the first-order terms are kept. An equivalent thermal variation is obtained for Eq. (11).

3.3 *The adjoint problem and the gradient equation*

Several steps are necessary to obtain the adjoint equations (Eq. (16) to Eq. (20)). The minimization must be realized under constrains (the direct problem). A Lagrange functional $L(\sigma)$ is written including the residual functional shown in Eq. (5) and the equations of the direct problem multiplied by Lagrange multipliers. Then, its variation $\delta L(\sigma)$ is developed from Eq. (15) and the system composed by Eq. (10) to (14). Finally, after integrations by parts, the vanishing of terms factorized by $\delta\theta$ or its derivations leads to Eq. (16) to (20).

$$-\rho(\theta) cp(\theta) \frac{\partial\psi}{\partial t} = \frac{1}{r}\frac{\partial}{\partial r}\left(r\lambda(\theta)\frac{\partial\psi}{\partial r}\right) + \frac{\partial}{\partial z}\left(\lambda(\theta)\frac{\partial\psi}{\partial z}\right) + \frac{1}{R}\sum_{m=1}^{M}\left[\theta(r,z,t;\sigma) - Y_m(t)\right]\delta(r-r_m)\delta(z-z_m) \quad (16)$$

$$\psi(r, \sigma, t) = 0 \quad \text{on [AB]} \quad (17)$$

$$-\lambda(\theta)\frac{\partial\psi}{\partial n} = h_{tot}\,\psi \quad \text{on [BD], [DE] and [EF]} \quad (18)$$

$$\lambda(\theta)\frac{\partial\psi}{\partial r} = 0 \quad \text{on [FA]} \quad (19)$$

$$\psi(r, z, t_{fin}) = 0 \quad \text{for } t=t_{fin} \quad (20)$$

From the definition of the standard inner product in an L^2 space in cylindrical coordinates, the variation of δJ can also by expressed by the following equation:

$$\delta J(\sigma;\delta\sigma)=\int_{t_0}^{t_{fin}}\int_{r_A}^{r_B}\nabla J\,\delta\sigma\, r\,dr\,dt \qquad (21)$$

After the vanishing of the different terms of the Lagrange functional, only one term is left:

$$\delta L(\sigma;\delta\sigma)=-\int_{t_0}^{t_{fin}}\int_{r_A}^{r_B}\lambda(\theta)\frac{\partial\theta}{\partial z}\frac{\partial\psi}{\partial z}(r,\sigma,t)\,\delta\sigma\, r\,dr\,dt \qquad (22)$$

From Eq. (21) and Eq. (22) the gradient functional ∇J can be expressed.

$$\nabla J(\sigma;\delta\sigma)=-\lambda(\theta)\frac{\partial\theta}{\partial z}\frac{\partial\psi}{\partial z}(r,\sigma,t) \qquad (23)$$

3.4 The stopping criterion

Here, the model ignores the measurement errors and a reduction of J to zero is thus possible. In a real application, experimental data (Y_m) contains measurement errors. Thus the discrepancy principle [9] is applied.

Let σ^* be the standard deviation of the measurement errors.

$$|\theta(r_m,z_m,t)-Y_m(t)|\approx\sigma^*\approx 0.0011 \qquad (24)$$

Including Eq. (24) in Eq. (5), the lower criterion can be written:

$$J^*=\sum_{m=1}^{M}\frac{r_m}{R}\int_{t_0}^{t_{fin}}\sigma^{*2}\,dt\approx 0.0027 \qquad (25)$$

The iterative procedure is then interrupted when $J(\sigma^k)=J^*$.

3.5 The algorithm

The three problems: direct, sensitivity, and adjoint are solved with the software COMSOL Multiphysics®. The automatic link created between Matlab© and this software is convenient to implement the complete inverse problem. The resolution steps are as follows:

step 1. Make an initial guess $\boldsymbol{\sigma(r,t)^0}$, **k=0**
step 2. Solve the direct problem $\theta(r,z,t)$ → store the solution $\theta(r_m,z_m,t)$ and $\theta_z(r,\sigma,t)$
step 3. Solve the adjoint problem $\psi(r,z,t)$ → store the solution $\psi_z(r,\sigma,t)$
step 4. Compute the gradient $\nabla J(r,z,t)$
step 5. Compute the conjugate coefficient β^{k-1}
step 6. Compute the direction of descent d^k
step 7. Solve the sensitivity problem $\delta\theta(r,z,t)$ → store the solution $\delta\theta(r_m,z_m,t)$
step 8. Compute the step size of descent γ^k
step 9. Compute the new estimate $\sigma(r,t)$
step 10. Check the stopping criterion J^*, if satisfied interrupt the procedure, otherwise $k=k+1$ and go back to step 2.

4. Results and discussion

The minimization of the functional (Eq. (15)) is realized with three methods of conjugate coefficient calculation: Fletcher and Reeves, Ribiere and Polak, Hestenes and Stiefel. A steepest descent method is also implemented. The efficiency of each method is shown in Fig. 4. The conjugate methods lead to divergences caused by the inertial effect associated with the approximation of the conjugate gradient. Nevertheless, the steepest descent method has the slowest convergence, but gives the best estimation and the estimated interface locations are shown in Fig. 5.

The errors are maximal at the initial time increments because the measurements are far from the interface and the sensitivity is the smallest. At the last time increments, the discrepancies are due to the fact that the adjoint variable (ψ) is imposed to zero at $t=t_{fin}$. Nevertheless, the interface has a low evolution for ending times. Therefore, the badly estimated curve can be ignored.

The results of estimation are not dependent on the initial guess. In Fig. 5, the interface locations are very close to zero. However, the positioning of the interface at $z=0$ (point B) must be known to reduce the effect of the gradient singularity. In the experimental case, this information is collected by an high-speed camera.

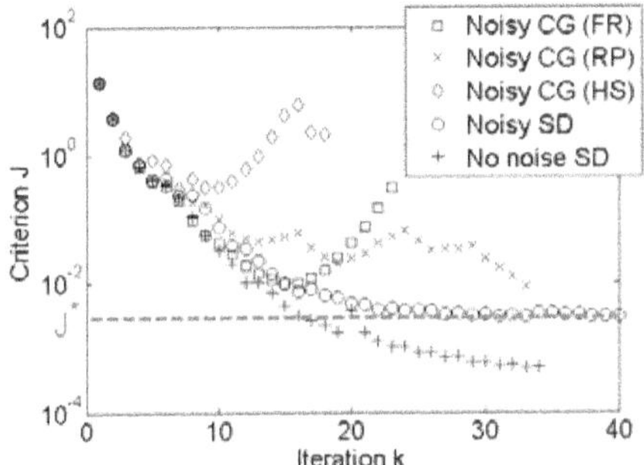

Figure 4: Criterion reductions.

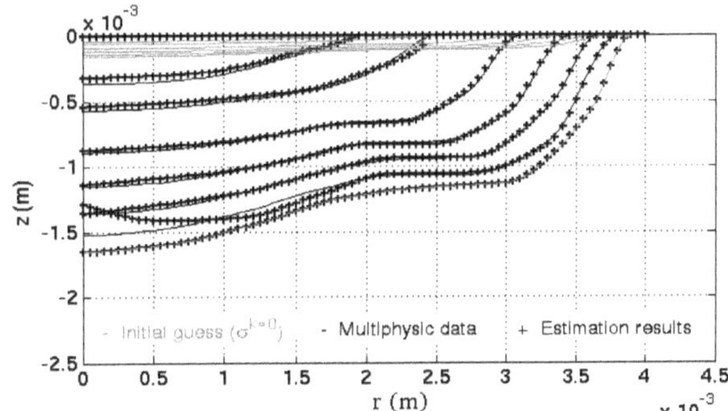

Figure 5: Estimated free boundary evolution.

5. Conclusion

A shape identification has been performed with a gradient method. A simplified modeling of a welding problem is validated by a comparison with a multipysic simulation. The inverse problem is solved with Matlab© and the three problems used to calculate the descent parameters are computed with COMSOL Multiphysics®. To validate the method, theoretical temperatures are used with realistic noises. Several ways to calculate the conjugate coefficient have been implemented and compared to the steepest descent method.

Finally, particular caution must be taken for some points:

- the way to impose the temperature on the free boundary (equivalent coefficient of convection),
- the decrease in quality of the moving mesh,
- the parameterization of the interface shape (give priority to reliable methods),
- the choice of the step times for the unknown function (avoid last time increments),

References

[1] Y. Wang, H.L. Tsai, "Impingement of filler droplets and weld pool dynamics during gas metal arc welding process", Int. J. Heat Mass Trans., Vol. 44, pp. ,2067-2080, 2001.

[2] M. Aissani et al., "Contribution à la modélisation du soudage TIG des tôles minces d'acier austénitique 304L par un modèle source bi-elliptique, avec confrontation expérimentale" , J. Phys. IV France, Vol. 124, pp., 213-220, 2005.

[3] D.D. Doan, "Modèle de source de chaleur pour la simulation du soudage avec et sans apport de matière", PhD thesis, Ecole centrale de Nantes, 2006.

[4] C.W. Hirt et al., "An Arbitrary Lagrangian-Eulerian Computing Method for All Flow Speeds", J. Comput. Phys., Vol. 135, pp., 203-216, 1997.

[5] J. M. Bergheau and R. Fortunier, Simulation numérique des transferts thermiques par éléments finis, hermes Sience, Paris, 2004.

[6] H. Orlande and N. Özisik, Inverse heat transfer, Taylor & Francis, 2000.

[7] J. Su and A. J. Silva Neto, "Two-dimensional inverse heat conduction problem of source strength estimation in cylindrical rods ", Applied Mathematical Modelling, Vol. 25, pp. 861-872, 2001.

[8] C.H. Huang et al., "A transient three-dimensional inverse geometry problem in estimating the space and time-dependent irregular boundary shapes", Int. J. Mass Heat Trans., Vol. 51, pp. 5238-5246, 2008.

[9] O.M. Alifanov, "Solution of an inverse problem of heat conduction by iteration methods", J. Engrg. Phys., Vol 26 (4), pp. 471-476, 1974.

Using Particle Swarm Optimization for Recovering the Heat Source Function in a Guarded Hot-Plate Apparatus

Obed Cortés-Aburto, Rita-Marina Aceves-Pérez, and Rafael Rojas-Rodríguez

Mechatronics and Automotive Systems Engineering Department,
Universidad Politécnica de Puebla, Tercer Carril del Ejido "Serrano" s/n,
San Mateo Cuanalá, Juan C. Bonilla, Puebla, México, C. P. 72640.
Email: obedca@uppuebla.edu.mx

Abstract

In this work we use Particle Swarm Optimization for solving an inverse heat conduction problem: Recovery of heat source in a Guarded-Hot-Plate Apparatus (GHPA). Polynomial functions are used for estimating heat source and number of parameters is from one to seven. Problem is one-dimensional in cylindrical coordinates. Results are compared with Levenberg-Marquardt Method. They are satisfactory for this kind of problem.

Key Words: PSO, GHPA, source problem

1. Introduction

Guarded-Hot-Plate Apparatus (GHPA) was formally adopted as a standard test method since 1945 by the American Society for Testing and Materials (ASTM). By 1977, more than 300 laboratories had been served, resulting in considerable improvement in the quality of thermal conductivity data on insulating and building materials reported in technical journals and handbooks. In many countries, there are universities and research centres dedicated to develop this kind of device. In Mexico, in Centro Nacional de Investigación y Desarrollo Tecnológico (CENIDET) was developed a device [1] for measuring thermal conductivities in insulating materials. In this work, temperature measurements taking during testing phase are used for inverse problem application of determining heat source term.

In recent years this problem was solved using Levenberg-Marquardt and Conjugate Gradient methods [2]. In that work, a polynomial heat source function was proposed and results obtained with this approach were analyzed. In this work, Particle Swarm Optimization is applied to same problem to analyze its performance. This paper shows the case for Guard of the GHPA because is more general. Formulation is in 1D cylindrical coordinates and transient analysis. First, a description of the problem is given. Next, a brief explanation of PSO and then, the way it is applied to inverse Heat Transfer problem. Finally, results are shown and discussed.

2. Direct and Inverse Problem

The guard is an annulus with a resistance $g(t)$ placed in $r_1 = 0.0983$ m, initial temperature $T_0 = 300.39$ K for time $t = 0$ s and for times $t > 0$ lose heat by convection to a medium with temperature $T_{amb} = 300.39$ K in both boundaries placed in $r = b = 0.0762$ m and $r = d = 0.1524$ m. A diagram of the Guard is shown in Fig. 1.

Mathematical formulation of these direct heat conduction problem is given by:

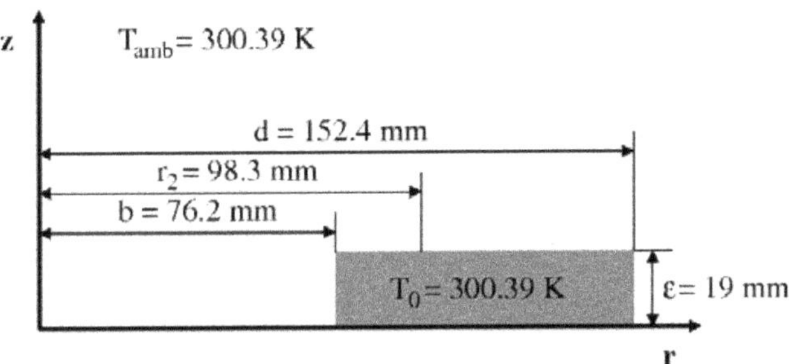

Figure 1: Physical model for the Guard.

$$\begin{aligned}
&\frac{\partial^2 T}{\partial r^2}+\frac{1}{r}\frac{\partial T}{\partial r}+\frac{g(t)\cdot\delta(r-r_2)}{2\pi k r}=\frac{1}{\alpha}\frac{\partial T}{\partial t} && b<r<d && t>0\\
&-k\frac{\partial T}{\partial r}+hT=hT_{\text{amb}} && r=b && t>0\\
&k\frac{\partial T}{\partial r}+hT=hT_{\text{amb}} && r=d && t>0\\
&T=T_0 && b\le r\le d && t=0
\end{aligned} \tag{1}$$

where $\delta(\cdot)$ is the Dirac delta function.

For this inverse problem, the time varying strength $g(t)$ of the heat source is regarded as unknown. The additional information obtained from transient temperature measurements taken at a location $r=r_{meas}$ at times t_i, $i=1,2,\ldots,I$, is the used for the estimation of $g(t)$. For comparison with the work of Cortés [2] it will be considered a polynomial approximation from one to five parameters, expressed in the following form:

$$g(t)=P_1+P_2t+P_3t^2+P_4t^3+P_5t^4 \tag{2}$$

The solution of this inverse heat conduction problem for the estimation of the unknown parameters is based on the minimization of the ordinary least squares norm given by:

$$S(\mathbf{P})=\sum_{i=1}^{I}\left[Y_i-T_i(\mathbf{P})\right]^2 \tag{3}$$

where Y_i are the measured temperature at time t_i, $T_i(\mathbf{P})$ are estimated temperature at time t_i, $\mathbf{P}^T$ is the vector of unknown parameters, N is the total number of unknown parameters and I is total number of measurements, where $I\ge N$.

The estimated temperatures $T_i(\mathbf{P})$ are obtained from the solution of the direct problem at the measurements location, $r=r_{meas}$, by using the current estimate for the unknown parameters. To minimize the least squares norm given by Eq. (3), it is needed to equate to zero the derivatives of $S(\mathbf{P})$ with respect to each of the unknown parameters.

The whole solution for these problems [2] is given by:

$$
\begin{aligned}
T(r,t) = {} & \left[bR_0(\beta,b)+dR_0(\beta,d)\right]\frac{hT_0}{\beta^2 kN}R_0(\beta,r)e^{-\alpha\beta^2 t} \\
& +\left[bR_0(\beta,b)+dR_0(\beta,d)\right]\frac{hT_{\mathrm{amb}}}{\beta^2 kN}R_0(\beta,r)\left(1-e^{-\alpha\beta^2 t}\right) \\
& +\frac{\alpha}{2\pi kN}R_0(\beta,r)R_0(\beta,r_2)\left\{\frac{P_1}{\alpha\beta^2}\left(1-e^{-\alpha\beta^2 t}\right)\right. \\
& +\frac{P_2}{\alpha^2\beta^4}\left[\alpha\beta^2 t-\left(1-e^{-\alpha\beta^2 t}\right)\right] \\
& +\frac{P_3}{\alpha^3\beta^6}\left[\alpha^2\beta^4 t^2-2\alpha\beta^2 t+2\left(1-e^{-\alpha\beta^2 t}\right)\right] \\
& +\frac{P_4}{\alpha^4\beta^8}\left[\alpha^3\beta^6 t^3-3\alpha^2\beta^4 t^2+6\alpha\beta^2 t-6\left(1-e^{-\alpha\beta^2 t}\right)\right] \\
& \left.+\frac{P_5}{\alpha^5\beta^{10}}\left[\alpha^4\beta^8 t^4-4\alpha^3\beta^6 t^3+12\alpha^2\beta^4 t^2-24\alpha\beta^2 t+24\left(1-e^{-\alpha\beta^2 t}\right)\right]\right\}
\end{aligned}
\tag{4}
$$

where

$$
\begin{aligned}
N &= \frac{\left(h^2+\beta^2 k^2\right)\left[d^2 R_0^2(\beta,d)-b^2 R_0^2(\beta,b)\right]}{2\beta^2 k^2} \\
R_0(\beta,r) &= \frac{J_0(\beta r)}{-\beta k J_1(\beta d)+hJ_0(\beta d)}-\frac{Y_0(\beta r)}{-\beta k Y_1(\beta d)+hY_0(\beta d)}
\end{aligned}
\tag{5}
$$

3. Particle Swarm Optimization

Natural creatures sometimes behave as a swarm. One of the main streams of artificial life research is to examine how natural creatures behave as a swarm and reconfigure the swarm models inside a computer. Kennedy and Eberhart [3] developed Particle Swarm Optimization (PSO) based on the analogy of swarms of birds and fish schooling. Each individual exchanges previous experiences in PSO. These research efforts are called swarm intelligence [4].

Kennedy and Eberhart (2001) developed PSO through simulation of bird flocking in a two-dimensional space. The position of each agent is represented by its x, y axis position and also its velocities are expressed by vx (the velocity of x axis) and vy (the velocity of y axis). Modification of the agent position is realized by the position and velocity information. This modification can be represented by the concept of velocity (modified value for the current positions). Velocity of each agent can be modified by the following equation:

$$
v_i^{k+1} = wv_i^k + c_1\mathrm{rand}_1\times\left(pbest_i - s_i^k\right)+c_2\mathrm{rand}_2\times\left(gbest - s_i^k\right) \tag{6}
$$

where v_i^k is velocity of agent i at iteration k, w is weighting function, c_j is weighting coefficients, rand is a random number between 0 and 1, s_i^k is current position of agent i at iteration k, $pbest_i$ is $pbest$ of agent i, and $gbest$ is $gbest$ of the group.

Namely, velocity of an agent can be changed using three vectors. The velocity is usually limited to a certain maximum value. PSO using Eq. (6) is called the Gbest model.

The weighting function, Eq. (7), is usually utilized in Eq. (6):

$$
w = w_{\max} - \frac{w_{\max}-w_{\min}}{\mathrm{iter}_{\max}}\times\mathrm{iter}, \tag{7}
$$

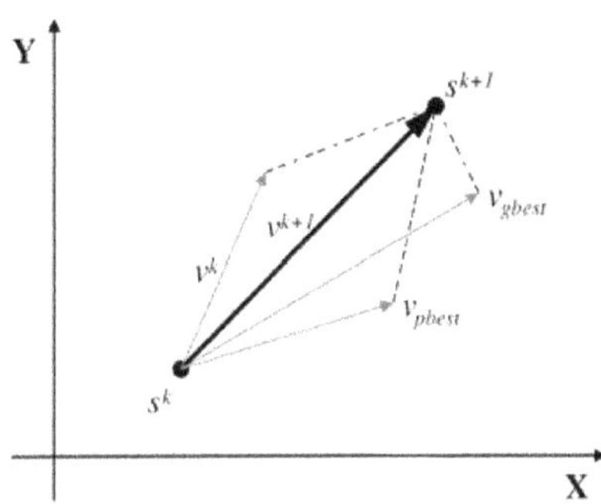

Figure 2: Concept of modification of a searching point by PSO.

where $w_{\max}$ is initial weight, $w_{\min}$ is final weight, $\mathrm{iter}_{\max}$ is maximum iteration number, and iter is current iteration number.

As shown below, for example, $w_{\max}$ and $w_{\min}$ are set to 0.9 and 0.4 [5]. Therefore, at the beginning of the search procedure, *diversification* is heavily weighted, while *intensification* is heavily weighted at the end of the search procedure such like simulated annealing (SA). Namely, a certain velocity, which gradually gets close to $pbest$ and $gbest$, can be calculated. PSO using Eq. (6) and Eq. (7) is called *inertia weights approach* (IWA).

The current position (searching point in the solution space) can be modified by Eq. (8):

$$s_i^{k+1} = s_i^k + v_i^{k+1} \tag{8}$$

Fig. 2 shows a concept of modification of a searching point by PSO. Shi and Eberhart [6, 7] tried to examine the parameter selection of the above parameters. According to their examination, the following parameters are appropriate and the values do not depend on problems:

$$c_i = 2.0, \qquad w_{\max} = 0.9, \qquad w_{\min} = 0.4.$$

4. Results

In the Hot-Plate of the GHPA, an electrical heat generation of 5 W was used. The temperature measurements were taken every 1 minute (624 time steps) at position $r_{\mathrm{meas}} = 0.0762$ m. The heat generation used in the Guard was 3 W. The measurements were taken every 30 seconds (345 time steps) at position $r_{\mathrm{meas}} = 0.1524$ m. The number of particles used for optimization was 20. Results are compared with Levenberg-Marquardt method (LMM) because it is a very stable and fast convergence method.

In Table 1, the ordinary least squares norm for LMM and PSO for Guard is shown. It can be seen that PSO has a best performance for all cases because its ordinary least squares norm is less than LMM. Table 2 shows the parameters estimated for each case with both methods.

In Fig. 3 the errors for cases with minimum errors in heat source estimation for each case with LMM and PSO for the guard are shown. Fig. 4 shows temperature field for the cases with minimum values of ordinary least squares norm for guard.

Table 1. Minimum ordinary least squares norm for guard.

Parameters	LMM	PSO
1	200.6026	200.5644
2	29.1051	28.4137
3	21.5268	13.5919
4	17.3220	13.5064
5	17.5800	13.3632

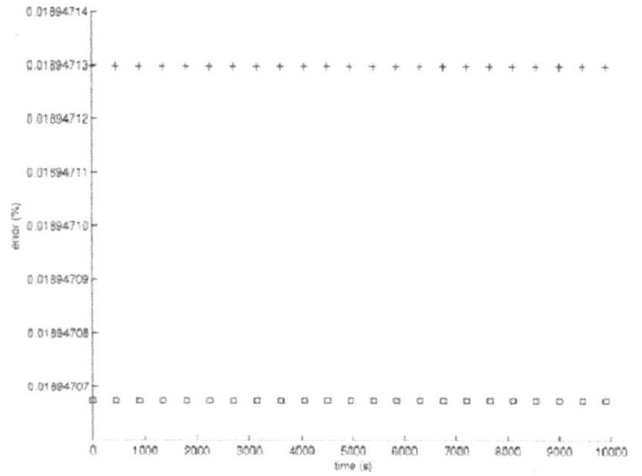

Figure 3: Errors for heat source estimation for LMM (□) and PSO (+) in the Guard.

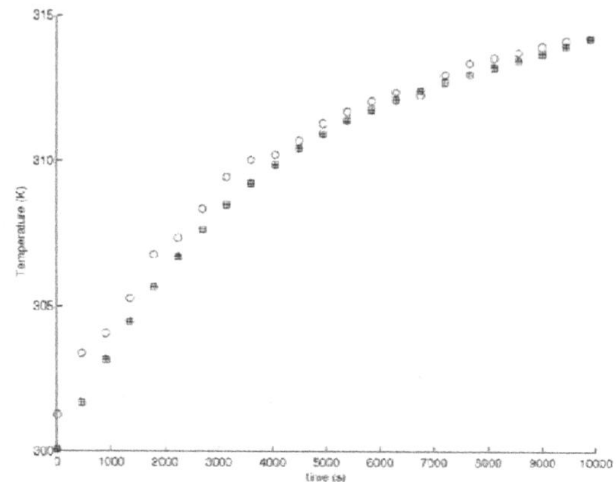

Figure 4: Temperature field for best ordinary least squares norm with LMM (□) and PSO (+) for Guard compared against measured temperature (○).

5. Conclusion

One strategy for inverse heat transfer problems has been applied. The Particle Swarm Optimization is a new method which was proposed, and it showed a good performance. This method was compared with Levenberg-Marquardt method, a method used for a long time for this kind of problem.

In the case of the guard PSO has shown a better performance than LMM. However, both methods are good enough to optimize the source term. It needs more applications for inverse problems to demonstrate PSO capability. At this time, we know of others works with PSO and results exhibit a good performance.

Table 2. Values estimated for the parameters of the heat generation function of guard for each case and method.

Case	Par.	LMM	PSO
I	1	390.4734	390.4734
II	1	467.4291	467.4386
	2	-0.01795	-0.01795
III	1	513.7841	515.4850
	2	-0.04634	-0.04733
	3	3.0314×10^{-6}	3.1345×10^{-6}
IV	1	499.5220	510.4633
	2	-0.03001	-0.04152
	3	-1.1185×10^{-6}	1.6476×10^{-6}
	4	2.8613×10^{-10}	1.0297×10^{-10}
V	1	500.4345	505.5068
	2	-0.03065	-0.03395
	3	-1.2106×10^{-6}	-1.3086×10^{-6}
	4	3.4898×10^{-10}	5.2527×10^{-10}
	5	-5.2327×10^{-15}	-2×10^{-14}

6. References

[1] Salazar R., "Diseño, construcción y caracterización de un equipo para medir conductividad térmica de materiales aislantes en el intervalo de temperatura de -75 a 250 °C", M. Sc. thesis, CENIDET, Cuernavaca, Morelos, México (1995).

[2] Cortés-Aburto, O., "Aplicación del Método de Levenberg-Marquardt y del Gradiente Conjugado en la estimación de la generación de calor de un Aparato de Placa Caliente con Guarda", M. Sc. Thesis, CENIDET, Cuernavaca, Morelos, México (2004).

[3] Kennedy J. and Eberhart R. "Particle swarn optimization". *Proceedings of IEEE International Conference on Neural Networks (ICNN'95)* Vol. 4 pp. 1942-1948, Perth, Australia (1995).

[4] Kennedy J. and Eberhart R. Swarm Intelligence. Morgan Kaufmann, San Mateo, CA (2001).

[5] Fukuyama Y. Fundamentals of Particle Swarm Optimization Techniques. In *Modern Heuristic Optimization Techniques.* Edited by K. Y. Lee and M. A. El-Sharkawi. Wiley & Sons (2008).

[6] Shi Y and Eberhart R. "A modified particle swarm optimizer", *Proceedings of IEEE International Conference on Evolutionary Computation (ICEC'98).* IEEE Press, pp. 69–73 Anchorage (1998).

[7] Shi Y. and Eberhart R. "Parameter Selection in Particle Swarm Optimization", *Proceedings of the Seventh Annual Conference on Evolutionary Programming*, MIT Press, San Diego (1998).

Star Shape Moving Source Reconstruction From Boundary Data

Nilson C. Roberty[a], **nilson@nuclear.ufrj.br**
Marcelo L. S. Rainha[b], **mrainha@nuclear.ufrj.br**
[a,b]**COPPE-UFRJ,Rio de Janeiro, RJ, BR**

Abstract.

In this work we consider computational and mathematical aspects related with the problem of reconstruction of an unknown characteristic star shape thermal source moving inside a domain. By introducing the concept of an Extended Dirichlet to Neumann map in the time space cylinder and the adoption of the anisotropic Sobolev-Hilbert spaces, we can treat the problem with methods similar to that used in the analysis of the stationary source reconstruction problem. Using Green formula we establish a reciprocity gap functional relating directly the boundary integral of Cauchy data with the domain integral of test functions in the unknown characteristic source support. Further, mean value based θ-scheme applied to the transient reciprocity gap equation leads to a model based on a sequence of modified Helmholtz equation solutions. The test functions are solutions of a homogeneous modified Helmholtz equation. For each modified Helmholtz equation the characteristic star-shape source function may be reconstructed uniquely from the Cauchy boundary data. Explicit solutions based on Fourier series for the equations involved in the model are presented. Numerical experiment related with the capture of a source inside a cubic box domain from boundary data are presented in two dimensional implementations. The problem of source centroid and shape determination is addressed.
Keywords: *source reconstruction, transient heat equation, extended Dirichlet to Neumann map, reciprocity gap, modified Helmholtz equation*

1. Introduction

The model to be presented in the conference is based on the modified Helmholtz Poisson equation that is obtained from the transient equation through the θ-scheme associated with an time finite differences discretization and averaging of the transients fields. Analysis of the related mathematical and computational work involved has been presented by the author in national conferences, [8], [9], [5], [6] and [7]. Theoretical aspects of the problem can be found in the paper [4].

This paper will be structured as follows: Some definitions and mathematical results extracted from [4] that are important to the understanding of the inverse problems are presented in the Section 2. There we introduce the concept of consistent Cauchy data and Extended Dirichlet to Neumann map. The inverse transient heat source problem is introduced in Section 3. An basic lemma about the Relative Extended Dirichlet to Neumann map and Reciprocity Gap Functional in the transient model is demonstrate. These concepts present original aspects which are numerically investigated in the present work. We show that the transient problem can be studied with aid of results demonstrated for the Modified Helmholtz Dirichlet Source problem found in reference [2]. The iterative source reconstruction scheme can be solved by the same methodology developed for stationary modified Helmholtz problems. Finally, the numerical results for two-dimensional reconstruction of sources are presented and discussed in Section 4. We conclude by pointing out the advances introduced by the present work.

2. Mathematical Formulation for the Direct Problem

By $\Omega \subset \Re^d$, $d = 1, 2, 3$ we denote a bounded space domain with Lipschitz boundary $\Gamma = \partial\Omega$, which means that Ω is locally on one side of its connected boundary. In the time-space $\Re^{d+1}$, we consider the time interval $I := (0, T)$, $T > 0$ to form the bounded cylinder $Q := I \times \Omega$, whose lateral time-space surface is $\Sigma := I \times \Gamma$. A section in this cylinder is $\Omega_t := \{t\} \times \Omega$ and the complete cylinder boundary is

$$\partial Q = \overline{\Sigma} \cup \Omega_0 \cup \Omega_T,$$

where Ω_0 and Ω_T are, respectively, the cylinders' bottom and top sections. At cylinder top and bottom there exist the corners $\Gamma_0 = \overline{\Omega_0} \cap \overline{\Sigma} \subset \Re^{d-1}$ and $\Gamma_T = \overline{\Omega_T} \cap \overline{\Sigma} \subset \Re^{d-1}$, respectively.

PROBLEM 2.1. *The direct transient heat source initial boundary value problem consists in finding $u(t,x)$ with $(t,x) \in Q$ given a boundary input $g(t,x)$ with $(t,x) \in \overline{\Sigma}$, an initial input $u_0(x)$ with $(t,x) \in \Omega_0$ and a source distribution $f(t,x)$ with $(t,x) \in Q$ that verifies the problem :*

$$(P_{u_0,g,f}) \begin{cases} \partial_t u - \Delta u = f & \text{in } Q, \\ u = u_0 & \text{in } \Omega_0, \\ u = g & \text{on } \Sigma. \end{cases} \tag{2.1}$$

and Dirichlet data compatibility condition, $u_0 = g$ at the time-space cylinder corner Γ_0.

3. Mathematical Formulation for the Inverse Source Problem

The two more important spaces for applications are $H^{2,1}(Q)$ and $H^{1,\frac{1}{2}}(Q)$, and for these space as has been shown by Lions and Magenes [1], the traces operators are not always onto.The solution of the non homogeneous problems (2.1.), $P_{u_0,g,f}$ is explicitly given by its Green's function:

LEMMA 3.1. *For regular data the Green's function exist, [3], and an explicit solution to $P_{u_0,g,f}$ is given by*

$$u(t,x) = \int_{\Omega_0} u_0(\zeta)G(t,x,0,\zeta)d\zeta + \int_\Sigma g(\tau,\zeta)\frac{\partial G(t,x,\tau,\zeta)}{\partial\nu_{(\tau,\zeta)}}d\sigma_{(\tau,\zeta)} + \int_Q f(\tau,\zeta)G(t,x,\tau,\zeta)d\zeta d\tau \tag{3.2}$$

for $(t,x) \in \overline{Q}$. By using problem's (2.1.) linearity we formally decompose the solution S in three parts

$$u = S[u_0,g,f] := S[u_0,0,0] + S[0,g,0] + S[0,0,f] \tag{3.3}$$

Since we need traces for treat non homogeneous problems (2.1.), $P_{u_0,g,f}$ with its associated inverse source problem, we will introduces the following definition:

DEFINITION 3.2. *By Consistent Cauchy data associated with problem (2.1.) we mean the functions:*

$$(u_0,g,u_T,g_\nu) \in (\gamma_0,\gamma,\gamma_T,\gamma_1)[H^{2,1}(Q)](\text{ resp. } (u_0,g,u_T,g_\nu) \in (\gamma_0,\gamma,\gamma_T,\gamma_1)[H^{1,\frac{1}{2}}(Q)]) \tag{3.4}$$
$$\subset H^{r-\frac{r}{2s}}(\Omega_0) \times H^{r-\frac{1}{2},s-\frac{s}{2r}}(\Sigma) \times H^{r-\frac{r}{2s}}(\Omega_T) \times H^{r-\frac{3}{2},s-\frac{3s}{2r}}(\Sigma)$$

where $(\gamma_0,\gamma,\gamma_T,\gamma_1)$ are respectively the traces operators at the bottom, top, lateral and normal lateral time space cylinder boundaries.

DEFINITION 3.3. *The Extended Dirichlet to Neumann map for the problem (2.1.) is defined by*

$$\Lambda^f_{\Omega,\Sigma}[(u_0,g)] = (\gamma_T,\gamma_1)\circ S[u_0,g,f] = (u|_{\Omega_T},\partial_\nu u|_\Sigma) \tag{3.5}$$

when $u = S[u_0,g,f]$ is solution of problem (2.1) with initial data $(u_0,g) = (u|_{\Omega_0},u|_\Sigma)$.

PROBLEM 3.4. *The inverse source problem $IP^f_{(u_0,g),(u_T,g^\nu)}$ is: To find $f \in H^{2,1}(Q)$ (resp. $f \in H^{1,\frac{1}{2}}(Q)$) such that*

$$(IP^f_{(u_0,g),(u_T,g^\nu)})\left\{\ (u_T,g^\nu) = \Lambda^f_{\Omega,\Sigma}(u_0,g)\right. \tag{3.6}$$

for all given data pair $(u_0,g)\times(u_T,g^\nu)$ corresponding to different solutions to the direct problem.

DEFINITION 3.5. *Consider two problems $P_{u_0,g,f}$ and $P_{u_0,g,0}$, one with source f and the other with zero source, but both with the same consistent initial time and Dirichlet data. By the Relative Extended Dirichlet to Neumann map we mean the application:*

$$(\Lambda^f_{\Omega,\Sigma} - \Lambda^0_{\Omega,\Sigma})[u_0,g] := \Lambda^f_{\Omega,\Sigma}[u_0,g] - \Lambda^0_{\Omega,\Sigma}[u_0,g]\,. \tag{3.7}$$

The following mathematical results can be use to implement a computational scheme for reconstruction of the star shape moving source:

LEMMA 3.6.

- *(i) The Relative Extended Dirichlet to Newman operator $\Lambda^f_{\Omega,\Sigma} - \Lambda^0_{\Omega,\Sigma}$ is an operator whose functional value depends only on the source function $f \in H^{2r,r}(Q)$, but is independent of the initial time and Dirichlet data (u_0,g).*
- *(ii) For all $v \in H^{2,1}_{-\partial_t-\Delta}(Q) = \{v \in H^{2,1}(Q)| -\partial_t v - \Delta v = 0\}$ the transient heat reciprocity gap equation is*

$$\int_Q fv dxdt = -\int_{\Omega_T}\Lambda^f_{\Omega,\bullet}[0,0]\gamma_T[v]dx - \int_\Sigma \Lambda^f_{\bullet,\Sigma}[0,0]\gamma[v]d\sigma_{(t,x)}. \tag{3.8}$$

- *(iii) Consider a partition of the time interval $[0,T]$ into N subintervals of length $\tau > 0$. Let $\{t_0,t_1,t_2,...,t_n,t_{n+1},...t_N\}$ be the knots of this partition, with $t_0 = 0$ and $t_N = T$. Let $H^2_\lambda(\Omega) := \{v \in H^2(\Omega) : -\Delta v + \lambda v = 0; \kappa^2 = \lambda = \frac{1}{\tau\theta}\}$ the space of Helmholtz functions. For $t_n < t < t_{n+1}, n = 0,1,N-1$ the mean value theorem can be used to approximate the equation (3.8)*

$$\int_\Omega f(t_{n+1},x)v(t_{n+1},x)dx \approx -\frac{\theta_2}{\theta_1}\int_\Gamma \Lambda^f_{\bullet,\Sigma}[0,0](t_{n+1},x)v(t_{n+1},x)d\sigma_x \tag{3.9}$$

where θ, θ_1 and θ_2 are same order numerical constants.

Proof:

- *(i) Let u_j, $j = 1,2,3,...$ be different solutions of problem 2.1 with the same source $f \in L^2(Q)$ and different initial time and Dirichlet data (u_{0_j}, g_j), $j = 1,2,3,...$, respectively. Let G be the Green function. Then, by applying the lateral boundary and final time traces operators on equation (3.2) for all solution of consistent data problems P_{f,u_{0_j},g_j}, $j = 1,2,3,...$, with the same source f, the Relative Extended Dirichlet to Newman operator will satisfies the equation*

$$(\Lambda^f_{\Omega,\Sigma} - \Lambda^0_{\Omega,\Sigma})[u_{0_j}, g_j] = \int_Q f(\tau,\varsigma)\left(G(T,x,\tau,\varsigma), \frac{\partial G(t,x,\tau,\varsigma)}{\partial \nu_{(t,x)}}\right) d\varsigma d\tau \text{ for all } j = 1,2,3,... \quad , \tag{3.10}$$

the Relative Extended Dirichlet to Neumann map will evaluates independent of the initial and Dirichlet data. Since this set includes the zero valued functions $(u_{0_j} = 0, g_j = 0)$ for some j,

$$(\Lambda^f_{\Omega,\Sigma} - \Lambda^0_{\Omega,\Sigma})[u_{0_j}, g_j] = \Lambda^f_{\Omega,\Sigma}[0,0] \tag{3.11}$$

will be the constant functional of acting on arbitrary data (u_{0_j}, g_j).

- *(ii) the second Green formula*

$$\int_Q ((\partial_t u - \Delta u)v - u(-\partial_t v - \Delta v))dxdt = \int_\Sigma (\gamma[u]\gamma_1[v] - \gamma_1[u]\gamma[v])d\sigma_{(t,x)}$$

$$-\int_{\Omega_T} \gamma_T[u]\gamma_T[v]dx + \int_{\Omega_0} \gamma_0[u]\gamma_0[v]dx \tag{3.12}$$

applied to problems $P_{u_0,g,f}$ with normal trace at the cylinder lateral boundary Σ, $\gamma_1[u] = g^\nu$, initial value at Ω_0, $\gamma_0[u] = u_0$ and $v \in H^{2,1}_{-\partial_t - \Delta}(Q) = \{v \in H^{2,1}(Q) | -\partial_t v - \Delta v = 0\}$ yield the following expression for reciprocity gap functional in the transient heat equation context:

$$\int_Q fvdxdt = \int_\Sigma (g\gamma_1[v] - g^\nu\gamma[v])d\sigma_{(t,x)} - \int_{\Omega_T} u_T\gamma_T[v]dx + \int_{\Omega_0} u_0\gamma_0[v]dx \tag{3.13}$$

or, by using the Extended Dirichlet to Neumann notation,

$$\int_Q fvdxdt = \int_\Sigma (g\gamma_1[v] - \Lambda^f_{\bullet,\Sigma}[u_0,g]\gamma[v])d\sigma_{(t,x)} - \int_{\Omega_T} \Lambda^f_{\Omega,\bullet}[u_0,g]\gamma_T[v]dx + \int_{\Omega_0} u_0\gamma_0[v]dx \tag{3.14}$$

Subtracting the Extended Dirichlet to Neumann map for the zero source problem $P_{u_0,g,0}$ with the same data, we obtain the Transient Heat Reciprocity Gap Equation (3.8).

- *(iii) Let us consider the Transient Heat Reciprocity Gap Equation (3.8) for the following time space cylinder*

$$Q = (t_{n+1}, t_{n+1} + \tau) \times \Omega$$

where τ is the time increment. Let us also consider the following test field $v \in H^{2,1}_{-\partial_t - \Delta}(Q) \wedge \gamma_{t_{n+1}+\tau}[v] = 0$. The equation in the definition of $H^{2,1}_{-\partial_t - \Delta}(Q)$ can be integrated in this time interval:

$$-\partial_t v - \Delta v = 0 \Rightarrow v(t_{n+1}, .) - \Delta \int_{t_{n+1}}^{t_{n+1}+\tau} v(t,.)dt = 0 \tag{3.15}$$

Since the interval is sufficiently small and the field is zero in its upper extremity, by the mean value theorem, we find an $0 \leq \theta \leq 1$ such that

$$\int_{t_{n+1}}^{t_{n+1}+\tau} v(t,.)dt = \theta\tau v(t_{n+1}, .). \tag{3.16}$$

Let us define $\lambda = \kappa^2 = \frac{1}{\theta\tau}$. By noting the definition of the space of Helmholtz functions $H^2_\lambda(\Omega) := \{v \in H^2(\Omega) : -\Delta v + \lambda v = 0; \kappa^2 = \lambda = \frac{1}{\tau\theta}\}$, we can see that it is an θ weight average of functions of the space $H^{2,1}_{-\partial_t - \Delta}(Q) \wedge \gamma_{t_{n+1}+\tau}[v] = 0$. The same averaging process can be done with the Transient Reciprocity Gap Equation (3.8)

$$\int_\Omega \int_{t_{n+1}}^{t_{n+1}+\tau} fvdtdx = -\int_\Gamma \int_{t_{n+1}}^{t_{n+1}+\tau} \Lambda^f_{\bullet,\Sigma}[0,0]\gamma[v]dtd\sigma_x \tag{3.17}$$

with

$$\int_{t_{n+1}}^{t_{n+1}+\tau} f(t,.)v(t,.)dt \approx \theta_1 \tau f(t_{n+1},.)v(t_{n+1},.)$$

and

$$\int_{t_{n+1}}^{t_{n+1}+\tau} \Lambda^f_{\bullet,\Sigma}[0,0]\gamma[v]dt \approx \theta_2 \tau \Lambda^f_{\bullet,\Sigma}[0,0](t_{n+1},.)v(t_{n+1},.)$$

we obtain equation (3.9).

REMARK 3.7. *The test field $v \in H^{2,1}_{-\partial_t-\Delta}(Q) \wedge \gamma_{t_{n+1}+\tau}[v] = 0$ is not empty, since with add of the time reversal operator defined for the the bounded cylinder $Q^1 := I \times \Omega^1$, $\Omega^1 \supset \Omega$,*

$$\kappa_T : H^{r,s}(Q^1) \to H^{r,s}(Q^1) \; ; \; v(t,x) \mapsto \kappa[v](t,x) = v(T-t,x)$$

an solution of the direct problem (2.1.) on Q^1 with zero source and initial value for with $u_0|_\Omega = 0$ can be converted to an such test function.

REMARK 3.8. *When the external domain Ω is a box $(0,1)^d \in R^d$, where $d = 1,2,3$ are the physical domain, and the Dirichlet boundary condition is homogeneous, $g = 0$, the transient heat problem $P_{u_0,0,f}$, (2.1.) has an explicit Fourier sine solution*

$$u(t,x_1,...,x_d) = \sum_{i=1}^{d}\sum_{n_i=1}^{N_i} c_{n_1...n_d}(t)\prod_{i=1}^{d}\sin(n_i\pi x_i) \tag{3.18}$$

where

$$c_{n_1...n_d}(t) = \exp(-t\pi^2\sum_{i=1}^{d}n_i^2)\int_0^1 ... \int_0^1 u_0(x_1,...,x_d)\prod_{i=1}^{d}\sin(n_i\pi x_i)dx_1...dx_d +$$

$$\int_0^t \exp(-(t-\tau)\pi^2\sum_{i=1}^{d}n_i^2)\int_0^1 ... \int_0^1 \chi_{\omega(\tau)}(x_1,...,x_d)\prod_{i=1}^{d}\sin(n_i\pi x_i)dx_1...dx_d d\tau \tag{3.19}$$

This solution can be convert into an element of $H^{2,1}_{-\partial_t-\Delta}(Q)$ with the straightforward application of the reversal operator

$$v(t,x_1,...,x_d) = \sum_{i=1}^{d}\sum_{n_i=1}^{N_i} c_{n_1...n_d}(T-t)\prod_{i=1}^{d}\sin(n_i\pi x_i) \tag{3.20}$$

REMARK 3.9. *By denoting $u^{n+1}(x)$ with $x \in \Omega$, the approximate solution at the time step t_{n+1}, the transient system (2.1) is approximated by the following θ-scheme, [4], sequence of stationary modified Helmholtz problems:*

$$(H^{n+1}_{\chi_\omega})\begin{cases} -\Delta u^{n+1} + \lambda u^{n+1} = f_0^n + f(t_{n+1}) & \text{in } \Omega, \\ u^{n+1} = 0 & \text{on } \Gamma, \\ \text{with } g^\nu(t_{n+1}) := \partial_\nu u^{n+1} & \text{on } \Gamma; \end{cases} \tag{3.21}$$

for $n = 0,1,2,...,N$. Here $\lambda = \frac{1}{\tau\,\theta}$ and $f_0^n = \frac{u^n+\tau(1-\theta)\Delta u^n+\tau\theta f(t_n)}{\tau\theta}$. An methodology for shape and centroid determination has been investigated in [2]. An new theorem proofing uniqueness of star shape sources reconstruction can be found there.

4. Numerical Reconstructions

We selected two numerical experiments to introduce this moving source reconstruction methodology. In the first problem a rectangle with centroid and sides varying according the equations

$$x_c = (.5 + .2 * \cos(t), .5 + .2 * \sin(t)) \text{ and } h = (0.10 + 0.05 \cdot |\sin(2t/3|, 0.15 - 0.05 \cdot |\sin(2t/3|) \,.$$

is moving inside the unit two dimensional box. Figure 1 shows the variation and centroid position for and entire period of circulation inside the square. Figure 2 shows the error associated with the size reconstruction. The utilized data are obtaining with the analytical solution (3.18) and the reconstruction is conducted with algorithm based on the reciprocity gap functional [2]. The reconstruction algorithm is based on equations (3.21). The second experiment shows the reconstruction of the moving shape

$$x_c = (0.5 \cdot \cos(2\pi t), 0.5 \cdot \sin(2\pi t)) \,;$$

$$r(\theta,t) = \frac{1.0 + 0.1 \cdot \cos(2\pi t) + 0.9 \cdot \cos(\theta) + 0.1 \cdot \sin(2\theta)}{1.0 + 0.75 \cdot \cos(\theta)};$$

also inside the unitary box. The source shape is parametrized with a Fourier series and the algebraic equation to be solved in the reconstruction process is deduced from the reciprocity gap equation (3.9), [2]. In the Figure 3 we show a reconstruction sequence of the moving star shape source for a cicle of times $t = \{0.0, 0.2, 0.4, 0.6, 0.8, 1.0\}$.

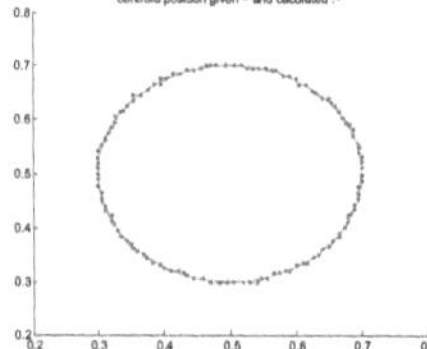

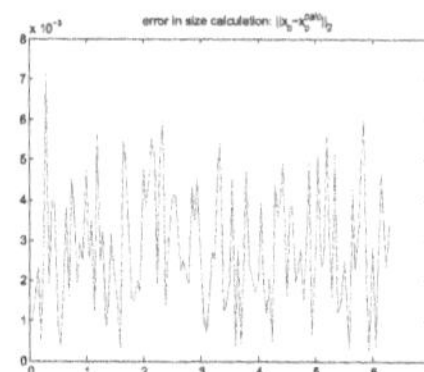

Figure 1. **a)Centroid reconstruction of a moving rectangle shape source b)Square norm error**

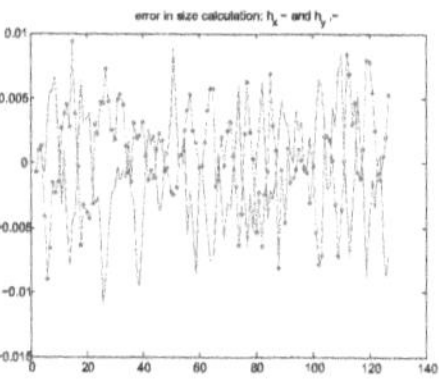

Figure 2. **Error in size calculation for the moving rectangle inside the unit box**

5. Conclusions and Further Work

The concept of Extended Dirichlet to Neumann map have permitted us to utilized results for reconstruction of stationary sources in the transient problem and the developing of numerical methods for moving characteristic source reconstruction in the transient heat model can be done by exploring the time discretization schemes such as the already introduced θ-scheme.

6. Acknowledgments

N.C. Roberty work is partially supported by the Brazilian agencies CNPq and Coopetec Foundation. M.L.S. Rainha are supported by CNPq.

7. References

[1]LIONS, J. L. AND MAGENES, E. - *Non-homogeneous boundary values problems and applications, 1972.*

[2]ROBERTY, N. C. AND RAINHA, M. L. S. -*Reconstruction of Star Shape Sources in the Modified Helmholtz Equation Model. Inverse Problems in Science and Engineering, accepted for publication, 2011.*

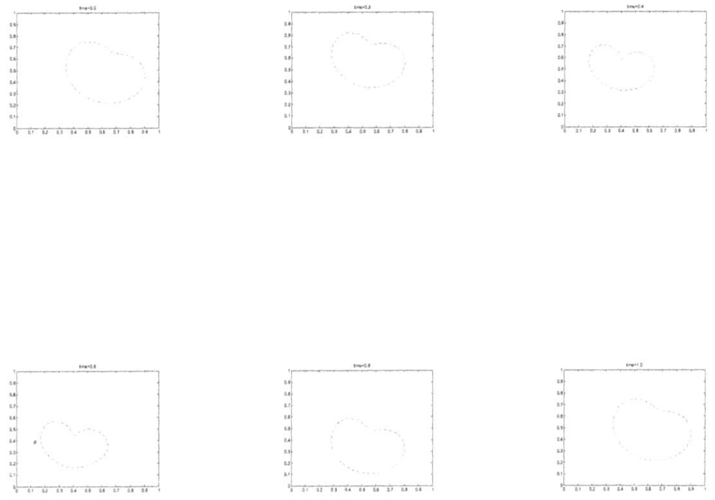

Figure 3. **Reconstruction of a moving star shape source with Fourier method($\tau = .05, \theta = 0.8, 5\%$ noise).**

[3]COSTABEL, M. -*Boundary integral operators for the heat equation. Integral Equations and Operator Theory, vol. 13, n. 4, pp. 498-552,1990.*

[4]ROBERTY, N. C. AND RAINHA, M. L. S. -*Moving heat source reconstruction from consistent boundary data. Mathematical Problems in Engineering, vol. 2010.*

[5]ROBERTY, N. C. AND SOUSA,D. M.- *Source reconstruction for the Helmholtz equation., 68 Seminário Brasileiro de Análise-SBA 2008, pp. 1-10, Universidade de São Paulo, 2008.*

[6]ROBERTY, N. C. AND RAINHA, M. L. S. -*Strong, variational and least squares formulations for the Helmholtz equation inverse source problem. 70 Semiário Brasileiro de Análise-SBA 2009 pp. 1-20, Universidade de São Paulo,2009.*

[7]*Roberty, N. C. and Rainha,M. L. S.- Star shape Sources reconstruction in the modified Helmholtz equation Dirichlet problem. Inverse Problems Design and Optimization Symposium, pp. 1-8, Universidade da Paraíba, 2010.*

[8]ROBERTY, N. C. AND ALVES, C. J. S.-*On the uniqueness of Helmholtz equation star shape sources reconstruction from boundary data. 65 Semiário Brasileiro de Análise-SBA 2007, pp. 141-150, Fapemig, São João del Rei, 2007.*

[9]ROBERTY, N. C. AND ALVES,C. J. S. -*Star-shape source reconstruction in the Helmholtz equations-frequency parameter limit.66 Seminário Brasileiro de Análise-SBA 2007, pp. 1-12, Universidade de São Paulo, 2007.*

Global Time Method for Inverse Heat Conduction Problem

Bryan S. Elkins[1], Majid Keyhani[1], and J. I. Frankel[1]*

[1] *Mechanical, Aerospace, and Biomedical Engineering Department*
The University of Tennessee, Knoxville, Tennessee 37996-2210, USA
**Corresponding Author: belkins1@utk.edu*

Abstract

Traditional space-marching techniques for solving the inverse heat conduction problem (IHCP) are highly susceptible to both measurement and round-off error. This problem is exacerbated if the problem requires small time steps to resolve rapid changes in the surface condition, since this can cause instability. The inverse technique presented in this paper utilizes a global time approach which eliminates the instability usually observed when using small time steps. It is demonstrated that a higher sampling rate (smaller time steps) in fact *improves* the inverse prediction. This is accomplished using a functional representation of the time derivative in the heat equation, and a physically-based regularization scheme. A Gaussian low-pass filter is used with an analytically determined optimum cutoff frequency. The filter delivers an analytical function which has smooth, bounded derivatives. The inverse technique is demonstrated to accurately resolve the transient surface thermal condition in the presence of noise.

Key Words: inverse, regularization, digital filter, global time

1. Introduction

The inverse heat conduction problem involves estimating the surface thermal condition without the use of surface-mounted sensors. Although the goal of the IHCP can be to resolve a variety of surface conditions – i.e., convection heat transfer coefficient, radiation heat transfer, surface heat flux, and temperature – the projection techniques are generally identical. Resolution of the surface condition always involves projecting interior data (an ill-conditioned procedure) to the surface. IHCPs are of particular interest when the surface is under a harsh thermal environment, such as high temperatures and/or high heat fluxes, which precludes the use of sensors on the surface. Applications of the IHCP include atmospheric reentry, solid rocket nozzles, hypersonic flow, quenching, and fire research. For a review of the available methods used to solve the inverse problem, see [1-6].

The chief difficulty associated with the inverse heat conduction problem is that it is ill-posed which complicates the resolution of the surface condition. The physics of conduction provide a venue for the explanation of this phenomenon. In the forward (direct) problem, diffusion damps out high frequency oscillation as the thermal front passes from the surface through the body. In the opposite direction (inverse), any high frequency oscillations present in the data (i.e., noise) are amplified as the surface condition is projected from the sensor site [1, 7]. Therefore, the IHCP is highly sensitive to measurement error and must be "regularized."

The present work is motivated by the global time and global space paper of Frankel and Keyhani [8]. In this paper, we present a global time and discrete space formulation of the inverse problem. A novel treatment of the temporal derivative in the heat equation is utilized; namely, the time derivative in the heat equation is not finite differenced, and the in-depth data are projected directly to the surface without any need to compute the temperature at intermediate spatial nodes. A Gaussian low-pass filter is employed for regularization. There are three main benefits to using this type of filter. First, this filter provides a global time regularization – i.e., all future and all past information is used in the regularization of the current time. Second, the output of this filter is an analytical function which can be analytically differentiated. Third, the filter ensures all time derivatives of the data to be smooth and bounded. The inverse technique presented in this paper is demonstrated to be robust in the presence of noise. Additionally, refining the data sampling rate *improves* the inverse prediction rather than causing instability.

2. Problem Description and Inverse Technique

Consider an isotropic, one-dimensional geometry with no internal generation and constant properties. The governing equation for this problem is given by the heat equation

$$\frac{\partial T}{\partial t}=\alpha\frac{\partial^2 T}{\partial x^2},\qquad 0\le x\le L,\ t\ge 0 \tag{1}$$

where $T = T(x,t)$ is temperature, t is the time variable, x is the spatial variable, α is thermal diffusivity, and L is the length of the (direct) spatial domain. Eq. (1) is subject to an initial condition

$$T(x,t=0)=T_i,\ \ 0\le x\le L \tag{2}$$

for the entire spatial domain. The surface (x=0) thermal condition is unknown

$$q(x=0,t)=-k\frac{\partial T}{\partial x}(x=0,t)=\text{unknown} \tag{3}$$

where k is the thermal conductivity. Heat flux and temperature data are provided at the embedded sensor site (x=d) such that

$$q(x=d,t)=q_0(t)=-k\frac{\partial T}{\partial x}(x=d,t),\quad d\le L,\qquad\qquad T(x=d,t)=T_0(t),\quad d\le L \tag{4-5}$$

where $q_0(t)$ is the heat flux data and $T_0(t)$ is the temperature data.

The goal of the inverse problem is to resolve the unknown surface (x=0) thermal condition using the provided sensor (x=d) temperature and heat flux data. This can be accomplished by first discretizing space such that $x_j = (N\text{-}j)\Delta x$ – i.e., $x_{j=0}$ corresponds to the sensor site (x=d) and $x_{j=N}$ corresponds to the surface (x=0). The spatial derivative of the heat equation in Eq. (1) is approximated using a central difference as

$$\frac{\Delta x^2}{\alpha}\frac{dT_j}{dt}(t)=T_{j+1}(t)-2T_j(t)+T_{j-1}(t),\quad j=1,2,\ldots,N \tag{6}$$

Traditionally, the time derivative is also approximated using a difference formula. However, if the heating rate and higher-time derivatives of temperature are directly measured [9] or can be obtained via post-processing, then time need not be differenced. We, therefore, define the operator, F, as

$$F=\frac{\Delta x^2}{\alpha}\frac{d}{dt}+2 \tag{7}$$

and solve Eq. (6) for T_{j+1} to obtain

$$T_{j+1}(t)=FT_j(t)-T_{j-1}(t),\qquad j=2,3,\ldots,N \tag{8}$$

An energy balance at the sensor site (j=0) yields the result

$$T_1(t)=\frac{F}{2}T_0(t)+\frac{\Delta x}{k}q_0(t) \tag{9}$$

where T_0(t) and q_0(t) are the sensor data. Using a process of repeated analytical substitution with Eqs. (8-9), the temperature at the Nth spatial node (i.e., at the surface) can be solved for in terms of the sensor temperature and heat flux explicitly. As an example, for N = 4 and N=7 the equations are

$$T_{N=4}(t)=\left(\frac{1}{2}F^4-2F^2+1\right)T_0(t)+\frac{\Delta x}{k}\left(F^3-2F\right)q_0(t) \tag{10}$$

$$T_{N=7}(t)=\left(\frac{1}{2}F^7-\frac{7}{2}F^5+7F^3-\frac{7}{2}F\right)T_0(t)+\frac{\Delta x}{k}\left(F^6-5F^4+6F^2-1\right)q_0(t) \tag{11}$$

respectively. For resolution of the surface temperature, N temporal derivatives of temperature data and N-1 temporal derivatives of the heat flux data are required. An energy balance at the surface yields

$$q_N(t)=\frac{k}{\Delta x}\left(\frac{F}{2}T_N(t)-T_{N-1}(t)\right) \tag{12}$$

which again can be found explicitly in terms of sensor temperature and heat flux only. As an example, Equation (12) for N = 4 and N = 7 becomes

$$q_{N=4}(t)=\frac{k}{\Delta x}\left(\frac{1}{4}F^5-\frac{3}{2}F^3+2F\right)T_0(t)+\left(\frac{1}{2}F^4-2F^2+1\right)q_0(t) \tag{13}$$

$$q_{N=7}(t)=\frac{k}{\Delta x}\left(\frac{1}{4}F^8-\frac{9}{4}F^6+\frac{13}{2}F^4-\frac{25}{4}F^2+1\right)T_0(t)+\left(\frac{1}{2}F^7-\frac{7}{2}F^5+7F^3-\frac{7}{2}F\right)q_0(t) \tag{14}$$

For resolution of the surface heat flux, N+1 temporal derivatives of the temperature data, and N temporal derivatives of the heat flux data are required. Therefore, if noise exists in the sensor data, the projection process is clearly ill-posed. Additionally, if the derivatives are obtained numerically, significant round-off error will accumulate as N increases, and again the process is ill-posed. However, if the data and the required number of temporal derivatives are bounded and smooth, the projection process is well-posed. This can be accomplished via a low-pass Gaussian filter.

3. Regularization via Digital Filtering

As discussed above, high frequency oscillations present in the data (noise) cause the inverse projection to become unstable. The reason for this is made clear through inspection of Eq. (13). For $N = 4$, five time-derivatives of the temperature data and four time-derivatives of the heat flux data are required to resolve the surface heat flux. Time differentiating noisy data produces an unbounded and unstable result. A physically meaningful way of stabilizing the inverse problem is to remove the non-physical high frequency energies from the signal by way of a low-pass digital filter. Consider the Gaussian low-pass filter given by

$$\hat{T}(t)=\left(\sum_{k=0}^{P}\exp\left(-\pi^2 f_c^2 (t-t_k)^2\right)\right)^{-1}\sum_{k=0}^{P}\tilde{T}(t_k)\exp\left(-\pi^2 f_c^2 (t-t_k)^2\right) \tag{15}$$

where f_c is the cutoff frequency in Hz, $\hat{T}$ is the filtered temperature and $\tilde{T}$ denotes the raw, discrete sensor data. Derivatives of Eq. (15) can be obtained analytically; the first time derivative is given by

$$\begin{aligned}\frac{d\hat{T}}{dt}(t) = 2\pi^2 f_c^2 \left(\sum_{k=0}^{M}\exp\left(-\pi^2 f_c^2 (t-t_k)^2\right)\right)^{-2}\Bigg[\sum_{k=0}^{M}(t-t_k)\tilde{T}(t_k)\exp\left(-\pi^2 f_c^2 (t-t_k)^2\right)\ \times \\ \sum_{j=0}^{M}\left(\frac{t-t_j}{t-t_k}-1\right)\exp\left(-\pi^2 f_c^2 (t-t_j)^2\right)\Bigg], \quad t\geq 0\end{aligned} \tag{16}$$

The output of Eqs. (15-16) are analytical functions which remove high frequency energies (noise) from the signal and ensure continuous, smooth temporal derivatives [7] as required by the inverse technique. Therefore, a carefully chosen cutoff frequency is the only parameter necessary to regularize the data, and make the problem well-posed.

The cutoff frequency of the filter can be determined using residual minimization in the least-squares sense. Traditional least squares would seek to minimize the difference between the filtered and noisy data – i.e. the residual. However, some authors [10, 11] have noted that the difference between the filtered data and the exact data should be minimized. First, let us define the noisy data to be

$$\tilde{T}(t) = T_{exact}(t) + \varepsilon(t) \tag{17}$$

where ε is the noise which is considered to be of a Gaussian distribution with a constant standard deviation, σ, and has a zero mean. Next, we define the residual as the difference between noisy and filtered data such that

$$R(t_k) = \tilde{T}(t_k) - \hat{T}(t_k) = T_{exact}(t_k) - \hat{T}(t_k) + \varepsilon(t_k) \tag{18}$$

Clearly, minimizing the difference between the filtered and exact data is the same principle as setting the residual equal to the noise. Therefore, we utilize the Euclidean norm and seek to minimize the function

$$\phi_1 = \left\|\tilde{T}(t_k) - \hat{T}(t_k)\right\|_2 - \left\|\varepsilon(t_k)\right\|_2, \quad t_{on} \leq t_k \leq t_{\max} \tag{19}$$

where t_{on} is the time at which the source at the surface is switched on, and t_{max} is the time at the final data sample. The function, ϕ_1, now contains the noisy data, filtered data, and the noise history. Additionally, lead data is commonly taken before the source is turned on. We use the mean of the lead data, T_i (initial condition), to define ϕ_2 as

$$\phi_2 = \left\|T_i - \hat{T}(t_k)\right\|_2, \qquad 0 \leq t_k \leq t_{on} \tag{20}$$

where T_i is again the initial condition. The cutoff frequencies which produce minimums of ϕ_1 and ϕ_2 provide a range of optimal cutoff frequencies from which the user can choose. It should be noted that the Gaussian low-pass filter of Eq. (15) is not overly sensitive to the choice of cutoff frequency; a change of 15-20% will not dramatically influence the filtered data.

At first, this procedure does not seem useful since in the real world both the exact data and the noise history are unavailable (apart from the initial condition). However, we utilize a procedure of estimating the noise outlined in [10]. A first order polynomial is fit to the noisy data, and the residual, R_1, between the noisy data and the first order least squares fit is calculated. Next, a higher order function, $\hat{R}_1$, is used to approximate R_1 via least squares. The error is then estimated as

$$\varepsilon_{est}(t_k) = \hat{R}_1 - R_1 \tag{21}$$

4. Results

In order to demonstrate the proposed technique, we present the classical Beck triangular surface heat flux problem [1]. The surface boundary condition is defined as

$$\bar{q}(\eta=0,t) = \begin{cases} 0, & 0 \le Fo \le 0.3 \\ \dfrac{Fo}{0.6} - 0.5, & 0.3 \le Fo \le 0.9 \\ -\dfrac{Fo}{0.6} + 2.5, & 0.9 \le Fo \le 1.5 \\ 0, & Fo \ge 1.5 \end{cases} \tag{22}$$

The problem has been made dimensionless via

$$Fo = \frac{\alpha t}{d^2}, \qquad \eta = \frac{x}{d} \tag{23-24}$$

$$\theta(\eta, Fo) = \frac{T - T_i}{q_S d/k}, \qquad \bar{q}(\eta, Fo) = \frac{q_x}{q_S}, \qquad \bar{f} = \frac{d^2}{\alpha} f \tag{25-27}$$

where Fo is dimensionless time, η is dimensionless space, θ is dimensionless temperature, $\bar{q}$ is dimensionless heat flux, and $\bar{f}$ is dimensionless frequency. Note that although the peak source heat flux, q_S, is unknown apriori, it is merely a scaling factor in Eqs. (25-26). Therefore, the actual value of q_S is unimportant for this analysis. The body is insulated at $x = L$, and the sensor is located at $x = d = L$.

For a dimensionless sampling rate of $\bar{f}_s$ = 1200, noise is simulated as in Eq. (17) using a normal distribution with a zero mean and a standard deviation of 0.01 for dimensionless temperature. Figure 1(a) shows the noisy data and a one-term least squares fit of the data. Figure 1(b) shows the residual, R_1, and an eighth order least squares fit of R_1 using Hardy multiquadric radial basis functions [10]. The estimated noise and actual noise can be seen in Figure 1(c).

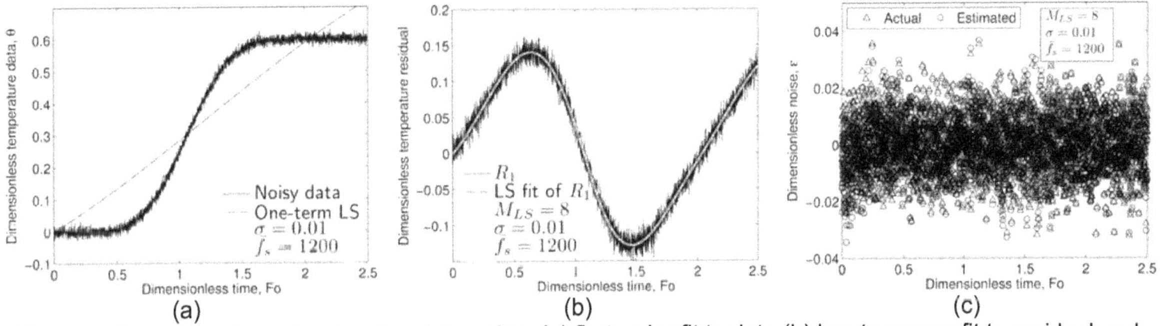

Figure 1: Technique for estimating the data noise: (a) first order fit to data (b) least squares fit to residual and (c) actual and estimated noise.

Once the noise history has been estimated, ϕ_1 and ϕ_2 from Eqs. (19-20) can be used to provide a range of appropriate cutoff frequencies. This can be seen in Figure 2(a) where the observed optimum dimensionless cutoff frequency range is 2.1 – 4.9. As noted above, the Gaussian low-pass filter used here is relatively insensitive to the actual choice of cutoff frequency. Figure 2(b) shows the exact and filtered temperature histories using cutoff frequencies of 2.1 and 4.9. Although there is a large difference between these two values of cutoff frequency, both filter outputs approximate the temperature data well. As a compromise between these two values, a dimensionless cutoff frequency of 3.5 was used to filter the data. The filter output was then passed to the inverse code; the results for the predicted surface temperature and heat flux histories can be seen in Figure (3). The peak value of heat flux is underestimated by only 7.0%.

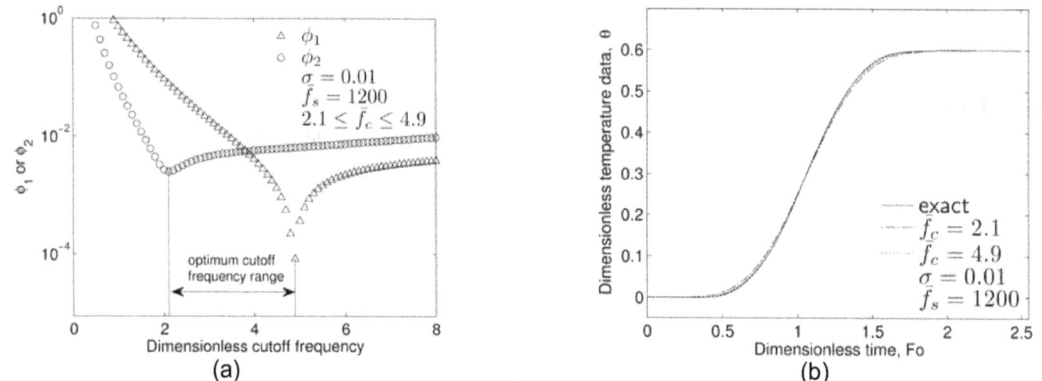

Figure 2: Gaussian low-pass filter exploration. (a) optimum cutoff frequency range and (b) insensitivity of Gaussian filter to change in cutoff frequency.

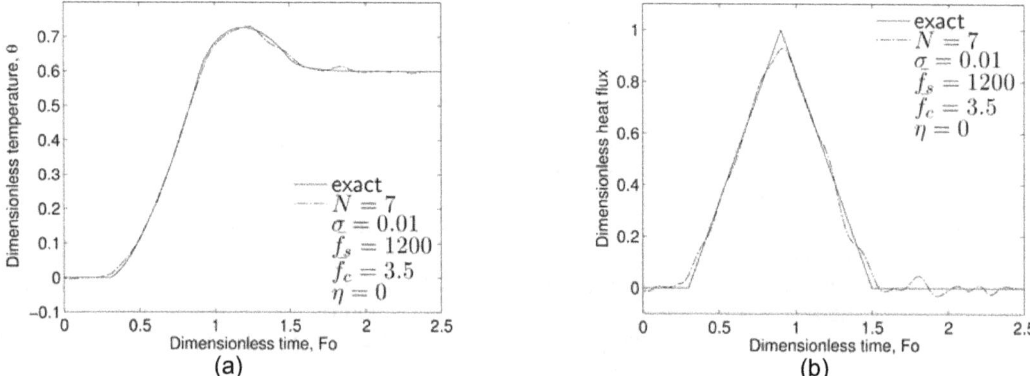

Figure 3: Inverse results using a normal distribution, σ = 0.01 and $\bar{f}_c$ = 3.5. (a) Inverse temperature results and (b) inverse heat flux results.

An investigation of the accuracy of the proposed inverse method as a function of the sampling rate was conducted. The sampling rate was varied from $100 \le \bar{f}_s \le 1200$. Five unique random error distributions were generated for each sampling rate and added to the data as in Eq. (17). The RMS of the surface inverse heat flux prediction was calculated for each dataset such that

$$q_{RMS} = \left[\frac{1}{M} \sum_{k=1}^{M} \left(q_{inverse}(t_k) - q_{exact}(x=0,t_k) \right)^2 \right]^{1/2} \tag{28}$$

where *M* is the number of time samples. The RMS error was averaged for the five runs at each sampling rate. Two different values for the order of the inverse scheme, *N*=4 and *N*=7, were used. The cutoff frequency was chosen as the midpoint of the optimum range from ϕ_1 and ϕ_2 for both values of *N*. Also, for *N*=7, $\bar{f}_c$ was held constant at $\bar{f}_c$ = 3.0 to investigate the effect of sampling rate independent of the cutoff frequency.

Table 1: Predicted heat flux RMS error as a function of the sampling rate, $\bar{f}_s$, projection order, *N*, and choice of cutoff frequency, $\bar{f}_c$; $\bar{f}_{c,mean}$ and $q_{RMS,mean}$ are average values resulting from five independent noise distributions.

	$N=4$, $\bar{f}_c = (\bar{f}_{c,\phi1} + \bar{f}_{c,\phi2})/2$		$N=7$, $\bar{f}_c = (\bar{f}_{c,\phi1} + \bar{f}_{c,\phi2})/2$		$N=7$, $\bar{f}_c = 3$	
$\bar{f}_s$	$\bar{f}_{c,mean}$	$q_{RMS,mean}$	$\bar{f}_{c,mean}$	$q_{RMS,mean}$	$\bar{f}_{c,mean}$	$q_{RMS,mean}$
100	2.56	0.0370	2.56	0.0368	3.00	0.0439
200	2.86	0.0352	2.83	0.0301	3.00	0.0357
400	3.11	0.0335	3.09	0.0273	3.00	0.0245
1200	3.41	0.0301	3.49	0.0212	3.00	0.0246

As expected, increasing N (more spatial nodes) improves the accuracy of the inverse results. However, in sharp contrast to other inverse methods [1], it is seen that the RMS error decreases as the sampling rate increases. These results hold whether the cutoff frequency is held constant at 3.0, or chosen as the midpoint of the optimum range. This suggests that optimum inverse results are obtained by sampling the data at the highest possible sampling rate, and using a high inverse projection order (N).

5. Conclusions

A robust global time inverse heat conduction method has been demonstrated. In stark contrast to traditional inverse methods, increasing the data density does not compromise stability of the inverse prediction. Indeed, the inverse results improve with faster data sampling. The time-derivative of the heat equation is not finite differenced. Instead, a functional representation of the higher-time derivatives of temperature are employed to project directly to the surface without the need to calculate the temperatures at intermediate spatial nodes. This is accomplished via a low-pass Gaussian filter with a physically based cutoff frequency which provides an analytical function as output. The proposed regularization scheme presented provides bounded, continuous, analytical time derivatives of the sensor data. Additionally, an analytical method for determining the optimum filter cutoff frequency range is demonstrated. The novel treatment of the temporal derivative in the heat equation, combined with the global time Gaussian low-pass filter provides the regularization required for stable, accurate results. Future studies with this concept should involve extension to two- and three-dimensional geometries.

6. Acknowledgements

The results presented here were partially supported under the NASA Cooperative Agreement NNX10AN35A.

7. References

[1] Beck, J. V., Blackwell, B., St. Clair, C. R., Jr., 1985, Inverse Heat Conduction, Wiley, New York.

[2] Kurpisz, K., and Nowak, A. J., 1995, Inverse Thermal Problems, Computational Mechanics Publications, Southampton, UK.

[3] Özisik, M. N., and Orlande, H. R. B., 2000, Inverse Heat Transfer, Taylor-Francis, New York.

[4] Beck, J. V., Blackwell, B., and Haji-Sheikh, A., 1996, "Comparison of some inverse heat conduction methods using experimental data," International Journal of Heat and Mass Transfer, 39(17), pp. 3649-3657.

[5] Beck, J. V., 2008, "Filter solutions for the nonlinear inverse heat conduction problem," Inverse Problems in Science & Engineering, 16(1), pp. 3-20.

[6] Taler, J., 1996, "A semi-numerical method for solving inverse heat conduction problems," Heat and Mass Transfer, 31(3), pp. 105-111.

[7] Frankel, J., 2007, "Regularization of inverse heat conduction by combination of rate sensor analysis and analytic continuation," Journal of Engineering Mathematics, 57(2), pp. 181-198.

[8] Frankel, J. I., and Keyhani, M., 1997, "A Global Time Treatment for Inverse Heat Conduction Problems," Journal of Heat Transfer, 119(4), pp. 673-683.

[9] Elkins, B. S., Huang, M., and Frankel, J. I., in review, "In-situ higher-time derivative of temperature sensors for heat transfer," International Journal of Thermal Sciences.

[10] Frankel, J. I., Keyhani, M., and Taira, K., 2004, "In-phase error estimation of experimental data and optimal first derivatives," AIAA Journal, 42(5), pp. 1017-1024.

[11] Murio, D., 2002, "Mollification and space marching," Inverse Engineering Handbook, CRC Press.

Inverse Analysis for Estimation of Inner Surface Temperature History of a Pipe from Outer Surface Temperature Measurement with Noise

***Yoshiyuki MATSUMOTO** [1], **Sayako ONCHI** [1] **Shiro KUBO** [1,2] **and Seiji IOKA** [1]*

[1]*Department of mechanical Engineering, Graduate School of Engineering*
Osaka University, 2-1, Yamadaoka, Suita, Osaka, 565-0871, Japan
[2]***Email: kubo@mech.eng.osaka-u.ac.jp***

Abstract

When high temperature slug flow runs through in a pipe, the temperature of a pipe changes, which may cause thermal fatigue fracture of the pipe. Therefore, it is necessary to know the temperature history in the pipe. However, it is not easy to measure the temperature in the pipe. Therefore establishment of the estimation method of the temperature history on the inner surface of a pipe is needed. In this paper, as a basic study on the estimation method of the inner surface temperature history, the estimation method of the temperature history on inner surface of a pipe from outer surface temperature measurement for a slug flow is proposed. The relationship between the outer surface temperature history and the inner surface temperature history is obtained analytically. Using the result of the mathematical analysis, the estimation method of the inner surface temperature history from the outer surface temperature history is proposed. In this method, the inner surface temperature history is estimated by amplifying the amplitude of the outer surface temperature history, and advancing the phase of the outer surface temperature history. The inner surface temperature history is estimated by using the proposed method. The effect of the measurement noise on the estimation is studied. It is found that the inner surface temperature history can be estimated from the outer surface temperature history by applying the inverse analysis method, even when the measurement noise is included in the outer surface temperature history.

Key Words: thermal stress, slug flow, measurement noise, heat conduction, pipe

1. Introduction

When high temperature fluid or low temperature fluid flows in a pipe, non-uniform temperature distribution may develop in the pipe. Then, the thermoelastic deformation and the transient thermal stress develop in the pipe. For example, transient thermal stress develops during start-up or shut-down of equipments. This transient thermal stress may be responsible for the low-cycle thermal fatigue fracture. It is also found that high cycle thermal fatigue is caused in the pipe due to thermal striping in the mixture of high temperature fluid and low temperature fluid. Therefore it is necessary to know the temperature distributions and the stress distributions in the pipe for the integrity assessment of the pipe. It is, however, difficult to measure the inner surface temperature directly. Therefore establishment of the estimation method of the temperature history on inner surface of pipe is needed.

Shao [1] studied the solution of a direct problem concerning temperature, displacement and thermal/mechanical stresses in a functionally graded circular hollow cylinder by using a multi-layered approach based on the theory of laminated composites. Tikhe et al. [2] studied an inverse thermoelastic problem of transient heat conduction in a finite circular plate with a given temperature distribution on the interior surface of a thin circular plate and determined the thermal defection on the outer curved surface of a thin circular plate. Kandil et al. [3] studied the distribution of the temperature and thermal stress of thick-walled cylinder under unsteady state by using numerical models.

The present authors [4-6] proposed an inverse method for determining the optimum history of the inner surface temperature and fluid temperature which reduces transient thermal stress considering the multiphysics.

In this paper, as a basic study on the evaluation of thermal fatigue estimation of a pipe, an estimation method of the temperature fluctuation on the inner surface of a plate due to slug flow from the temperature on the outer surface by using the inverse analysis method is studied. Furthermore, the applicability of the proposed method to the estimation of the inner surface temperature from the outer surface temperature with a measurement noise is studied.

2. Inverse analysis method of the inner surface

We consider a plate shown in Fig. 1. When high temperature fluid and low temperature fluid flow alternately on the inner surface of the plate, the inner surface temperature changes in time, and the inside temperature of the plate is in nonstationary state. We obtain the relation between the outer surface temperature and the inner surface temperature. Then we propose an estimation method of the inner surface temperature history.

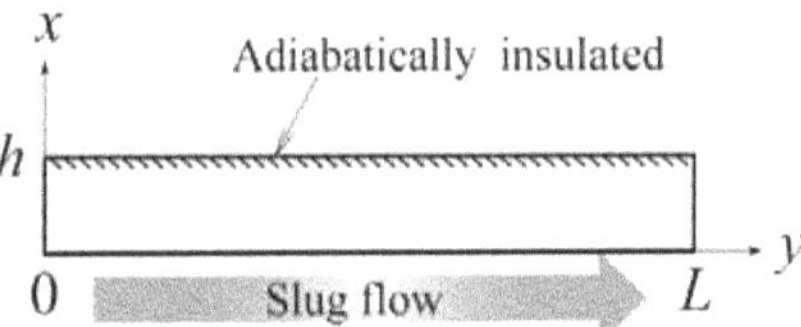

Figure 1 Model used for thermal analysis.

2.1. Mathmatical analysis

The plate shown in Fig. 1 is subjected to the temperature change on the inner surface and the outer surface is insulated. Slug flow runs on the inner surface at a velocity of v. We study the relation between the inner surface temperature history and the outer surface temperature history analytically. The temperature of the plate at location (x,y) and at time t is denoted by $\phi(x,y,t)$. The heat equation is written as follows:

$$\frac{\partial \phi}{\partial t} = \kappa^2 \left(\frac{\partial^2 \phi}{\partial x^2} + \frac{\partial^2 \phi}{\partial y^2} \right), \tag{1}$$

where κ^2 is the thermal diffusivity.

We assume that the temperature on the inner surface changes is expressed by a sinusoidal function of y whose wave length is $l=2\pi/c$, and moves in the y-direction at a velocity of v. The cycle period T_p of this sinusoidal wave is written as

$$T_p = \frac{l}{v}. \tag{2}$$

Then the boundary condition on the inner surface $x = 0$ can be written as follows:

$$\phi(x, y, t) = A\sin(-cy + vct - fx) \quad \text{on} \quad x = 0. \tag{3}$$

For Eq. (3), we introduce a particular solution as follows:

$$\phi = A(x)\sin(-cy + vct - fx). \tag{4}$$

Here, we assume that $A(x)$ can be denoted as follows:

$$A(x) = a\exp(-\alpha x) \tag{5}$$

From Eqs. (1), (4) and (5), we obtain the following two equations:

$$\alpha = \sqrt{\frac{c^2 + \sqrt{c^4 + v^2c^2/\kappa^4}}{2}}, \tag{6}$$

$$f = \frac{vc}{2\kappa^2\alpha}. \tag{7}$$

Then, the inner surface temperature and the outer surface temperature can be expressed as

$$\phi(0, y, t) = a\sin(-cy + vct), \tag{8}$$

$$\phi(h, y, t) = a\exp(-\alpha h)\sin(-cy + vct). \tag{9}$$

These equations indicate that the temperature on the outer surface is reduced by a ratio R in amplitude and is delayed by a phase-lag Δp compared to the temperature on the inner surface. The reduction ratio R of the amplitude of the outer surface temperature against the inner surface temperature is written as

$$R = \left|\frac{\phi(h, y, t)}{\phi(0, y, t)}\right| = \exp\left(-h\sqrt{\frac{c^2 + \sqrt{c^4 + v^2c^2/\kappa^4}}{2}}\right). \tag{10}$$

The phase-lag Δp of the outer surface temperature against the inner surface temperature can be expressed as

$$\Delta p = fh = \frac{vch}{2\kappa^2\alpha}. \tag{11}$$

Then the time-lag Δt can be written as

$$\Delta t = \Delta p \times \frac{2\pi}{T_p} = \frac{chl}{\pi\kappa^2\sqrt{8\left(c^2 + \sqrt{c^4 + v^2c^2/\kappa^4}\right)}}. \tag{12}$$

The factor c depends on wave length l. From Eqs. (10) and (12), it is seen that the reduction ratio of amplitude R and time-lag Δt depend on the wave length l and velocity v of the wave. Therefore, it is seen that these factors depend on the frequency of the wave.

2.2. Inverse analysis method

By using the reduction ratio R and the time lag Δt obtained in the subsection 2.1, the inner surface temperature history can be estimated from the outer surface temperature history. The temperature history at a point on the outer surface ϕ_{out} can be expressed in terms of the Fourier series expansion as

$$\phi_{\text{out}} = c_0 + \sum_j a_j \sin\frac{2\pi vj}{l}t + \sum_j b_j \cos\frac{2\pi vj}{l}t. \tag{13}$$

We obtain the reduction ratio R_j and the time-lag Δt_j for each frequency component from Eq. (10) and (12). By amplifying the magnitude of the outer temperature by the amplification ratio $M_j=1/R_j$ and by advancing the time by the time-lag Δt_j, we can estimate the inner surface temperature history ϕ_{in} as follows:

$$\phi_{\text{in}} = c_0 + \sum_j M_j a_j \sin\left(\frac{2\pi v}{l}j(t - \Delta t_j)\right) + \sum_j M_j b_j \cos\left(\frac{2\pi v}{l}j(t - \Delta t_j)\right). \tag{14}$$

However, if M_j is very high, the noise included in the outer surface temperature may deteriorate the estimated inner surface temperature. Then we set the upper limit of M_j to be 1000 considering that the representing value of the noise is order of 0.1%. In other words, M_j is set as

$$M_j = \min\left(\frac{1}{R_j}, 1000\right) \quad (15)$$

Setting a limit to M_j, we can ensure the appropriateness of the solution of this inverse analysis.

3. Numerical example of the inner surface temperature estimation

3.1. Estimation of the inner surface temperature history

The inner surface temperature history is estimated from the outer surface temperature history by using the method described in the subsection 2.2. We consider the following two inner surface temperature:

Case1

$$\phi_{\text{in}} = 10\sin\frac{2\pi}{l}(vt-y)+200 , \quad (16)$$

where l=0.1[m], v=0.0005[m/s].

Case2

$$\phi_{\text{in}} = 10\sin\frac{2\pi}{l_1}(v_1t-y)+10\sin\frac{2\pi}{l_2}(v_2t-y)+10\sin\frac{2\pi}{l_3}(v_3t-y)+200 , \quad (17)$$

where l_1=0.1[m], l_2=0.05[m], l_3=0.02[m], v=0.0005[m/s]. The material of the plate of the analysis model is SUS316, and the geometry condition and the material property are shown in Table 1.

When the inner surface temperature is given by Eqs. (16) or (17), the outer surface temperature is calculated by using the finite element method (F.E.M.). From the outer surface temperature obtained by F.E.M. analyses, the inner surface temperature is estimated by the inverse analysis method described in the subsection 2.2.

Estimated inner surface temperature histories for Cases 1 and 2 are shown in Figs. 2(a) and 2(b), respectively. The dotted lines and the dashed lines in Fig. 2 show the estimated inner surface temperature and the actual inner surface temperature given in F.E.M. analysis, respectively. The chained lines show the outer surface temperature obtained by the F.E.M. analysis.

The amplitude of the estimated inner surface temperature is about 1.7 times larger than that of inner surface temperature given in F.E.M. analyses. On the other hand, it is found that the phases of the estimated inner surface temperature agree well with those of the given inner surface temperature, even for the case that the temperature history is expressed by the multiple frequency components.

To improve the accuracy of the amplitude of the estimation of the inner surface temperature history, we introduce a correction factor r. The correction factor r is defined by the ratio between the reduction ratio of the amplitude obtained by F.E.M. analysis R and that obtained by the mathematical analysis R_m as follows:

$$r = R / R_m . \quad (18)$$

The correction factors were calculated for some combinations of wave velocity v and wave length l. The correction factor r is plotted in Fig. 3 as a function of v and l. When a combination of wave velocity v and wave length l is given, the correction factors are estimated by 2D linear interpolation by using the nearest three point of grid of Fig. 3. Modified reduction ratio R^* is expressed as follows:

$$R^* = rR_m . \quad (19)$$

Using the modified reduction ratio, the inner surface temperature for Cases 1 and 2 can be estimated. Inner surface temperature histories estimated by using the correction factor for Cases 1 and 2 are shown in Figs. 2(a) and 2(b), respectively. The solid lines in Fig. 2 show the inner surface temperature estimated by using the correction factor. It is found that the amplitude of the estimated inner surface temperature agrees well with that of the given inner surface temperature, and that the temperature history on the inner surface is reconstructed well.

Table 1 Thermal and mechanical properties

Plate thickness	h [m]	0.01
Plate length	L [m]	0.1
Heat diffusivity	κ^2 [m^2/s]	4.46×10^{-6}

3.2. Effect of a measurement noise on the inner surface temperature estimation

To examine the influence of the measurement noise on the estimation result of the inner surface temperature, the inner surface temperature history is estimated from the outer surface temperature which include a measurement noise. Using the outer surface temperature history with a noise, the inner surface temperature is estimated for Cases 1 and 2. The measurement noise is set to be within 0.1%, 0.15% and 0.3% of the outer surface temperature obtained by F.E.M. analysis. Inner surface temperature histories estimated by using the outer surface temperature history with an noise for Cases 1 and 2 are shown in Figs. 4(a) and 4(b), respectively. From Fig. 4, it is seen that when the measurement noise included in the outer surface is within 0.1%, the good estimation of the inner surface temperature can be made.

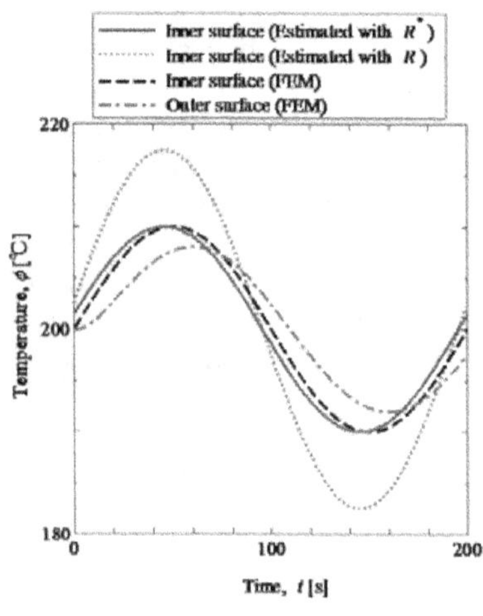

(a) Case 1 (v=0.0005[m/s], l=0.1[m])

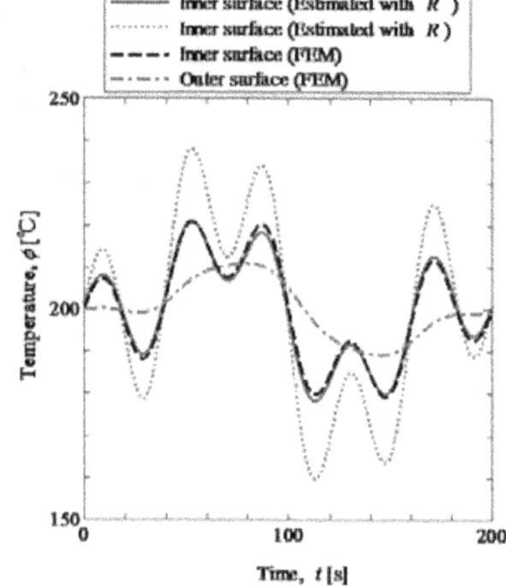

(b) Case 2 (v=0.0005[m/s], l_1=0.1[m], l_2=0.05[m], l_3=0.02[m])

Figure 2 Temperature history.

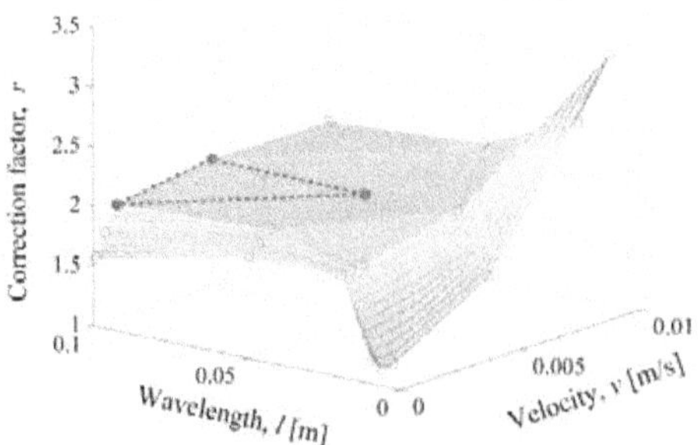

Figure 3 Influence of wave length and velocity on correction factor r.

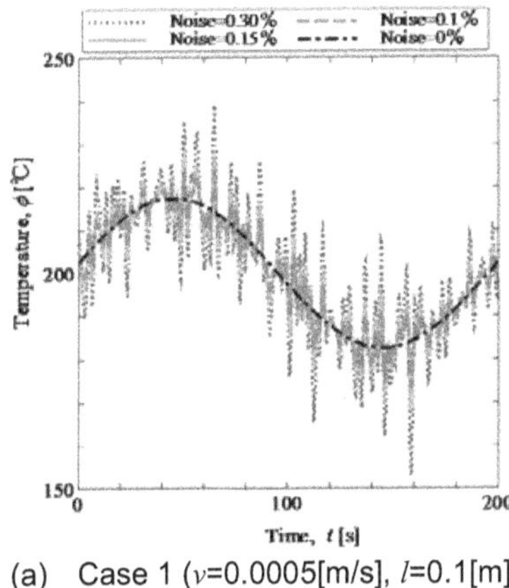

(a) Case 1 (v=0.0005[m/s], l=0.1[m])

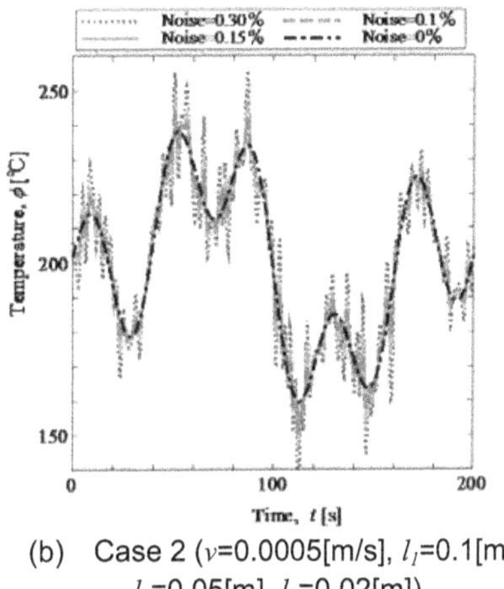

(b) Case 2 (v=0.0005[m/s], l_1=0.1[m], l_2=0.05[m], l_3=0.02[m])

Figure 4 Effect of noise on estimated inner surface temperature history.

4. Conclusions

The estimation method of the temperature history on the inner surface of a pipe from outer surface temperature measurement for a slug flow was proposed. Numerical simulations using the finite element method were conducted to demonstrate the effectiveness of estimation method of the inner surface temperature history. It was found that the inner surface temperature can be estimated by using the proposed inverse analysis method. The effect of measurement noise on the estimation of inner surface temperature was examined. When the measurement noise included in the outer surface temperature was within 0.1%, the good estimation of the inner surface temperature history can be obtained.

5. References

[1] Shao Z. S. , "Mechanical and thermal stress of a functionally graded circular hollow cylinder with finite length," Int. J. Pressure Vessels and Piping, Vol.82, pp.155, 2005.

[2] Tikhe A. K. and Deshmukh K. C., Inverse hest condition problem in a thin circular plate and its thermal defection, Applied Mathmatical Modeling, Vol.30, pp.554, 2006.

[3] Kandil A. A., El-kady A. and El-Kafrawy A., "Transient Thermal Stress Analysis of Thick-Walled Cylinders," Int. J. Mechanical Science, Vol.37, pp.721, 1995.

[4] Ishizaka T., Kubo S. and Ioka S., "An Inverse Method for Determining Thermal Load History Which Reduces Transient Thermal Stress," paper # PVP2006-ICPVT-11-93618, *Proceedings of PVP2006-ICPVT-11*, ASME 1

[5] Kubo S., Uchida K., Ishizaka T. and Ioka S., "Determination of the Optimum Temperature History of Inlet Water for Minimizing Thermal Stresses in a pipe by Multiphysics Inverse Analysis," Journal of physics: Conference Series, *Proceedings of 6th International Conference on Inverse Problems in Engineering Theory and Practice (ICIPE2008))* 135 1 2008. (Published online)

[6] Ishizaka T., Kubo S. and Ioka S., "Multiphysics Inverse Analysis Method for Reducing Transient Thermal Stress in a Thic-Walled Pipe," *Proceedings of the ASME 2009 Pressure Vessels and Piping Division Conference*, ASME 1, 2009.

Identification of defects and inclusions from thermal data by using FEM design sensitivity analyses based on adjoint variable method

Toshiro Matsumoto[1], Toru Takahashi[1], and Guang Wang[1]

[1] Department of Mechanical Science and Engineering
Nagoya University, Furo-cho, Chikusa-ku, Nagoya, 464-8603, Japan
t.matsumoto@nuem.nagoya-u.ac.jp

Abstract

We present a numerical approach for identifying the location and size of defects and inclusions in solids from temperature in steady state measured on specific parts of the surface of the solids. Defects and inclusions are modeled as an appropriate distribution of thermal conductivity, and their locations and sizes are identified through inverse calculations of the equivalent thermal conductivity distribution. The inverse problem is solved by searching the distribution of the thermal conductivity that minimizes a cost function defined by an integral of the square of the error between the measured temperature and that calculated by FEM for an assumed distribution of the thermal conductivity. A gradient method such as conjugate gradient method is utilized for the searching process. The sensitivity of the cost function with respect to the thermal conductivity of an arbitrary element is obtained by adjoint variable method. The adjoint system is defined so that the sensitivity coefficients of nodal temperatures with respect to the thermal conductivity of each element vanish in the expression of the variation of the cost function. The derived adjoint system is governed by the same governing equation as that for the temperature, with the boundary condition of Dirichlet, Neumann, and Robin types. The final expression for the sensitivity of the cost function results in a simple integral of a dot product of the gradients of the adjoint variable and the temperature for the element whose thermal conductivity is the design variable, hence the computation cost spent for calculating sensitivities is quite low. The effectiveness of the present approach is demonstrated in part through a numerical example for the case in which measured temperature values are given on specific part of the boundary.

Key Words: Heat conduction, Defect and inclustion, Adjoint variable method, FEM

1. Introduction

When a defect is included in the domain of a solid, temperatures and heat fluxes measured on its boundary become different from those measured for the solid without defects. Because the heat hardly transfers through defects, we can consider the defects as the media with null thermal conductivity. Therefore, if we can identify the distribution of the thermal conductivity including null ones from the measurement data, the size, shape and location of the defects in the media can be identified. This idea can also be extended to the identification of the inclusions.

The distribution of thermal conductivities is identified so that the discrepancy between the measured boundary data and the numerical data calculated for the assumed distribution of the thermal conductivity become minimum. The objective function appropriately defined with this discrepancy is minimized by searching the correct distribution of the thermal conductivity, hence gradient methods such as the steepest descendent method and the conjugate gradient method, etc. can be used for the searching process. The gradient methods require sensitivities of the objective function in its minimization process. The sensitivities are related to the sensitivities of temperatures and heat fluxes with respect to the thermal conductivity distribution. The thermal conductivity can also be discretized like the temperature. The calculations of the temperature sensitivities with respect to the discretized thermal conductivity values are very costly when the so-called direct differentiation method is used. Hence, adjoint methods[1, 2, 3, 4] can be used to eliminate the unknown sensitivities of the temperature and heat flux. Huang and Chin[3] used the

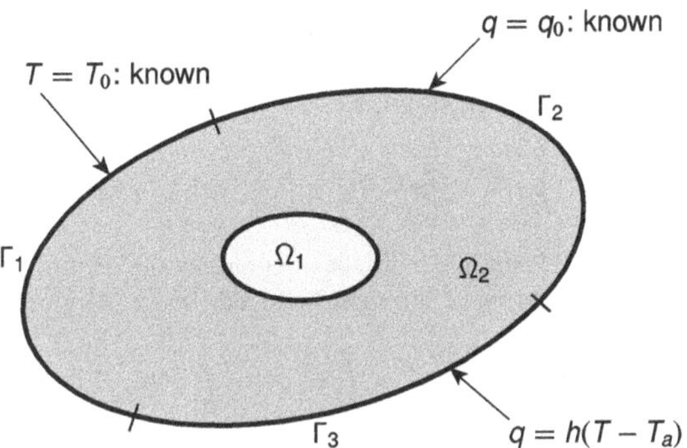

Figure 1: A domain with an inclusion and boundary conditions

conjugate gradient method and the adjoint method to obtain the thermal conductivity distributions of inhomogeneous materials from the measurement data of transient thermal data.

In this paper, we attempt to show a scheme for identifying defects and inclusions by solving thermal conductivity identification problem using steady-state heat conduction data. We assume that the temperature on some part of the boundary is measured and all the other boundaries are subjected to Neumann or Robin type boundary condition. We also assume that we can obtain several different measurement data under different heating conditions[5] to increase the accuracy of the results. The problem is solved based on minimizing an objective function defined as the boundary integral of the square residuals between the measured temperature and the calculated temperature under an assumed thermal conductivity distribution. The thermal problem is solved by means of FEM, and the thermal conductivity is discretized as piecewise constant within each finite element. An adjoint problem is derived so that the unknown temperature and heat flux sensitivities are eliminated from the variation of the objective function.

2. Formulation

Consider a domain consisting of different materials as shown in Fig.1. The thermal conductivity of Ω_2 is assumed to be different from that of Ω_1. The governing differential equation of steady-state heat conduction problem for isotropic material is

$$\nabla(-\kappa \nabla T) + b = 0, \tag{1}$$

where T denotes the temperature, κ the thermal conductivity, and b the heat source. The boundary conditions are assumed as

$$T = T_0 \qquad \text{on } \Gamma_1, \tag{2}$$

$$q = -\kappa \frac{\partial T}{\partial n} = q_0 \qquad \text{on } \Gamma_2, \tag{3}$$

$$q = h(T - T_a) \qquad \text{on } \Gamma_3, \tag{4}$$

where n is the outward normal direction, h is the heat transfer coefficient, and T_a is the ambient temperature. In the present study, we assume $b = 0$ for simplicity, hence instead of Eq.(1), we use

$$\nabla(-\kappa \nabla T) = 0. \tag{5}$$

We suppose that on some part of the boundary Γ_m the temperature can be measured. Then, the objective function is given as

$$J = \int_{\Gamma_m} (T - T_m)^2 \, d\Gamma, \tag{6}$$

where T_m is the measured temperature, and T the temperature calculated by assuming the thermal conductivity distribution.

By introducing Eq.(5) as an equality constraint, we can augment the objective function J, as follows:

$$J = \int_{\Gamma_m} (T - T_m)^2 \, d\Gamma + \int_{\Omega} \lambda \nabla (-\kappa \nabla T) \, d\Omega, \tag{7}$$

where Ω is the entire domain, and λ the Lagrange multiplier, serving as the adjoint variable later.

Integrating by parts the second integral of the right-hand side of Eq.(7) gives

$$J = \int_{\Gamma_m} (T - T_m)^2 \, d\Gamma + \int_{\Gamma} \lambda q \, d\Gamma + \int_{\Omega} \kappa \nabla \lambda \cdot \nabla T \, d\Omega. \tag{8}$$

By taking a variation of Eq.(8), we have

$$\begin{aligned}
\delta J &= \int_{\Gamma_m} 2(T - T_m)\,\delta T \, d\Gamma + \int_{\Gamma} \lambda\, \delta q \, d\Gamma + \int_{\Omega} \delta\kappa \nabla \lambda \cdot \nabla T \, d\Omega + \int_{\Omega} \kappa \nabla \lambda \cdot \nabla (\delta T) \, d\Omega \\
&= \int_{\Gamma_m} 2(T - T_m)\,\delta T \, d\Gamma + \int_{\Gamma} \lambda\, \delta q \, d\Gamma + \int_{\Omega} \delta\kappa \nabla \lambda \cdot \nabla T \, d\Omega - \int_{\Gamma} \left(-\kappa \frac{\partial \lambda}{\partial n}\right) \delta T \, d\Gamma + \int_{\Omega} \nabla (-\kappa \nabla \lambda)\, \delta T \, d\Omega \\
&= \int_{\Gamma_m} 2(T - T_m)\,\delta T \, d\Gamma + \int_{\Gamma_1} \lambda\, \delta q \, d\Gamma + \int_{\Gamma_2} \lambda\, \delta q \, d\Gamma + \int_{\Gamma_3} \lambda\, h \delta T \, d\Gamma \\
&\quad - \int_{\Gamma_1} \mu\, \delta T \, d\Gamma - \int_{\Gamma_2} \mu\, \delta T \, d\Gamma - \int_{\Gamma_3} \mu\, \delta T \, d\Gamma + \int_{\Omega} \delta\kappa \nabla \lambda \cdot \nabla T \, d\Omega + \int_{\Omega} \nabla (-\kappa \nabla \lambda)\, \delta T \, d\Omega.
\end{aligned} \tag{9}$$

Suppose Γ_m is devided into three parts belonging to Γ_1, Γ_2, and Γ_3, respectively, i.e., $\Gamma_m = \Gamma_{m1} \cup \Gamma_{m2} \cup \Gamma_{m3}$ and $\Gamma_{m1} \in \Gamma_1$, $\Gamma_{m2} \in \Gamma_2$, $\Gamma_{m3} \in \Gamma_3$. Since $\delta T = 0$ on Γ_1 on which the temperature is prescribed as the boundary condition, and $\delta q = 0$ on Γ_2, Eq.(9) can be written as follows:

$$\begin{aligned}
\delta J &= \int_{\Gamma_{m2}} \left[2(T - T_m) - \mu\right] \delta T \, d\Gamma - \int_{\Gamma_2 \setminus \Gamma_{m2}} \mu\, \delta T \, d\Gamma \\
&\quad + \int_{\Gamma_{m3}} \left[2(T - T_m) + \lambda h - \mu\right] \delta T \, d\Gamma - \int_{\Gamma_3 \setminus \Gamma_{m3}} (\mu - h\lambda)\, \delta T \, d\Gamma \\
&\quad + \int_{\Gamma_1} \lambda\, \delta q \, d\Gamma + \int_{\Omega} \delta\kappa \nabla \lambda \cdot \nabla T \, d\Omega + \int_{\Omega} \nabla (-\kappa \nabla \lambda)\, \delta T \, d\Omega.
\end{aligned} \tag{10}$$

In Eq.(10), δT on Γ_2 and Γ_3, and in Ω, and also δq on Γ_1 are unknown. Therefore, in order to eliminate them from δJ, we consider an adjoint field satisfying the following differential equation and the boundary conditions:

$$\nabla (-\kappa \nabla \lambda) = 0 \qquad \text{in } \Omega \tag{11}$$

$$\lambda = 0 \qquad \text{on } \Gamma_1 \tag{12}$$

$$\mu = -\kappa \frac{\partial \lambda}{\partial n} = \begin{cases} 2(T - T_m) & \text{on } \Gamma_{m2} \\ 0 & \text{on } \Gamma_2 \setminus \Gamma_{m2} \end{cases} \tag{13}$$

$$\mu = \begin{cases} h\left[\lambda - \{-2(T - T_m)/h\}\right] & \text{on } \Gamma_{m3} \\ h\lambda & \text{on } \Gamma_3 \setminus \Gamma_{m3} \end{cases} \tag{14}$$

Using Eqs.(11) to (14) in Eq.(10), we finally obtain

$$\delta J = \int_{\Omega} \delta\kappa \nabla \lambda \cdot \nabla T \, d\Omega. \tag{15}$$

3. Discretization of thermal conductivity

In order to calculate the sensitivity of the objective function, we have to discretize the thermal conductivity distribution. The simplest case is the employment of piecewise constant interpolation. Suppose the domain is divided into elements and the thermal conductivity is constant within each element. Then, Eq.(15) becomes

$$\delta J = \sum_{i=1}^{N} \delta \kappa_i \int_{\Omega_i} \nabla \lambda \cdot \nabla T \, d\Omega, \tag{16}$$

where N denotes the number of elements and Ω_i the domain corresponding to elemen i. Thus, we obtain

$$\frac{\partial J}{\partial \kappa_i} = \int_{\Omega_i} \nabla \lambda \cdot \nabla T \, d\Omega. \tag{17}$$

Because the differential equation for the adjoint field λ is the same as that for T, and the boundary conditions are also of the same type, the distribution of λ can be calculated efficiently. Then, all the sensitivities of J with respect to the thermal conductivities of the elements are very efficiently calculated by using Eq.(17).

4. Inverse analysis

Both the distributions of T and λ are calculated using FEM. The thermal conductivity is also discretized in accordance with the same mesh as used for FEM analyses. The distribution of the thermal conductivity of the medium with defects and inclusions are calculated by using any nonlinear programming method, starting from the initial guess for its distribution.

The measurement data of temperature are supposed to be obtained under several different heating conditions, i.e., different types of boundary conditions. For each heating condition, we are able to estimate the distribution of the thermal conductivity. In this study, we simply average the results. The computation procedure for identifying the distribution of the equivalent thermal conductivity of the medium with defects is given below.

Step 1: Register all the element thermal conductivities to the design variable set.

Step 2: Do FEM analysis and calculate J. Stop if $J < \varepsilon$.

Step 3: Go to *Step 2* if all the thermal conductivities are smaller than that of the matrix material, otherwise go to *Step 4*.

Step 4: Remove the element thermal conductivities greater than that of the matrix material from the design variable set. Update the design variable set and go to *Step 2*.

Step 5: Remove the maximum design variable from the design variable set. Update the design variable set and go to *Step 2*.

When we do not know if the thermal conductivities of the inclusions are greater or smaller than that of the matrix material, we use the following procedure:

Step 1: Register all the element thermal conductivities to the design variable set.

Step 2: Do inverse analysis for the current set of desing variables.

Step 3: Remove the element thermal conductivities closest to that of the matrix material from the design variable set. Update the design variable set and go to *Step 2*.

5. Numerical example

As a defect identification example, we consider a rectangular domain with a square shaped defect as shown in Fig.2. The square domain is divided into 100 linear rectangular elements with totally 121 nodes. The thermal conductivity of each element is assumed as constant in the element. The temperature is prescribed on some part of the boundary and all the remaining parts are assumed to be subjected to heat transfer condition with the heat transfer coefficient is 0.1 W/m^2K and the ambient temperature is 30°C. We use the measurement data obtained for four different types of heating conditions shown by Type 1 to 4 in Fig.2. In this study, the measurement data are generated by FEM analyses and no measurement errors are taken into account for simplicity.

In Fig.3(a) to Fig.3(f) are shown the original defect location and thermal conductivity distribution obtained in each iterative step. The elements in white are those removed from the design variable set. It was found that by removing the thermal conductivities greater than that of the matrix material, and that of the maximum value from the design variable set, the area, in which the defect is located, is getting focused.

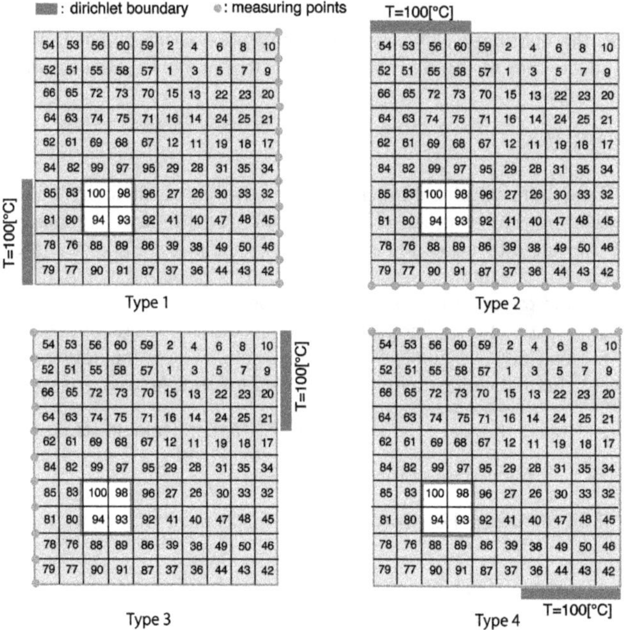

Figure 2: A domain with an inclusion, its FEM meshes, and heating conditions. All the boundaries except those where temperature is specified is subjected to the heat transfer boundary condition.

6. Concluding remarks

An inverse method has been presented for identifying defects and inclusions in media by using FEM based design sensitivity analysis. The defects and inclusions are modeled as specific distributions of thermal conductivity, so that they are identified by solving inverse problems to obtain distributions of thermal conductivities from boundary measurement data. The inverse analysis is based on the minimization of the objective function defined by a boundary integral of the square of the residual between the calculated temperature and the measured temperature. The thermal conductivity is also discretized appropriately based on the FEM mesh. By defining an adjoint variable satisfying the same differential equation as the original one for the temperature, the sensitivities of the objective function with respect to each of the thermal conductivities of the finite elements can be calculated very efficiently. The effectiveness of the present method has been demonstrated in part through a numerical example for two-dimensional defect identification example.

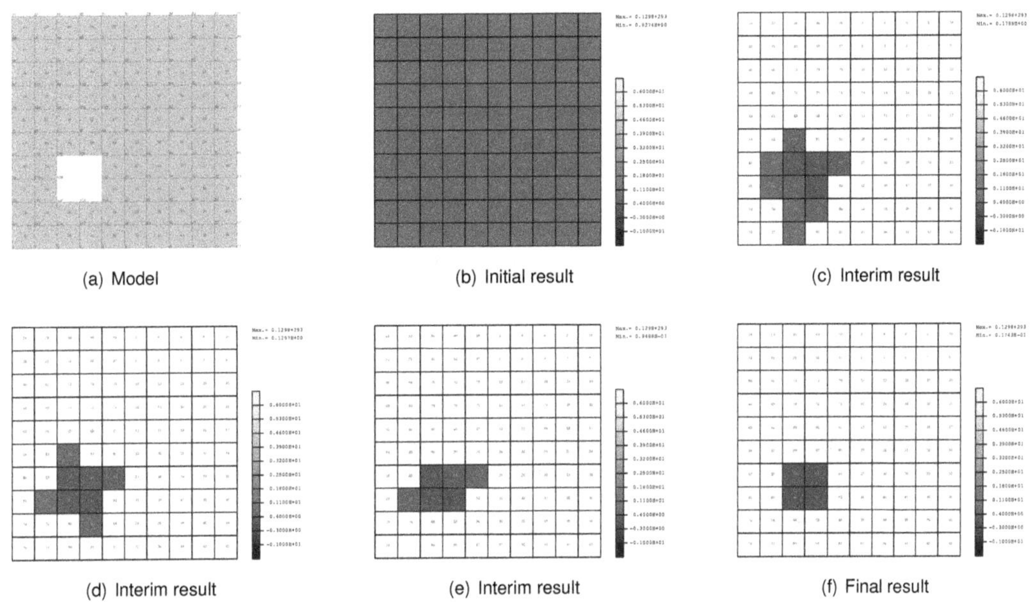

(a) Model (b) Initial result (c) Interim result

(d) Interim result (e) Interim result (f) Final result

Figure 3: Defect model and distributions of thermal conductivity obtained in iterative steps.

7. References

[1] Haug, E.J., Choi, K.K. and Kombov, V., Design Sensitivity Analysis of Structural Systems, Academic Press, New York, 1986.

[2] Cho, S. and Choi, J.Y., "Efficient topology optimization of thermo-elasticity problems using coupled field adjoint sensitivity analysis method," Finite Elements in Analysis and Design, Vol.41, pp. 481–1495, 2005.

[3] Huang, C. and Chin, S.C., "A two-dimension inverse problem in imaging the thermal conductivity of a non-homogeneous medium," Int. J. Heat Mass Transfer, Vol.43, pp. 4061?-4071, 2000.

[4] Zabaras, N. and Yang, G.Z., "A function optimization formulation and implementation of an inverse natural convection problem," Comput. Meth. Appl. Mech. Eng., Vol.144, pp. 245–274, 1997.

[5] Sakagami, T., Kubo, S., Hashimoto, T., Yamawaki, H. and Ohji, K., "Quantitative measurement of two-dimensional inclined cracks by the electric-potential CT method with multiple current applications, " JSME Int. J., Ser.I, Vol.31-1, pp. 76–86, 1988.

Temperature and heat flux errors in thin film thermometry

Keith A. Woodbury[1], *Jonathan W. Woolley*[2]

[1]*Mechanical Engineering*
The University of Alabama, Tuscaloosa, AL 35487-0276
Email: keith.woodbury@ua.edu
[2]*AEGIS Corporation*
Albuquerque, NM
Email: j.w.woolley@gmail.com

Abstract

Thin platinum resistance thermometers (herein called thin film sensors) a re often used in applicatio ns where rapid measurement s of surface temperature are required. These gages are typically vap or deposited onto a non-conducting substrate surface and electrically connected with small wi res through access holes to the surface. The time response of the gage i s measured in millis econds and surface temperature data obtained with this g age is often combined with a pseudo-inverse heat conduction algorithm to provide info rmation about the surfa ce heat flux. Howeve r, the thermal mass of th e connecting wires, thou gh small in ab solute terms, is large compared to that of the thin film, and the capacitive effect of this mass gives ri se to distortions in the temperature fie ld in the are a of the gage, resulting in a small error in the sensed temperature. This temperature error, when used in the inversion for heat flux, also results in an error.

In this re port, a detailed model of a particular thin film gage i s used to com pute the respon se of the sensor to supposed heating conditions. The response of the sensor and the undisturbed surface temperature are compared to estimate the temperature error. Finally, the error in the computed heat flux is determined. A simple approxim ate technique based on superposition is applied to reduce the error in the estimated heat flux.

Key Words: temperature measurement bias

1 Introduction

It is well establish ed that thermocouples imbedded within a low co nductivity material give rise to distortions in the temperature field surrounding the junction [1-10]. This distortion of the temperature field ultimately results in a tem perature bias error. Typically, this bias error results in a sensed temperature lower than th e temperature of the su rroundings. This is because the pre sence of the hig h conductivity thermocouple lead wires offers a preferential path for heat removal from the bead and cooling of th e region around the bead.

Most recently, Woolley [10], condu cted an com prehensive investigation into temperature m easurement errors through high-fidelity modeling of thermocouple installations. His findings affirm previously reported results [1-3] for thermo couples mounted perpendicular to the heated su rface. However, he also demonstrated that temperatu re field distortion and me asurement error will also occur when thermocouples are installed parallel to the heated surface, although this effect is con siderably smaller than for the perpendicular case.

What is le ss established is the effect of inst rumentation on other type s of conta ct temperature measurements: RT D's, thermistors, etc. Since th ese types of instrumentation also rely on metalli c conductor wires to transmit signals to a recording or readout device, it is reasonable to presume that these devices might also be subject to bias errors.

In this paper, the results of a numerical investigation into the possible temperature field disturbance associated with a specific thin film thermometer design and installation are reported. The numerical model details are presented in the next section, followed by the results of a computation to determine the temperature field in the region surrounding the sensor when excited by a constant heat flux history. Further computations are performed using a triangular time-varying heat flux, and the bias error is evaluated from these simulations. A "pseudo-inverse" inverse heat conduction solution is used to determine the heat flux using these biased results, both without and with simulated noise. Finally, a simple approach to improve the heat flux estimates is used to obtain an improved heat flux estimate.

2 Sensor Model

A photograph of the subject thin film sensor is shown in Fig. 1(a) and the corresponding mesh model is seen in Fig. 1(b). The serpentine conductor between the two circular connections is actually the thin film of platinum, which was vapor-deposited and laser-etched on the surface. The body of the object on which the sensor is mounted is Macor and is about 0.25 inch thick. The fill in the main of the circular connecting posts is a silver epoxy which gives electrical continuity between the platinum film and copper wires beneath the surface. Note that the scale of the sensor is small with an overall length dimension of about 0.180 inch and the distance between centerlines of the connections about 0.135 inch.

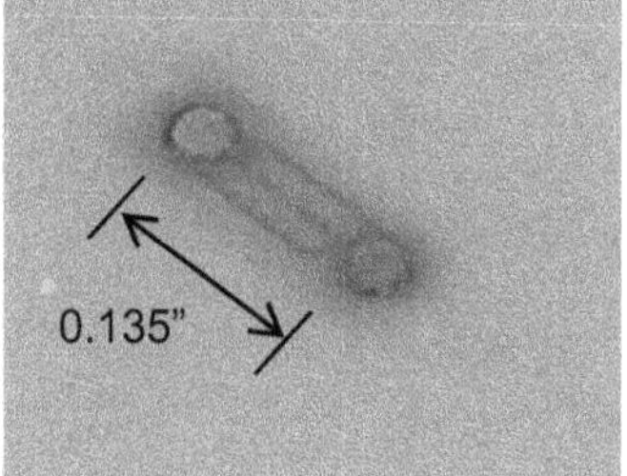

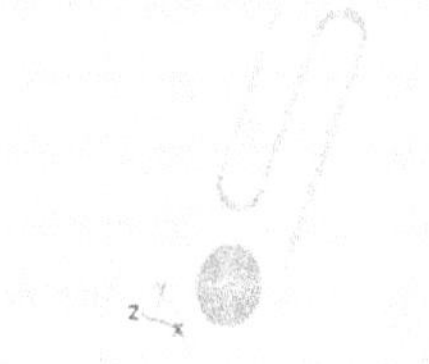

Figure 1: (a) Photograph of sensor; (b) grid model of sensor

A numerical model of sensor was created by isolating the sensor from the surroundings on a small plug of Macor as illustrated in Fig. 2(a). This plug is about 1 inch long and 0.75 inch wide and about 0.25 inch thick. This isolation entails the assumption that the heat flow into the surface far away from the sensor is strictly one-dimensional, and therefore the boundary along the edge of the "plug" is adiabatic. Also shown in Fig. 2(a) are the copper lead wire models which are connected to the thin film sensor via drilled holes in the Macor and the silver epoxy potting material. These wires are about 0.005 inch diameter and are seen in a cross sectional view in Fig. 2(b). The blue material in Fig. 2(b) is the silver epoxy. It should be noted that perfect thermal contact between all interfaces was assumed.

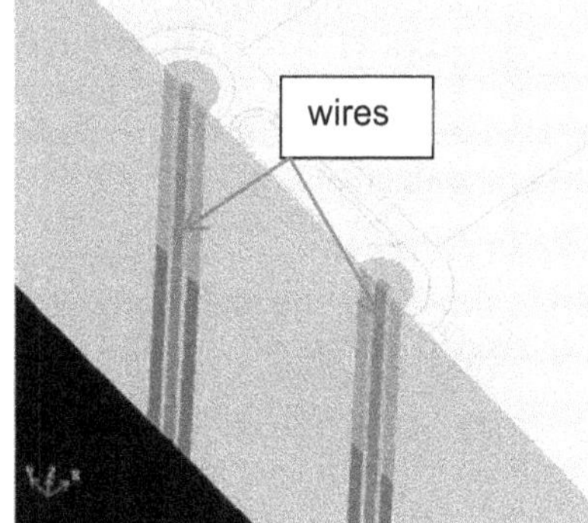

Figure 2: (a) Final grid; (b) split view of model showing wires and epoxy

An initial grid with about 1.7 million calculation volumes was used and an integration time refinement study performed. The grid density was then increased to about 3 million volumes and integration time refinement performed. It was determined that the higher density grid requires a smaller integration time to accurately resolve the solution. The final grid and an integration time step of 0.0025 seconds was employed for the following calculations.

3 Demonstration of Temperature Bias

To demonstrate the bias error, a constant heat flux of 1000 W/m^2 was imposed on the top surface of the Macor plug. All other surfaces were treated as adiabatic. To determine the sensor temperature, the volume averaged temperature along the length of the serpentine platinum wires was computed. This is used as the sensed temperature as the actual sensor temperature is that corresponding to the average resistance of the thin film on the substrate.

The source of the bias error can be seen in the centerline cross section temperature contours of Fig. 3(a). The depression in the temperature field on the surface in the area of the connection terminals can clearly be seen.

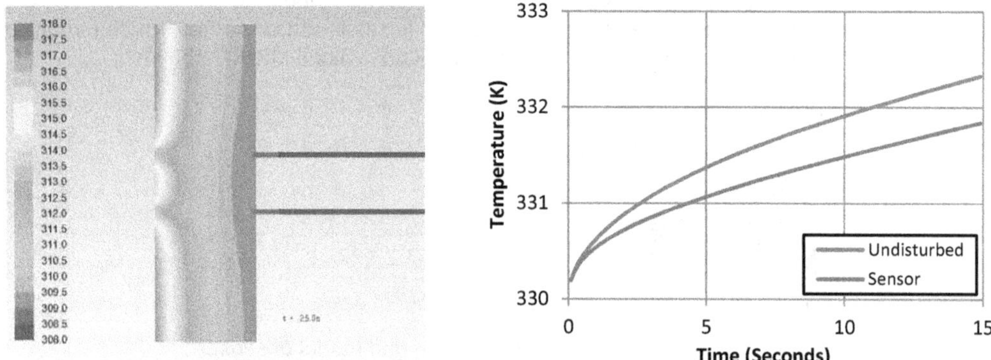

Figure 3: (a)Typical temperature contours; (b) temperature histories

The evolution of this bias error over time can be found by comparing the sensed temperature to the "undisturbed" temperature. The undisturbed temperature was extracted from the numerical model by tracking the temperature of the surface of the Macor at a point near the edge of the plug. The sensed temperature and the undisturbed temperature are shown in Fig. 3(b) for a fifteen second heating with a 1000 W/m^2 heat flux. Note that the error at the end of the heating is about 0.5C.

4 Effect of Bias on the IHCP solution

The determination of surface heat fluxes from measured surface temperatures is often called a "pseudo-inverse" problem. This is because the temperature field in the body can, theoretically, be determined using the measured temperature of the surface as a boundary condition. The heat flux into the domain can then be found from the temperature field using the Fourier Law. Nevertheless, inverse methods, such as Beck's sequential function specification method (SFSM) [11] can be used to invert the measured surface temperature to obtain the surface heat flux. Use of an inverse method offers the possibility to regularize the solution to combat noise in the measured surface temperatures.

Briefly, Beck's SFSM method computes heat flux components sequentially in time. Regularization is afforded by assuming, at each time step, that the heat flux is constant over the next several ("r") time steps. This introduces a bias into the computation, and a balance must be struck between this bias and the stability afforded by the use of future times in the selection of the "r" parameter. The algorithm for Beck's SFSM is summarized in Eq. (1) below:

$$\hat{q}_M = \frac{\sum_{i=1}^{r}\left(Y_{M+i-1} - \hat{T}_{M+i-1}|_{q_{M\ldots}=0}\right)\varphi_i}{\sum_{i=1}^{r}\varphi_i^2} \quad (1)$$

Here, $\varphi_i = \varphi(t_i)$ is the value of the sensitivity function at time t_i. The sensitivity function is the response of the domain at the temperature measurement point to a step input in surface heat flux. This sensitivity function is typically found by determining the temperature response at the sensor location due to a constant heat flux and computing the derivative of the temperature history to the heat flux. The term $\hat{T}_{M+i-1}\big|_{q_{M\ldots}=0}$ is the calculated temperature at the sensor location based on all the heat flux components up to but excluding the present ("M") time.

A triangular heat flux pulse, illustrated in Fig. 4(a), was used as input to the Macor plug and sensor. From this computation, the sensed temperature and the undisturbed temperature histories (not shown) were extracted. These temperatures were used directly as input to solve the corresponding IHCP using Beck's method, and the results are seen in Fig 4(b). The upper (blue square) symbols are the results of the estimation using the undisturbed data, while the lower (red diamond) symbols are the results obtained using the sensed temperature data. Note that the upper curve overlays the imposed heat flux, while the lower curve under-predicts the heat flux by about 20% at the peak.

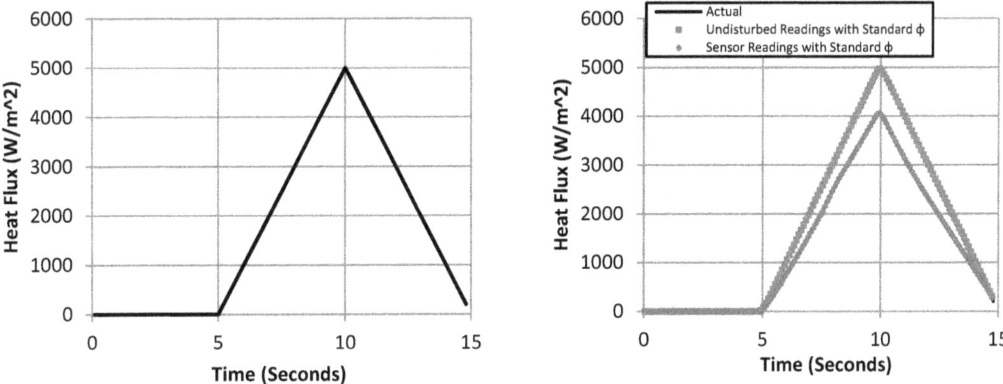

Figure 4: (a) imposed heat flux; (b) recovered heat flux (errorless data)

To demonstrate the effect of noise on the data to the solution of the pseudo-inverse problem, random noise in the range of +/- 0.5C was added to both the sensed and undisturbed temperature histories. These simulated measurements were then used as input to Beck's method. The results, obtained with a value of r=8, are seen in Fig. 5. Fig. 5(a) shows the recovered heat flux if the "undisturbed" temperature with noise is used as data, and Fig. 5(b) shows the result if the sensed temperature with noise is used. Note that a relatively low level of random noise (+/- 0.5C) causes significant excursions in the heat flux predictions. This is a manifestation of the inverse nature of the problem, even though data are obtained at the surface of the heat disturbance. Use of an inverse solver allows some smoothing of the solution through regularization of the problem.

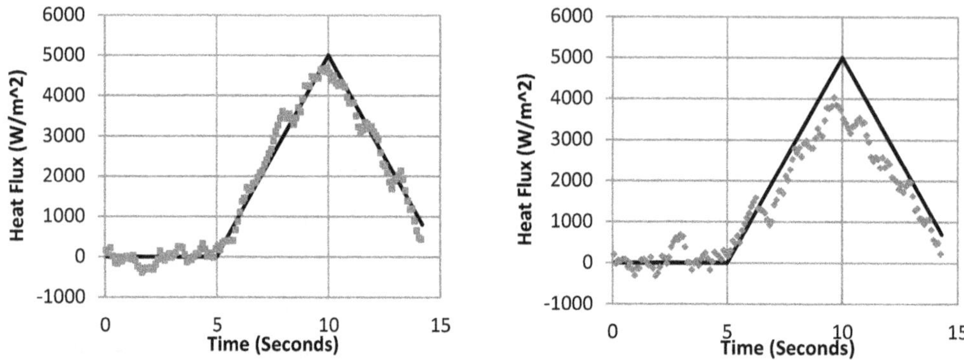

Figure 5: (a) results with noisy "undisturbed" data; (b) results with noisy "sensed" data

5 Improved Estimates through Sensor Modeling

The IHCP solution can be substantially improved if an accurate detailed model of the sensor dynamics is available. The sensitivity coefficients, φ_i, in Eq. (1) can be computed from the detailed model by imposing a constant heat flux and monitoring the sensor output. This nu merical technique for determining the φ function can be combined with the traditional Beck's meth od of Eq. (1) to obtai n much improved estimates.

Fig. 6(a) shows the sensitivity functions obtained from the numerical model. The upper (blue) line is the "undisturbed" sensitivity function and corresponds to the φ function used for all the p receding IHCP solutions in this paper. The lower (red) line shows the temperature rise of the sensor due to a 1000 W/m^2 heat flux imposed on the surface. When this latter φ function is used as the sensitivity function in the IHCP and noised data from the thin film sensor are used, the heat fluxes in Fig. 6(b) are obtained. These results are dramatically improved over the previous estimates using the sensed data (cf. Fig. 5(b)).

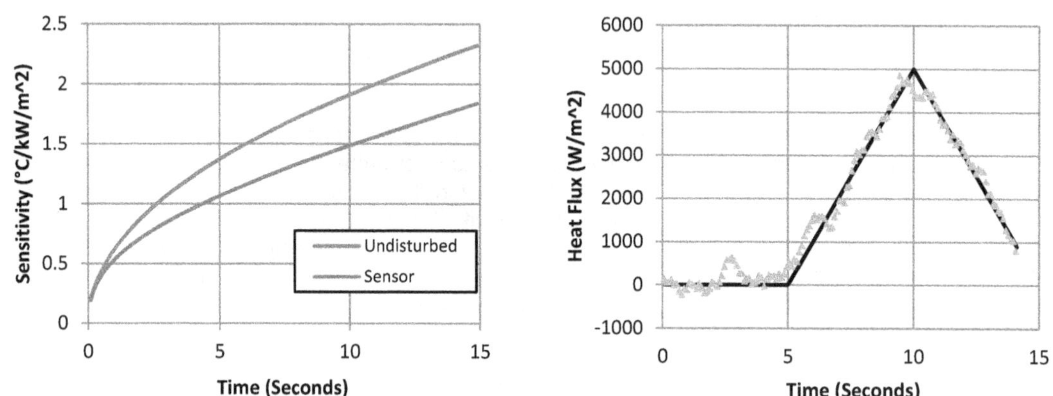

Figure 6. (a) sensitivity coefficient functions; (b) results with noised sensor data using sensor sensitivity

6 Conclusions

The presence of lead wires in a low conductivity domain may result in temperature disturbances in the regions of interest leading to biased te mperature readings. Use of these bia sed temperatures in th e IHCP can, owing to the inverse nature of the problem, result in large errors (~20%) in the estimated heat flux. Even if the problem is pseu so-inverse, and noise is present in the data, the regula rization in an

IHCP solution will help smooth the solution. Detailed sensor modeling can be used to determine sensitivity coefficients that take account of the biasing sensor dynamics. Incorporation of these more accurate sensitivity coefficients in the solution results in dramatic improvement in the estimated heat flux history.

7 References

[1] Beck, J.V., "Thermocouple Temperature Disturbances in Low Conductivity Materials," Journal of Heat Transfer, vol. 84C, pp. 124-132, 1962

[2] Pfahl, Jr., R.C., and Dropkin, D., "Thermocouple Temperature Perturbations in Low-Conductivity Materials," American Society of Mechanical Engineers – Papers, Paper 66-WA/HT-8, 1966

[3] Attia, M.H., and Kops, L., "Distortion in Thermal Field around Inserted Thermocouples in Experimental Interfacial Studies," Journal of Engineering for Industry, vol. 108, pp. 241-246, 1986

[4] Attia, M.H., and Kops, L., "Distortion in Thermal Field around Inserted Thermocouples in Experimental Interfacial Studies-Part II: Effect of the Heat Flow Through the Thermocouple," Journal of Engineering for Industry, vol. 110, pp. 7-14,1988

[5] Park, J.E., Childs, K.W., Ludtka, G.M., and Chu, W., "Correction of Errors in Intrinsic Thermocouple Signals Recorded During Quenching," AIChE Symposium Series - Heat Transfer - Minneapolis, vol. 87, n. 283, pp. 309-318, 1991

[6] Attia, M.H., and Kops, L., "Distortion in Thermal Field around Inserted Thermocouples in Experimental Interfacial Studies-Part III: Experimental and Numerical Verification," Journal of Engineering for Industry, vol. 115, pp. 444-449, 1993

[7] Attia, M.H., Cameron, A., and Kops, L., "Distortion in Thermal Field around Inserted Thermocouples in Experimental Interfacial Studies-Part IV: End Effect," Journal of Manufacturing Science and Engineering, vol. 124, pp. 135-145, 2002,)

[8] Tszeng, T.C., and Saraf, V., "A Study of Fin Effects in the Measurement of Temperature Using Surface-Mounted Thermocouples," Journal of Heat Transfer, vol. 125, pp. 926-935, 2003

[9] Gupta, A., "Effect of Deterministic Thermocouple Errors on the Solution of the Inverse Heat Conduction Problem," M.S. Thesis, Department of Mechanical Engineering, The University of Alabama, 2004

[10] Woolley, J. W., "Accounting for Transient Temperature Measurement Error with a High Fidelity Thermocouple Model and Application to Metal/Mold Interfacial Heat Flux Estimation",Ph.D. Dissertation, Department of Mechanical Engineering, The University of Alabama, 2009

[11] Beck, J. V., Blackwell, B., and St. Clair, C., *Inverse Heat Conduction: Ill-posed Problems*, John-Wiley, New York, 1985

Reconstruction of heat transfer coefficient at nucleate and film boiling using Proper Orthogonal Decomposition

Ziemowit Ostrowski[1], *Ryszard Bialecki*[1], *Marc Muster*[2] *and Roman Weber*[2]

[1]*Insitute of Thermal Technology*
Silesian University of Technology, Gliwice, Poland
Email: ziemowit.ostrowski@polsl.pl, ryszard.bialecki@polsl.pl
[2]*Institute for Energy Process Engineering and Fuel Technology*
Clausthal University of Technology, Clausthal, Germany
Email: muster@ievb.tu-clausthal.de, roman.weber@ievb.tu-clausthal.de

Abstract

Proper Orthogonal Decomposition (POD) combined with Radial Basis Functions interpolation is used to retrieve from transient temperature measurements the heat transfer coefficients at boiling, both for nucleate and film regimes. The object under consideration is a hot metal ball immersed in water. It is shown, that POD can be seen as a generalized separation of variables technique with eigenfunctions evaluated numerically by a statistical analysis of temperature fields obtained by solving a sequence of direct problems. The regularization of the technique comes from the truncation of the resulting series and filtering out higher frequencies of the solution. The amplitude of the eigenvalues are obtained by fitting the measurements results with the POD-RBF model.

Key Words: inverse, POD, RBF, heat transfer coefficient retrieval

1. Introduction

1.1. Boiling

The importance of heat transfer at boiling lies mainly in the high heat fluxes arising in this heat convection mechanism leading to significant reduction of investment costs of heat transfer equipment. Steam, a frequently used heat carrier, is produced through boiling. The popularity of this medium comes from its high energy density resulting from the high latent heat of evaporation. Another advantages of using steam is its low price and low compression costs as this process can be carried out in liquid phase. All this features cause that boiling is a very important mode of heat transfer.

Mathematical description of boiling is still in its infancy, as the phenomenon involves nonequilibrium overheating, bubble formation controlled by the presence of active centers on the boiling surface, the roughness thereof and surface tension-pressure interplay. All these complexities make a reliable CFD simulation almost impossible, leaving as the only option for describing the rate of heat transferred by boiling, to experimental identification.

1.2. Inverse analysis

When designing an experimental setup for inverse analysis it is convenient to let the rig work in transient mode. The reasons for this are twofold:

- the time of the experiment can be made short, as steady state need not be reached,
- the influence of the boundary conditions on the solution is limited. Moreover, they need not to be stabilized, which might be a source of an additional error.

Though inverse analysis can readily produce temperature and heat flux on the boiling surface, these kind of BCs are of limited interest in practice. The reason for this is that both temperature and heat flux vary strongly in time. Moreover, the results cannot be generalized to other thermal loads. Robin's condition does not suffer from such drawbacks. The heat transfer coefficient (HTC) arising in this BC only weakly depends on both time and temperature. Not only can the experiments be carried out at room temperature, but also the constancy in time introduces additional regularization in the inverse procedure. Moreover, there is a well established theory of similarity, which defines the way, the HTC can be generalized to other fluids.

1.3. Filtration of higher frequencies

Analytical solutions of heat conduction problems in domains of finite dimensions have a form of Fourier series

$$u(\mathbf{r},\tau)=\sum_{i=1}^{\infty}a_i(\tau)\Phi_i(\mathbf{r}) \tag{1}$$

Where u is the temperature, $\mathbf{r}$ stands for vector coordinate, τ is time, a_i are amplitudes and Φ_i orthogonal eigenfunctions of appropriate Sturm Liouville problem. The higher is the index of the eigenfiunctions, the higher is the frequency of its oscillation (Fig. 1).

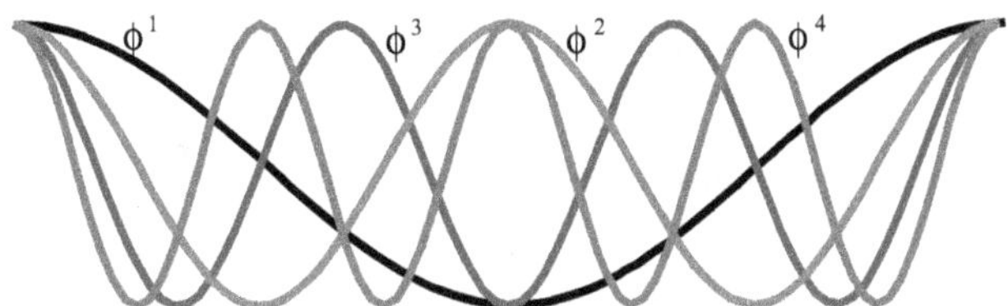

Figure 1: Plot of first four eigenfunctions of a sample Sturm Liouville problem arising in heat conduction.

Expression (1) can be obtained analytically by separation of variables or integral transforms. If the time is not too short, Fourier series converge rapidly, so that only few first, low frequency terms, are needed to reproduce the temperature field. This feature is attractive in the context of the inverse problems, as the truncation of the series means filtering out the numerical noise.

Analytical methods can be applied only to very simple shapes and linear problems. However, there is a way of evaluating the eigenfuctions numerically using a statistical technique, known as Proper Orthogonal Decomposition (POD) [1].

The idea of this method is to excite the system using different input parameter and process the output of the BVP to extract the correlations between the resulting fields. As the object under consideration is always the same physical object, its responses (i.e. temperature fields) are not independent. It has been shown that POD produces in this case approximate eigenfunctions of the Sturm Liouville problem [2].

To be of use in the context of the inverse analysis, Eq. (1) should be modified to accommodate the dependence of the solution from the parameters to be retrieved. Assume that these parameters gathered in an excitation vector $\mathbf{k}$ are the approximation coefficients of the unknown HTC distribution. The temperature field can then be written in a form of

$$u(\mathbf{r},\tau,\mathbf{k})=\sum_{i=1}^{\infty}a_i(\tau,\mathbf{k})\Phi_i(\mathbf{r}) \tag{2}$$

As already mentioned, POD gives an approximation of the the eigenfunctions. To employ this technique in the inverse analysis, a technique of evaluating the amplitudes $a(\tau,\mathbf{k})$ should be devised. The proposed inverse technique uses Radial Basis Functions (RBFs) for this purpose. Some earlier papers on this subject [3,4,5] applied POD to other inverse problems and did not point out the direct association of the method with eigenfunction expansion.

2. Proper Orthogonal Decomposition

To be able to handle problems in arbitrary geometries, numerical methods of solving the BVP are used. This brings Eq. (2) to a discretized form of

$$\mathbf{u}(\mathbf{r},\tau,\mathbf{k})=\mathbf{\Phi}\cdot\mathbf{a} \tag{3}$$

Where $\mathbf{u}$ is the vector of temperatures at nodal points of the applied numerical mesh, $\mathbf{\Phi}$ is the modal matrix whose columns store the discretized eigenfunctions (POD modes) of the problem and $\mathbf{a}$ is a vector of amplitudes. Modal matrix is orthogonal ie $\mathbf{\Phi}^T\cdot\mathbf{\Phi}=\mathbf{I}$ with $\mathbf{I}$ standing for the unit matrix.

The first step of the procedure is to solve a sequence of direct heat conduction problems using subsequent values of the excitation vector $\mathbf{k}$. Each solution is stored in a vector $\mathbf{u}$, hereafter referred to as a *snapshot.* A collection of snapshots is gathered *snapshot matrix* $\mathbf{U}$ whose column store subsequent snapshots.

POD can be seen as a decomposition of the snapshot matrix into a product of POD basis and a matrix of amplitudes. Columns of the latter are the vectors of amplitudes entering Eq. (2). The decomposition is an exact operation and can be written as

$$\mathbf{U}=\mathbf{\Phi}\cdot\mathbf{A} \tag{4}$$

The dimensionality (number of POD modes) in Eq. (4) is that of the snapshot matrix (i.e. number of snapshots). The theory of POD shows that the dimensionality of the modal matrix can be significantly reduced, if only there is a correlation between the snapshots. This brings Eq. (4) to a truncated form

$$\mathbf{U} = \bar{\mathbf{\Phi}} \cdot \bar{\mathbf{A}} \tag{5}$$

Where the overbar stands for truncated matrices. Similarly to the full POD matrix, its truncated counterpart is also orthogonal

$$\bar{\mathbf{\Phi}}^T \cdot \bar{\mathbf{\Phi}} = \mathbf{I} \tag{6}$$

The POD basis can be constructed using different formulae. The one used in this paper is that of Karhunen-Loeeve approach that resorts from the least square condition stating that

$$\| \mathbf{U} - \bar{\mathbf{\Phi}} \cdot \bar{\mathbf{A}} \| \to \min \tag{7}$$

The central proposition of the POD technique states that among all orthogonal bases of the same dimensionality, the POD basis is the best possible one in terms of the energy carried by subsequent modes. For a given number of the modes, no other basis can contain more energy [1].

Calculation of the POD basis requires a solution of an eigenvalue problem defined as

$$\mathbf{D}\mathbf{v}^j = \lambda_j \mathbf{v}^j \tag{8}$$

Where $\mathbf{D} = \mathbf{U}^T \cdot \mathbf{U}$ is a positive definite symmetrical matrix, $\mathbf{v}^j$ stands for eigenvector and λ_j for an eigenvalue. It can be shown [1] that the latter can be interpreted as a measure of the energy carried by the *j*th POD mode. If the snapshots are strongly correlated, subsequent eigenvalues rapidly decay. A criterion of truncating the POD basis is the fraction of the neglected energy.

$$\gamma = \frac{\sum_{j=1}^{M} \lambda_j}{\sum_{j=1}^{N} \lambda_j} \tag{9}$$

Where N is the total number of generated snapshots (number of columns in the snapshot matrix $\mathbf{U}$) and N stands for the number of columns in the truncated POD matrix $\bar{\mathbf{\Phi}}$.

Once the eigenvectors are known, the *j*th POD mode being the columns of the POD matrix can be determined from Sirovich [6] equation

$$\phi^j = \mathbf{U} \cdot \mathbf{v}^j \tag{10}$$

The determination of the amplitude matrix starts with selecting a proper Radial Basis Functions (RBF) $g(\mathbf{k})$. Several types of these functions exists. In this study, thin plate spline [7] has been used. This function is defined as

$$g_i(\mathbf{k}) = (|\mathbf{k} - \mathbf{k}_i|/r)^2 * \ln(|\mathbf{k} - \mathbf{k}_i|/r) \tag{11}$$

with r standing for the smoothing parameter. In the present study the value of the parameter has been taken as 0.5, $\mathbf{k}$ denotes the vector of excitation and $\mathbf{k}_i$ is the center of the RBF being the excitation vector used to generate the *i*th snapshot. It is recommended to used normalized components of the excitation vector.

Collecting the RBF in a matrix $\mathbf{G}$ whose entries are defined as $\{\mathbf{G}\}_{ij} = g_j(\mathbf{k}_i)$ and requiring that the truncated amplitude matrix can be expressed as a product of the RBF matrix and a coefficient matrix $\mathbf{B}$, one arrives at an equation

$$\bar{\mathbf{A}} = \mathbf{B} \cdot \mathbf{G} \tag{12}$$

The amplitude matrix can readily be determined from Eq. (5) by multiplying it by transposed truncated POD matrix. The result reads

$$\bar{\mathbf{A}} = \bar{\mathbf{\Phi}}^T \cdot \mathbf{U} \tag{13}$$

Combining Eqs. (12) and (13) yields the coefficient matrix $\mathbf{B}$ whose entries can be evaluated by solving a set of linear equations

$$\mathbf{G}^T \cdot \mathbf{B}^T = \bar{\mathbf{A}}^T \tag{14}$$

with multiple right hand sides. In many cases, a better option is to solve the system using the Singular Value Decomposition. Once the coefficient matrix $\mathbf{B}$ is known, any nodal temperature vector can be expressed in a form

$$\mathbf{u}(\mathbf{k}) = \bar{\mathbf{\Phi}} \cdot \mathbf{B} \cdot \mathbf{g}(\mathbf{k}) \tag{15}$$

Denoting as $\mathbf{C}$ the product of the known POD and coefficient matrices one arrives finally at

$$\mathbf{u}(\mathbf{k}) = \mathbf{C} \cdot \mathbf{g}(\mathbf{k}) \tag{16}$$

Equation (16) can be seen as a Reduced Order Model of the temperature field induced by excitation vector $\mathbf{k}$. This equation can be used to retrieve the unknown excitation vector by fitting the representation to the measured temperatures. This is equivalent to a solution of the nonlinear programming problem

$$\min_{w.r.t.\mathbf{k}} \sum_{k=1}^{k} [\hat{u}_k - \sum_{j=1}^{M} C_{jk} g_j(\mathbf{k})]^2 \tag{17}$$

Where $\hat{u}_k$ stands for the reading of the *k*th sensor. Index $k = 1, 2, \ldots K$ runs over temperatures measured by all sensors at all time instants.

The solution of Eq. 17 can be accomplished using any nonlinear optimization solver. Genetic Algorithm solver has been used in this study. The reason for selecting this algorithm was that it does not show a tendency to stuck in local minima.

3. Experiment

A nickel ball equipped with three thermocouples has been heated up to about 920°C and dropped into 20°C water. The temperature readings of three thermocouples mounted close to the surface of the ball were recorded. Sketch of the test rig, temperature sensors locations and surfaces where the HTC was retrieved are shown in Fig. 2.

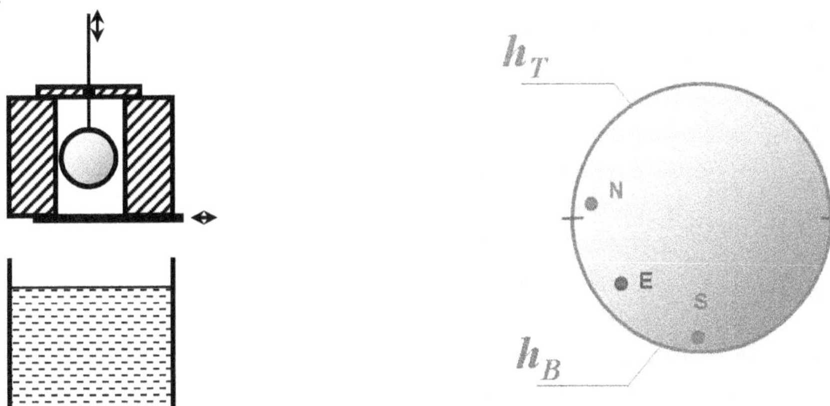

Figure 2: The test rig (left) and location of temperature sensors, surfaces where the HTC was retrieved (right).

The experiment has been recorded using a video camera, four excerpts are shown in Fig. 3. As the initial temperature of the ball was very high, film boiling developed at the beginning of the phenomenon (Fig. 3(a)). After a certain time, the temperature of the ball dropped and the steam blanket separating the metal from liquid collapsed (Fig. 3(c)&(d)), so that the boiling regime turned into nucleate (Fig. 3(d)). Analysis of the recorded pictures revealed that the boiling crisis occurred later on the upper hemisphere. The times the boiling crisis happened has been found by analyzing the recorded pictures of the phenomenon.

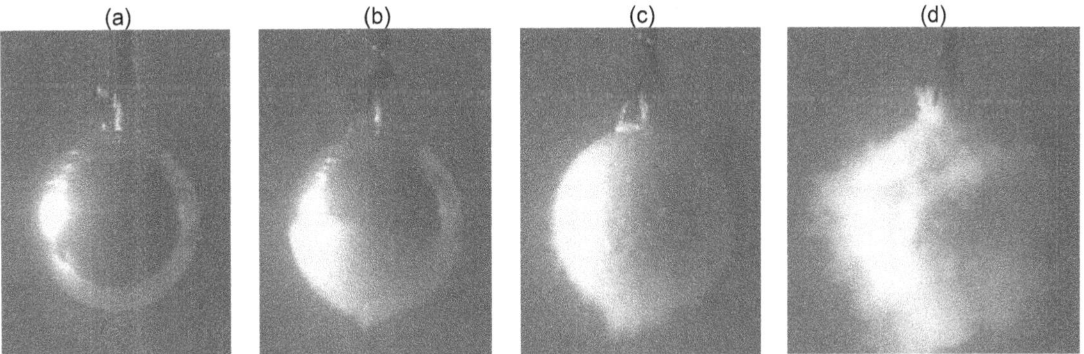

Figure 3: Camera recorded experiment (excerpts from the movie):
film boiling (a), steam blanket separating the metal from liquid collapsing (b) & (c) and nucleate boiling (d).

4. Results of inverse analysis

The goal of the study was to retrieve the values of the HTC for nucleate and film boiling, separately for the top and bottom part of the ball, which means that four values of the HTC are to be retrieved.

The excitation vector consisted of five components $\{h_B^F, h_B^N, h_T^F, h_T^N, \tau\}$ where h stands for the HTC, subscript B and T denote bottom and top hemispheres of the ball while F and N refer to film and nucleate boiling regime, respectively. τ denotes time the snapshot has been generated and is treated as a known parameter. The total number of generated snapshots was 6912, the length of the snapshot was 1303 and correspond to the number of nodes used by the numerical solver to generate the snapshot. The measured temperatures were recorded every 5 seconds starting from 5 till 60 seconds. HTC were sampled at following values, for both bottom and top hemispheres:

- for nucleate boiling: between 200W/m^{2}K and 800W/m^{2}K with a step of 200W/m^{2}K,
- for film boiling: between 1 000W/m^{2}K and 16 000 W/m^{2}K with a step of 3 000W/m^{2}K.

The time of the boiling crisis have been assessed by inspection of the recorded movie and taken as 33s for the top and 32s for the bottom hemisphere.

The plot of the first eigenvalues is shown in the left part of Fig. 4. Rapid decay of the eigenvalues indicates strong correlations between snapshots. The neglected energy fraction was 0.9 E-12, which truncated the POD basis to 20 vectors. The resulting HTC at bottom and top hemispheres for nucleate and film boiling are shown in the right part of Fig. 4.

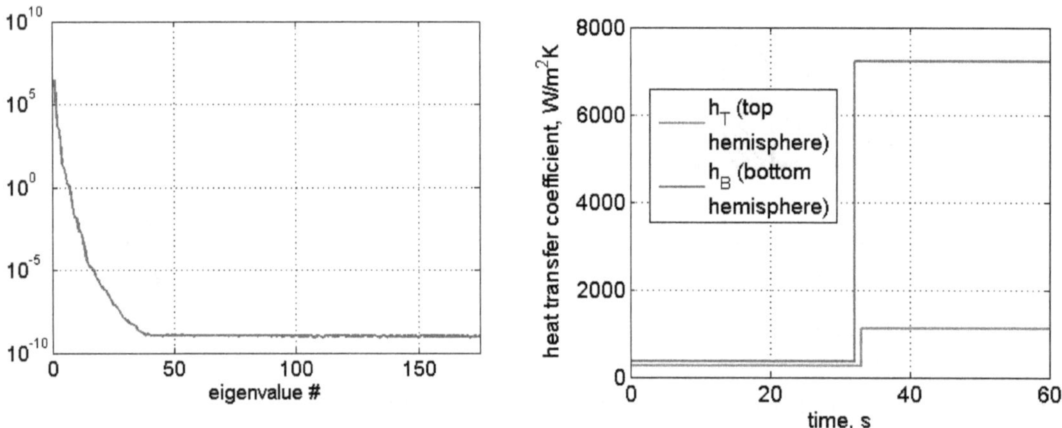

Figure 4: Eigenfunctions of the POD problem (left),
and retrieved values of the HTC for bottom and top of the ball in film and nucleate boiling regime (right).

The accuracy of the retrieved results have been assessed by an *a-posteriori* analysis. The retrieved values of the HTC have been taken as boundary conditions of an appropriate direct problem. The evaluated temperatures have been compared with the measured temperature. Fig. 5 shows result of such comparison of these values for all thermocouples versus time.

5. Conclusions

The POD-RBF technique has a similar functionality as neural network. A deeper analysis shows that the POD technique has some common roots with separation of variables. The extension of the latter for the purposes of the inverse analysis relies on including in the amplitudes of the eivenfunctions the dependence on parameters to be retrieved. The property of filtering out the higher order frequencies of the eigenfunctions introduces some regularization in the inverse solver. The weak point of the proposed technique is its numerical cost associated with

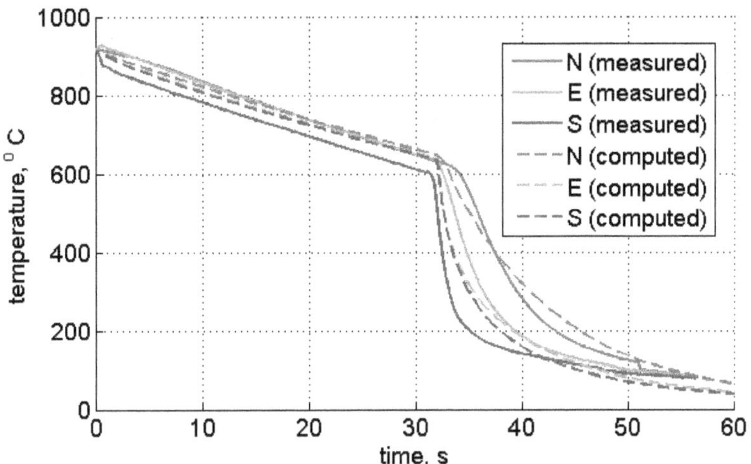

Figure 5: *A-posteriori* analysis. Comparison of measured (solid lines) and computed (broken lines) temperatures. Codes of thermocouples location as in Fig. 2 (right).

the generation of numerous snapshots. At the present stage of development, the proposed technique still bases on intuition rather than a sound mathematical reasoning. Numerous successful applications of RBF-POD in inverse analysis show that the technique can be seen as robust and general technique of solving inverse problems.

Acknowledgement

The research of Z.O. and R.B. has been supported by a grant of the Polish Ministry of Research and Higher Education. This financial assistance is gratefully acknowledged herewith.

6. References

[1] Liang, Y.C., Lee, H.P., Lim, S.P., Lin, W.Z. Lee, K.H.,and Wu C.G, "Proper Orthogonal Decomposition and its Applications - Part I: Theory", Journal of Sound and Vibration, Vol. 252, No. 3, pp. 527-544, 1995.

[2] Białecki, R.A., Kassab, A.J. and Fic, A., "Proper Orthogonal Decomposition and modal analysis for acceleration of transient FEM thermal analysis", International Journal for Numerical Methods in Engineering, Vol. 62, No. 6, pp. 774-797, 2005.

[3] Ostrowski, Z., Białecki, R.A. and Kassab, A.J., „Solving inverse heat conduction problems using trained POD-RBF network inverse method", Inverse Problems in Science and Engineering Vol. 16, No. 1, pp. 39-54, 2008.

[4] Buljak, V. and Maier, G., "Proper Orthogonal Decomposition and Radial Basis Functions in material characterization based on instrumented indentation", Engineering Structures, Vol. 33, No. 2, pp. 492–501, 2011.

[5] Larson, R.S. and Jones, M.R., "Reduced-order modeling of time-dependent reflectance profiles from purely scattering media", Journal of Quantitative Spectroscopy & Radiative Transfer, Vol. 109, No. 2, pp. 201-209, 2008.

[6] Sirovich, L., "Turbulence and the dynamics of coherent structures", Quarterly of Applied Mathematics, Vol. 45, No. 3, pp. 561-571, 1987.

[7] Buhmann, D.M., Radial basis functions, Cambridge University Press, 2003.

Accelerated Detection of Rapid Heat Flux Changes in Real Time Remote Prediction of Heat Source Temperature

Marcin Janicki, Zbigniew Kulesza, and Andrzej Napieralski

Department of Microelectronics and Computer Science,
Technical University of Lodz, Wolczanska 221/223, 90-924 Lodz, Poland,
E-mail: janicki@dmcs.pl

Abstract

This paper discusses the possibility of accelerating the detection of rapid heat flux changes, when estimating the source heat flux from remote temperature sensor measurements. Obviously, due to the temperature lagging, in such cases there is an inevitable delay in the heat flux change detection, however the modifications proposed in this paper to the sequential function specification algorithm allow significant acceleration of the detection, which is crucial in certain practical real time applications. The theoretical considerations are illustrated with the presentation of the experimental results obtained both on simulated and measured data.

Key Words: heat conduction, remote temperature sensor network, real time heat flux estimation.

1. Introduction

Many modern electronic applications require continuous monitoring of their power dissipation and heat source temperature. Considering that the temperature distribution in semiconductor chips is usually particularly non-uniform, distributed sensor networks and reliable algorithms for prediction of heat source temperature have to be used. Although the theoretical bases for the solution of such problems are already well established, new hardware and software solutions for such applications still need to be developed.

There are many methods developed expressly to solve inverse heat transfer problems. The excellent overview of these methods can be found in [1]. Here, taking into account the particular real time application, we focus on the sequential Function Specification Method (FSM) described in [2]. This method is particularly suitable for digital filter applications, thus it was adapted and further modified by the authors as described in [3]-[4]. The next section describes the particular problem considered in this paper. Then, the inverse algorithms used by the authors in experimental part are presented in detail. Finally, the results of numerical simulations and temperature measurements performed for a real structure are discussed providing important conclusions and indications for further practical applications of the proposed method

2. Problem Statement

The particular inverse problem concerned in this paper consists in real time estimation of hot spot temperature in electronic structures. Such structures, as shown in Fig. 1, typically contain a certain number of heat sources and can be additionally equipped with a network of temperature sensors. The temperature distribution in a structure is usually modelled by the second order partial differential heat equation, which can be solved for a given circuit thermal model using virtually any analytical or numerical solution method [5]-[6].

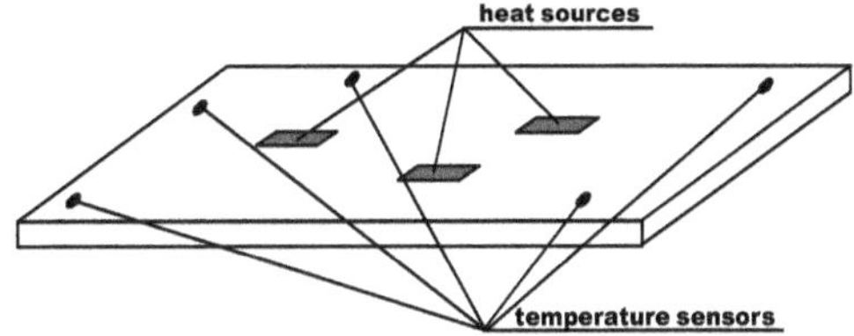

Figure 1: Semiconductor die with heat sources and temperature sensors.

Here, the goal of the solution is to compute the thermal influence coefficients, which link the density of power dissipated in the heat sources to the resulting temperature rise in a structure. Then, the temperature values at selected locations T can be found multiplying the source heat flux (power density) vector Q by the matrix of sensitivity coefficients A as shown in Eq. 1:

$$T = A * Q \tag{1}$$

Owing to the relatively simple geometry of the considered problem, the sensitivity coefficients in most cases can be computed, as outlined in [7]-[8], using the analytical Green's function solution method. This method has been already successfully applied in the thermal analysis of electronic circuits [9]. The detailed descriptions of the layered thermal model used by the authors and the respective heat equation solution method can be found in [10].

3. Inverse Algorithms

The simplest solution to measure the temperature of heat sources in an electronic structure is to place sensors directly where the heat is generated. However, in practice this solution might not be possible and then the heat source temperature has to be estimated based on remote sensor measurements solving an inverse heat conduction problem. The inverse problems of this kind are usually extremely ill-posed because of lagging and damping of a temperature response. Therefore, some special algorithms have to be used to produce robust estimates of heat source temperature.

The inverse algorithm considered in this paper is based on the function specification method, also known as the sensitivity coefficient method, proposed by Beck and adapted by the authors for the multi-source real time temperature estimation of electronic circuits. Extending the concept of the pseudo-inverse Moore-Penrose rectangular matrix, this sequential algorithm uses the temperature data from multiple sensors and averages them in time to produce the heat source power estimates. During the algorithm operation the density of power dissipated in the heat sources Q is estimated, as shown in Eq. (2), multiplying the measured sensor temperature values T by a kind of pseudo-inverse matrix of the thermal influence coefficients. The detailed derivation of the algorithm and the exact construction of this matrix are presented in [2].

$$Q = (A^T * A)^{-1} * A^T * T \tag{2}$$

The numerical simulations of electronic circuit thermal model, presented in [4], demonstrated that in the considered case the nature of the inverse problem ill-conditioning is dependent on the cooling conditions. Namely, with the forced water cooling the temperature profile is steep and when using remote temperature sensors one might expect problems due to the damping of the thermal response. On the contrary, with still air cooling the temperature is much more uniform throughout the chip, hence the damping of the thermal response is not so important, but the poor conditioning of the problem comes rather from the fact that the power dissipation in a heat source causes similar response in all sensor locations. Then, it is virtually impossible to estimate the power dissipation in the individual sources, but their temperature can be predicted quite accurately, even using remote sensors. Therefore, here we will discuss only the earlier case, when a circuit is water cooled, which is anyway more natural for high power electronic applications.

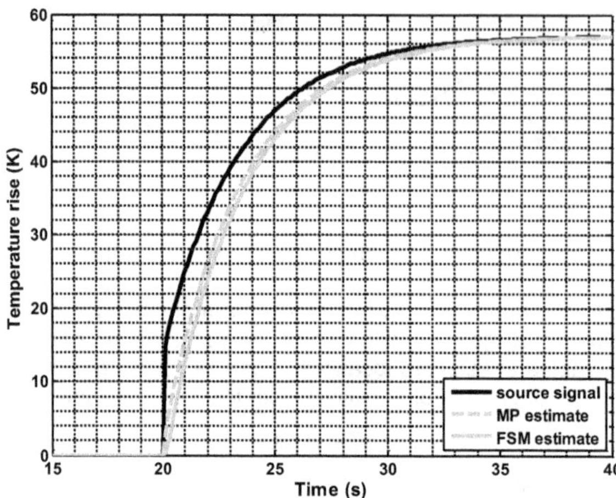

Figure 2: Simulated heat source heating curve and its estimates for exact data.

Thus, as stated previously, the main problem considered here is the lagging of the thermal response. This is caused by the fact that the heat generated in a source diffuses relatively slowly, especially compared to the electrical signals, and the delay in the sensor responses ranges typically from tens of microseconds to hundreds of milliseconds. This problem is illustrated in Fig. 2 on the simulated data generated for the integrated circuit thermal model discussed in the following section. This figure shows the actual heating curve of a heat source (solid black line) as well as its Moore-Penrose (dashed grey line) and FSM (solid grey line) estimates obtained based on temperature measurements from 16 remote sensors.

As can be seen, there is an observable delay, exceeding even 1 s, between the source signal and its estimates. This delay is caused by the limited speed of the heat diffusion process and, in the case of the FSM estimate, by the averaging of several consecutive measurements. Although, the delay might seem relatively small, but already after this short time the temperature rise in the heat source reaches almost half of its steady state value, hence the source temperature can be seriously underestimated; even by 50 %. This might cause serious problems in the detection of unexpected thermal transients and consequently lead to the destruction of a circuit.

Thus, for faster detection of rapid thermal transients, we propose a modified version of the inverse algorithm in which, as shown in Eq. (3), the algorithm uses additionally the difference in time between consecutive sensor measurements. The coefficient α, as discussed in the next section, can be a simple constant or it can depend on the distance $\boldsymbol{r}$ between a heat source and a sensor.

$$Q = (A^T * A)^{-1} * A^T * (T + \alpha(\boldsymbol{r}) * \Delta T) \tag{3}$$

4. Experimental Results

All the simulations and measurements presented in this section concern the thermal test integrated circuit pictured in Fig. 3. The detailed description of this circuit was already given in [11], so here only the most important of its features will be highlighted. This circuit contains two matrices: a matrix of 9 large heat sources, marked in the figure with the black rectangles, and a matrix of 25 temperature sensors; 16 remote ones indicated by the white circles and 9 additional ones located in the middle of each heat source. The differential analog output sensors signals are available at the circuit pins, where they can be read out and processed further by an external measurement system. The circuit itself is also equipped with internal analog-to-digital converters and digital filters able to process the data on-chip. The approximate area of the semiconductor die is 4.2 mm x 5.3 mm.

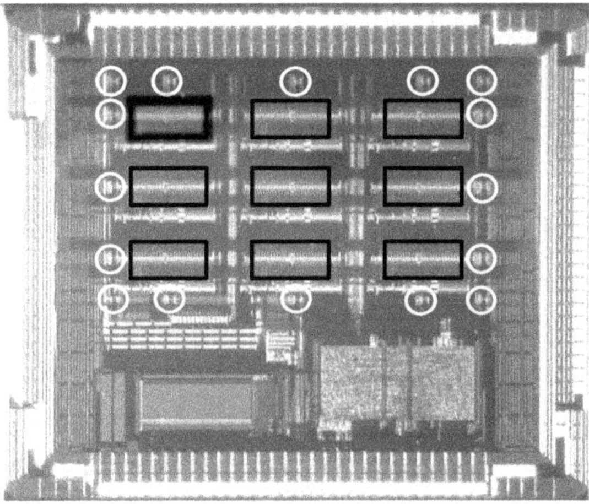

Figure 3: Integrated circuit photograph showing heat source and temperature sensor locations.

First, the operation of the inverse algorithms was tested on the simulated exact data, which were generated using the thermal model of the circuit for the sensor temperature sampling intervals of 1 ms and assuming that suddenly the power of 1 W is dissipated in the top left source indicated in the figure by the wide border. For the estimation, only the temperature measurements from the 16 remote sensors were used. The results obtained for the first second of the heating process are shown in Fig. 4. As can be seen, the standard Moore-Penrose and FSM heat source temperature estimates, indicated by the grey lines, indeed are delayed by almost a second and the heat source is underestimated by 40 %. On the other hand, the results obtained for the modified algorithm with the constant value of the coefficient α (the dashed black line) contain a characteristic spike, whose height depends on the coefficient value. If the coefficient value depends on the distance between the source and a sensor, i.e. the value decreases gradually with the distance, the spike disappears (see the dotted black line). Nevertheless, the proposed modification of the algorithm in both cases accelerates the detection of the sudden change in the hat flux by at least 10 times and the source temperature estimation error is also greatly reduced in the first phase of the heating process. The presence of the spike indicates that the proposed modifications might cause the noise amplification.

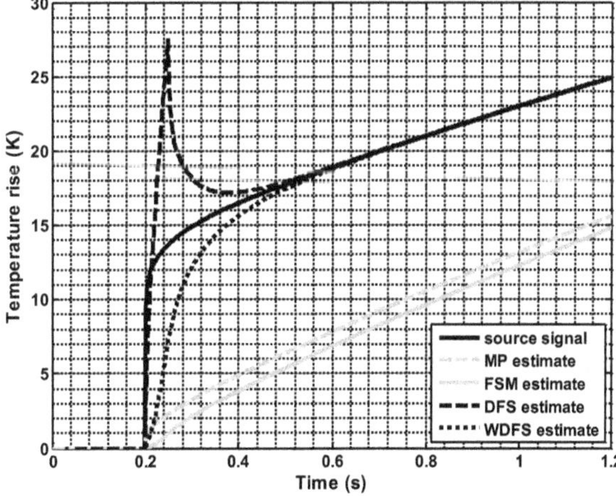

Figure 4: Simulated operation of the modified algorithm with exact data.

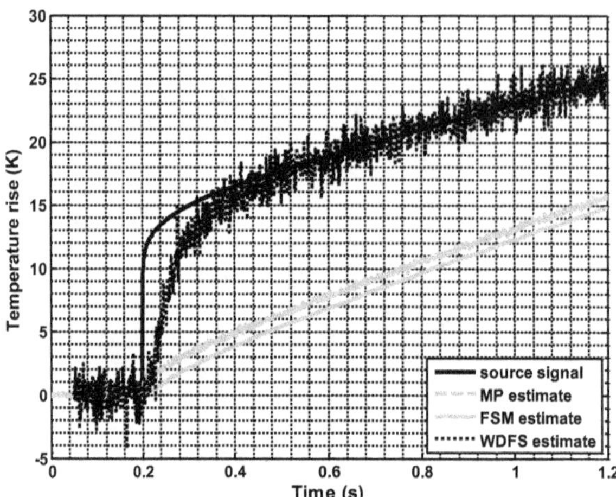

Figure 5: Simulated operation of the original algorithm for noisy data.

Next, the operation of all the algorithms was verified also on simulated noisy input data. The noise was assumed to be additive with the zero mean and the standard deviation of 0.5 K, which is a reasonable value in real cases. The results obtained for the noisy data are presented Fig. 5. As can be seen in the figure, for the standard version of the algorithm the noise deviation is reduced below 0.3 K, whereas for the modified one it is more than doubled, thus confirming the hypothesis on the possibility of noise amplification. For the reasons of figure clarity, the temperature estimate in the case of the modified algorithm with the constant value of the parameter α was removed from the figure. Depending on a particular application, the value of this parameter can be used to adjust the estimation delay or the quality of the estimate, i.e. its standard deviation.

Finally, the real time operation of all the algorithms was investigated on the measured data. During the experiments, sensor temperature measurements were taken by the dedicated system, described in [12], capable of processing in parallel all the differential sensor inputs sampled with frequencies up to 1 MHz each with the 12-bit effective resolution. The inverse algorithms in the form of digital filters were implemented in an FPGA chip. The estimation results obtained for the original FSM algorithm and the modified one with the constant value of α are shown in Fig. 6 (note that the time scale is logarithmic). The standard deviation of the noise present in the measured values equalled 1.5 K, which is quite a high value. As can be seen, in spite of the high noise level, the algorithm is able to produce a reasonable estimate for the source temperature and the detection time of the change in the power dissipation is at least halved. This result is promising and proves the feasibility of the practical realisation of real time circuit temperature monitoring systems based on remote sensor measurements. More detailed description of the measurement system and the practical results can be found in [13].

5. Conclusions

This paper analyzed in detail the operation of the modified function specification algorithm adapted for fast detection of rapid heat flux changes in real time remote prediction of heat source temperature. The algorithm was tested both on simulated and measured data. The experiments showed that the proposed modification indeed accelerated significantly the detection of rapid heat flux changes but at the expense of noise amplification. However, this might be acceptable in some practical applications when the speed of detection is more important than the quality of estimates. Ideally, some trade-off between the accuracy and the noise amplification should be reached when adjusting the parameters of the inverse algorithm.

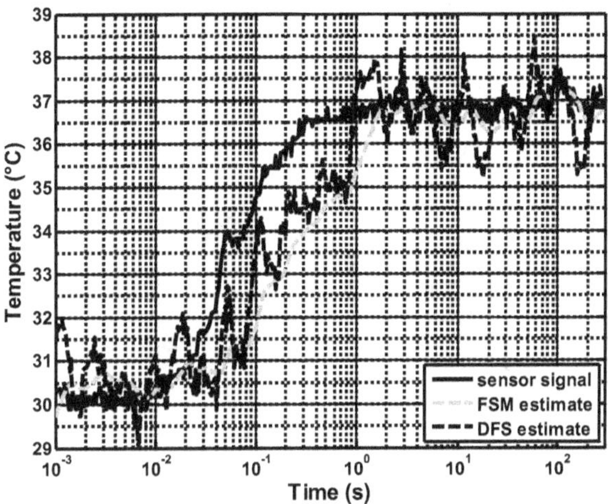

Figure 6: Operation of the algorithm on measured data.

Acknowledgements

This work was supported by the internal university grant No. Dz.St. K-25/2011/1.

References

[1] Woodbury, K.A., ed.., Inverse Engineering Handbook, CRC Press, 2003.

[2] Beck, J.V., Blackwell, B., and St. Clair, C.R., Inverse Heat Conduction - Ill-posed Problems, Wiley, New York, 1985.

[3] Janicki, M., Zubert, M., and Napieralski, A., "Application of inverse problem algorithms for integrated circuit temperature estimation," Microelectron. J., Vol. 30, pp. 1099-1107, 1999.

[4] Janicki, M., Napieralski, A., "Real time temperature estimation of heat sources in integrated circuits with remote temperature sensors," J. Phys. Conf. Ser., Vol. 124, 012027, 2008.

[5] Carslaw, H.S., and Jaeger, J.S., Conduction of Heat in Solids, Clarendon, Oxford, 1947.

[6] Ozisik, M.N., Heat Conduction, Wiley, New York, 1993.

[7] Beck, J.V., Cole, K., Haji-Sheikh, A., Litkouhi, B., Heat Conduction Using Green's Functions, Hemisphere, Washington, 1992.

[8] Haji-Sheikh, A., and Beck, J., "Temperature solution in multi-dimensional multi-layer bodies," Int. J. Heat Mass Transfer, Vol. 45, pp. 1865-1877, 2002.

[9] Gerstenmaler, Y., and Wachutka, G., "Time dependent temperature fields calculated using eigenfunctions and eigenvalues of the heat conduction equation," Microelectron. J., Vol. 32, pp. 801-808, 2001.

[10] Janicki, M., De Mey, G., and Napieralski, A., "Thermal analysis of layered electronic circuits with Green's functions," Microelectron. J., Vol. 38, pp. 177-184, 2007.

[11] Szermer, M., Kulesza, Z., Janicki, M., Napieralski, A., "Test ASIC for Real Time Estimation of Chip Temperature," *Proceedings of NSTI Nanotechnology Conference and Trade Show Nanotech,* June 1-5, 2008, Boston, MA, USA.

[12] Janicki, M., Kulesza, Z., Pietrzak, P., and Napieralski, A., "Multichannel System for Real Time Registration of Electronic Circuit Temperature Response," *Proc. of 17th Conference on Mixed Design of Integrated Circuits and Systems MIXDES*, June 24-26, 2010, Wroclaw, Poland.

[13] Janicki, M., Kulesza, Z., and Napieralski, A., "Distributed Network of Remote Sensors for Real Time Prediction of Hot Spot Temperature Values," *Proceedings of IEEE Sensors Conference*, November 1-4, 2010, Waikoloa, HI, USA.

Determination of Anisotropic Properties in Flash Diffusivity Experiments

Robert L. McMasters[1] and Ralph B. Dinwiddie[2]

[1]Department of Mechanical Engineering

Virginia Military Institute

Lexington, Virginia, USA 24450

mcmastersrl@vmi.edu

[2]High Temperature Materials Laboratory

Oak Ridge National Laboratory

Oak Ridge, Tennessee 37831

dinwiddierb@ornl.gov

Abstract

This work involves the measurement of thermal diffusivity with the laser flash method in anisotropic material using only a spot temperature measurement to record the temperature history. A two-dimensional model is fitted to the temperature data by nonlinear regression to obtain the thermal parameters. Two values for thermal diffusivity are computed through this process, one radial and one axial. Additionally, the magnitude of the laser flash and the Biot number, a dimensionless term proportional to the convection coefficient, are found simultaneously with the two thermal diffusivity values. The results are presented comparing four different models, each with an increasing level of sophistication. A model involving simple isotropic properties is compared with a model involving isotropic properties and penetration of the flash beyond the heated surface. The other two models investigated include the two-dimensional anisotropic model, both with and without penetration of the flash. The performance of these models is compared quantitatively by looking at the standard deviation of the residuals in each case.

Keywords: *Anisotropic, Flash, Diffusivity, Two-Dimensional, Penetration, Laser.*

1. Introduction

The laser flash method has been used in measuring thermal diffusivity for many years. The early mathematical models used in the analysis of these experiments were reasonably simple. Assumptions included an infinitesimally short flash duration, isotropic thermal properties, surface heating of the material, one-dimensional heat transfer and, in many cases, negligible convection from the surface [1]. With the growth of computing power through the years, mathematical models applied to these experiments have become more sophisticated.

The material used in this research is Alton foam, which is a carbon bonded carbon fiber material. It is made from strands of carbon fiber bonded in a carbon slurry which is pressed as the material is cured and hardened. As part of this process, the fibers of carbon tend to align perpendicularly to the direction in which they are being pressed. As such, the thermal conductivity can tend to be greater along the axis of the carbon fibers than in the direction perpendicular to the fibers. The anisotropic conductivity exhibited by this material can be a desirable property for thermal protection systems in aerospace and automotive applications. In such cases, the objective is to limit heat transfer transversely through the material. However, transferring heat laterally along the surface of the material can help dissipate the heat, which aids the overall objective.

The flash method of determining thermal diffusivity is convenient because of the simplicity of the experimental method, the small sample size required, and the short duration of the experiment. These advantages allow multiple samples to be tested in batches. Other experimental methods have been used for determination of anisotropic thermal diffusivity such as those described by Amazouz, et al [2] which utilized more than one temperature history, each measured at a different location. These temperature histories were then analyzed using the method of moments in order to arrive at the two values of thermal diffusivity. Similarly, Graham, et al [3] employed a mathematical model which made use of a temperature ratio between the temperatures which were measured. By contrast, the present research uses only conventional flash diffusivity experimental results generated from a single temperature detector. Since most laser flash thermal diffusivity instruments only measure temperature in one location, it would be convenient to have an analysis method that would generate results for anisotropic materials using this equipment. The purpose of the present research is to achieve that objective. The differential equation being solved in this anisotropic analysis is

$$\rho c \frac{\partial T}{\partial t} = k_a \frac{\partial^2 T}{\partial x^2} + \frac{k_r}{r}\frac{\partial}{\partial r}\left(r \frac{\partial T}{\partial r} \right) \tag{1}$$

where ρc is volumetric heat capacity (J/m^3), k_a is the thermal conductivity in the axial direction and k_r is the thermal conductivity in the radial direction (W/m-K). The boundary conditions subsequent to the flash are

$$k_a \left. \frac{\partial T}{\partial x} \right|_{x=0} = h\left(T(r,0,t) - T_\infty\right) \tag{2}$$

$$-k_a \left. \frac{\partial T}{\partial x} \right|_{x=L} = h\left(T(r,L,t) - T_\infty\right) \tag{3}$$

$$-k_r \left. \frac{\partial T}{\partial r} \right|_{r=r_o} = h\left(T(r_o,x,t) - T_\infty\right) \tag{4}$$

In these equations, h is the convection coefficient (W/m^2-K), T_∞ is the ambient temperature (K), and r_o is the outer radius of the sample. The flash heating is modeled as an initial condition at the surface.

2. Results

The results from the four analysis methods in this experiment are shown in Table 1. The first column of this table shows the results for the axial thermal diffusivity for each of the four analysis models. Since the sample is disc-shaped, the axial diffusivity is in a direction perpendicular to the heated surface. Table 1 also shows the standard deviation of the residuals for each analysis method. The residuals are simply the error between the measured values from the experiment and the calculated values from the mathematical model. As can be seen in this table, the standard of deviation of the residuals is significantly reduced through the use of the anisotropic model. More importantly, it can be seen that there is a significant difference between the transverse and radial thermal diffusivities of this material, nominally an order of magnitude.

The magnitude of the flash and the Biot Number are also computed in this analysis out of necessity, but they are not parameters of interest. The thermal diffusivity values are the parameters which are sought and the reason for which the experiment is performed. Still, a minimum of three parameters must be found simultaneously even in the simplest model assuming isotropic behavior. In the most sophisticated model, the bottom analysis listed in Table 1, five parameters must be computed simultaneously. The photon men free path, measured in millimeters, is computed because of the porous surface of the material. As the laser flash reaches the sample surface, a portion of the energy is absorbed at $x = 0$, but some of the flash energy penetrates beyond the surface and heats the material deeper inside. In this portion of the mathematical model, the distribution of the flash energy is assumed to follow an exponential pattern in accordance with Beer's Law, as discussed in Reference [4]. Using this model, another parameter must be computed, related to the depth of flash penetration, which is the mean free path of a photon.

Figure 1 shows a graphical plot of the residuals as a function of time, comparing the last two models summarized in Table 1. Both of these models account for flash penetration with one accounting for anisotropic conductivity and the other assuming isotropic conductivity. As can be seen in Figure 1, there is a noticeable improvement in the residuals when using the model which takes into account the anisotropy of the material.

Table 1. A summary of the results from analyzing one typical experiment using four different analysis models. Each model is successively more sophisticated, containing additional parameters, but generates improved results.

Axial Diffusivity (mm^2/s)	Flash Magnitude	Biot Number	Radial Diffusivity (mm^2/s)	Photon Mean Free Path (mm)	Std. Dev. of Residuals (°C)
0.5184	9.7413	0.1408	NA	NA	0.077279
0.5440	8.086	0.0463	3.570	NA	0.060564
0.4496	6.4086	0.2219	NA	0.1998	0.017783
0.4847	5.1425	0.0625	4.424	0.1933	0.014201

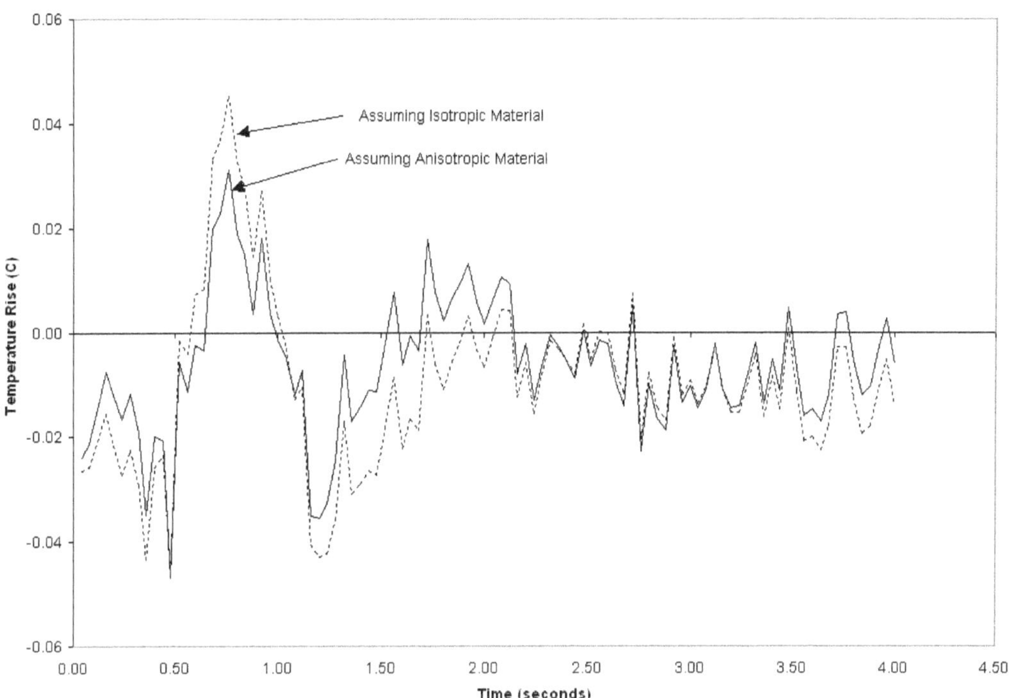

FIGURE 1. Graph of residuals comparing the results of two models used on the same experimental data: one assuming isotropic material and the other assuming anisotropic material.

The validity of the improvement shown here in the conformance of the mathematical model to the measured data can be supported using statistical tools such as the "F Test" which employs the "F" statistic. In this case, comparing the first two models, it can be shown that the addition of the anisotropic parameter reduces the sum of squares of the errors between the two results by 0.230 degrees C^2. Dividing this number by the square of the standard deviation of the residuals in the higher-order model gives a number of 62.8 (dimensionless). This is significantly greater than 3.95, which is the value of the "F" statistic at a 95% confidence level for the number of degrees of freedom in this model. This outcome signifies that the addition of the anisotropic parameter is statistically significant at a 95% confidence level. Likewise, in comparing the third and fourth models, both of which account for the penetration of the flash into the porous material, the addition of the anisotropic term is shown to be statistically significant. The difference of the squares of the errors divided by the square of the standard deviation in this case is 56.8 which also exceeds the 3.95 criterion value.

3. Conclusions

The inclusion of anisotropic thermal conductivity in the mathematical model, which generates two diffusivity values for the material, provides a better fit between the model and the experimental measurements. Moreover, the inclusion of this aspect of the model is shown to be statistically significant at a 95% level of confidence. This is true when comparing the isotropic model to the anisotropic model, when accounting for flash penetration, as well as when flash penetration is not accounted for. The anisotropic model is therefore given validation to be used in this experiment and the radial conductivity is shown to be approximately an order of magnitude larger than the axial conductivity.

4. References

[1] Parker, W., Jenkins, B., and Abbott, G., “Flash method of determining thermal diffusivity, heat capacity and thermal conductivity”, *Journal of Applied Physics*, Vol. 32, No. 9, pp. 1679-1684, 1961.

[2] Amazouz, M., Moyne, C., and Degiovanni, A., “Measurement of the thermal diffusivity of anisotropic materials”, *High Temperatures-High Pressures*, Vol. 19, pp. 37-41,1987.

[3] Graham, S., McDowell, D. and Dinwiddie, R., “In-plane thermal diffusivity measurements of orthotropic materials”, *Thermal Conductivity*, Vol. 24, pp. 241-252, 1999.

[4] McMasters, R., Beck, J., Dinwiddie, R. and Wang, H., “Accounting for Penetration of Laser Heating in Flash Thermal Diffusivity Experiments”, *ASME Journal of Heat Transfer*, Vol. 121, pp. 15-21, 1999.

Estimating of Material's Thermal and Radiative Properties

A.V. Nenarokomov[1], O.M. Alifanov[1], D.M. Titov[1]

[1]Moscow Aviation Institute (Technical University), Dept. of Space System Engineering
4 Volokolamskoe Sh., Moscow, 125993, Russia
nenar@mai.ru

ABSTRACT

In many practical situations it is impossible to measure directly such characteristics of analyzed materials as thermal and radiation properties. The only way, which can often be used to overcome these difficulties, is indirect measurements. This type of measurements is usually formulated as the solution of inverse heat transfer problems. Such problems are ill-posed in mathematical sense and their main feature shows itself in the solution instabilities. That is why special regularizing methods are needed to solve them. The experimental methods of identification of the mathematical models of heat transfer based on solving of the inverse problems are one of the modern effective solving manners. The goal of this paper is to estimate thermal and radiation properties of advanced materials using the approach based on inverse methods (as example: thermal conductivity $\lambda(T)$, heat capacity $C(T)$ and emissivity $\varepsilon(T)$). New metrology under development is the combination of accurate enough measurements of thermal quantities, which can be experimentally observable under real conditions and accurate data processing, which are based on the solutions of inverse heat transfer problems. In this paper, the development of methods for estimating thermal and radiation characteristics is carried out for thermally stable high temperature materials. Such problems are of great practical importance in the study of properties of materials used as non-destructive surface coating in objects of space engineering, power engineering etc.

Keywords: thermal conductivity, heat capacity, emissivity

1. INTRODUCTION

In the modern engineering systems we deal with structures operating in the conditions of intensive, often extreme thermal effects. The general tendency in the development of technology is connected with the increase of the number of responsible, thermally loaded engineering objects. For such systems the support of thermal conditions is one of the most important aspects of design, determining the main design solutions.
The modern approaches to the design of structures assume broad application of mathematical and physical simulation methods. But mathematical simulation is impossible if there is no true information available on the characteristics (properties) of objects analyzed. In the majority of cases in practice the direct measurement of materials' thermal properties, especially of complex composition, is impossible. There is only one way which permits to overcome these complexities - the indirect measurement. Mathematically, such an approach is usually formulated as a solution of the inverse problem: through direct measurements of system's state (temperature, component concentration, etc.) define the properties of a system analyzed, for example, the thermophysical properties. Violation of cause-and-effect relations in the statement of these problems results in their correctness in mathematical sense (i.e., the absence of existence and/or uniqueness and/or stability of the solution). Hence to solve such problems we develop special methods usually called regularized.

In estimating properties of modern structural, thermal-protective and thermal-insulating materials - as temperature-dependent - the most effective are methods based on solution of the coefficient inverse heat conduction problems. The most promising direction in further development of research methods for non-destructive composite materials using the procedure of inverse problems is the simultaneous determination of a combination of material's thermal and radiation properties (thermal conductivity $k(T)$, heat capacity $C(T)$ and integral emissivity $\varepsilon(T)$). Such problems are of great practical importance in the study of properties of composite materials used as non-destructive surface coating in objects of space technology, power engineering etc (Artyukhin E.A., Ivanov G.A. and Nenarokomov A.V. [1], Artyukhin E.A. and Nenarokomov A.V. [2], D.Baillis, M.Raynaud and Sacadura J.F. [3], Ho C.-H. and Ozisik M.N. [4], Li H.Y. [5], Li H.Y. and Ozisik M.N. [6], Li H.Y. and Yang C.Y. [7]). The experimental equipment and the method developed could be applied for determination of material's three characteristics; the availability of two specimens of the material allows us to provide uniqueness of the solution. The mathematical model of heat transfer in specimen is

$$C(T)\frac{\partial T}{\partial \tau} = \frac{\partial}{\partial x}\left(k(T)\frac{\partial T}{\partial x}\right),\ x\in(X_0,X_1),\ \tau\in(\tau_{\min},\tau_{\max}] \tag{1}$$

$$T(x,\tau_{\min}) = \mathrm{T}_0(x),\ \ \mathrm{x}\in[X_0,X_1] \tag{2}$$

$$-\beta_1 k(T)\frac{\partial T(X_0,\tau)}{\partial x} + \alpha_1 T(X_0,\tau) = q_1(\tau),\quad \tau\in(\tau_{\min},\tau_{\max}] \tag{3}$$

$$-k(T)\frac{\partial T(X_1,\tau)}{\partial x} = \varepsilon_{eff}(T)\sigma\left(T_h^4(\tau)-T^4(X_1,\tau)\right),\quad \tau\in(\tau_{\min},\tau_{\max}] \tag{4}$$

Where

$$\varepsilon_{eff}(T) = \frac{\varepsilon(T)\varepsilon_h(T_h)}{\varepsilon(T)+\varepsilon_h(T_h) - \varepsilon_h(Th)\varepsilon_h(T_h)}$$

In model (1)-(4) the quantities $C(T)$, $\lambda(T)$and $\varepsilon(T)$ are u nknown. If emissivity of the he ater (ε_h) is known a-priori, and its temperature is measured, the heat flux from the heater can be calculated as irradiation with known (measured by thermocouple) temperature of the heater $T_h(\tau)$ and a-priory known (theoretically) emissivity of heater $\varepsilon_h(T)$. In pre sented paper a case is considered, when $\varepsilon_h(T)$ is not kno wn a-priori and should be estimated. At this p aper the emi ssivity of the he ater is considered as additional (forth) estimating functions. Therefore the accuracy of the inverse problems, considered bellow, will not d epend to th e a-priory information about the radiation properties of the heater's material.

The results of temperature measurements inside the specimen are a ssigned as necessary additional information to solve an inverse problem

$$T^{\exp}(x_m,\tau) = f_m(\tau),\quad \mathrm{m}=\overline{1,\mathrm{M}} \tag{5}$$

With the pre sented statement of inverse problem, the data mea sured in the single experiment are not sufficient for simultaneous estimating of four therm al and radiative characteristics (thermal conductivity, heat capacity and two emissivity), because data by values of the heat flux applied to a specimen are also needed. To solve this p roblem the da ta of several N (two or m ore) similar experiments with equal mat erial specimen and different heating regimes were processed simultaneously.

The experimental equipment and th e method d escribed below could be applie d for estimating of material's three ch aracteristics. This paper is concerned with modification of the app roach, presented at (Nenarokomov A.V. and Alifanov O.M. [8]).

2. INVERSE PROBLEM ALGORITHM

In the inverse problem E qn.(1)- Eqn.(5) it is ne cessary first of all to indi cate as a tem perature range $[T_{\min},T_{\max}]$ of the unkn own functions, which is g eneral for all experiments, and for whi ch the inverse probl em analysis has a uniqu e solution. For $T_{\min}$ the minimu m value of initial temperature is u sed. Of much g reater importance is a correct sampling of value $T_{\max}$. Proceeding from the necessity to provide uniqueness of solution, it seems p ossible to sample, for $T_{\max}$, a minimum among maximum temperature values gained on the thermocouple positioned on the heate d surface at every testing specimen. The same should be d one for the heater temperature range $[T_{h,\min},T_{h,\max}]$.

Suppose then that the unknown characteristics are given i n their parametric form. With this p urpose introduce in the interval $[T_{\min},T_{\max}]$ three uniform difference grids with the number of nodes $N_i,\ i=1,2,3$

$$\omega_i = \{T_k = T_{\min} + (k-1)\Delta T,\ \ \mathrm{k}=\overline{1,\mathrm{N_i}-1}\},\ \ \mathrm{i}=\overline{1,3} \tag{6}$$

Approximating the unknown functions on grids Eqn. (6) using the cubic B-splines

$$C(T)=\sum_{k=1}^{N_1} C_k \varphi_k^1(T)$$
$$k(T)=\sum_{k=1}^{N_2} k_k \varphi_k^2(T)$$
$$\varepsilon(T)=\sum_{k=1}^{N_3} \varepsilon_k \varphi_k^3(T) \qquad (7)$$

where C_k, $k=1,\ldots,N_1$, k_k, $k=1,\ldots,N_2$, ε_k, $k=1,\ldots,N_3$ - parameters.

Let's introduce in the interval $\left[T_{h\min}, T_{h\max}\right]$ uniform difference grids with the number of nodes N_4

$$\omega_4=\left\{T_k=T_{h\min}+(k-1)\Delta T_h,\ \ k=\overline{1, N_4-1}\right\} \qquad (6a)$$

and approximate the unknown function $\varepsilon_h(T)$ on grids (6a) using the cubic B-splines

$$\varepsilon_h(T)=\sum_{k=1}^{N_4} \varepsilon_{hk} \varphi_k^4(T) \qquad (7a)$$

where ε_{hk}, $k=1,..,N_4$ - parameters.

As a result of approximation, the inve rse problem is reduced to the sea rch of a vecto r of unknown parameters $\overline{p}=\{p_k\}$, $k=1,\ldots,N_P$, with dime nsions $N_P=N_1+N_2+N_3+N_4$. Writing down a me an-square error of the design and experimental temperature values in points of thermal sensors positioning, than the residual functional will depend to four functions

$$J(\overline{p})=J(C(T),k(T),\varepsilon(T),\varepsilon_h(T))=\sum_{n=1}^{N}\sum_{m=1}^{M_n}\int_{\tau_{\min}^m}^{\tau_{\max}^m}\left(T^n(x_m^n,\tau)-f_m^n(\tau)\right)^2 d\tau \qquad (8)$$

where $T^n(x,\tau)$ is determined from a solution of the boundary-value problem Eqn. (1)- Eqn. (4) for n-th experiment using the approximations of Eqn. (7). It is assumed here that the conditions of uniqueness of the inverse problem solving are satisfied. Bellow to simplify the notation of equations index n will be excluded.

Proceeding from the principle of iterative regularization (Nenarokomov A.V. and Alifan ov O.M. [8], Alifanov O.M., Artyukhin E.A. and Rumyantsev S.V. [9], Alifanov O.M. [10]), the unknown vector can be determined through minimization of functional Eqn. (8) by gradient methods of the first order prior to a fulfilment of the condition

$$J(\overline{p})\le\delta_f \qquad (9)$$

where $\delta_f=\sum_{m=1}^{M}\int_{\tau_{\min}}^{\tau_{\max}}\sigma_m(\tau)d\tau$ - integral error of temperature measurements $f_m(\tau)$, $m=\overline{1,M}$, and σ_m - measurement variance.

To construct an iterative algorithm of the inverse problem solving a conjugate gradient method was used. The greatest difficulties in reali zing the gra dient methods are connected with calculation of the minimize d functional gradient. In the approach being developed the methods of cal culus of vari ations are used. Here an analytic expression for the minimized functional gradient can be obtained

$$J'_{C_k} = -\int_{\tau_{\max}}^{\tau_{\max}} \int_{X_0}^{X_1} \psi(x,\tau)\cdot\varphi_k^1(T)\frac{\partial T}{\partial \tau}\,dxd\tau$$
$$k=\overline{1,N_1} \tag{10}$$

$$J'_{k_k} = -\int_{\tau_{\max}}^{\tau_{\max}} \int_{X_0}^{X_1} \psi(x,\tau)\left(\frac{\partial^2 T}{\partial x^2}\cdot\varphi_k^2(T)+\left(\frac{\partial T}{\partial x}\right)^2\cdot\frac{\partial\varphi_k^2}{\partial T}\right)dxd\tau -$$
$$-\beta_1\int_{\tau_{\max}}^{\tau_{\max}} \psi(X_0,\tau)\frac{\partial T}{\partial x}\cdot(X_0,\tau)\varphi_k^2(T(X_0,\tau))d\tau +$$
$$+\int_{\tau_{\max}}^{\tau_{\max}} \psi(X_1,\tau)\frac{\partial T}{\partial x}(X_1,\tau)\cdot\varphi_k^2(T(X_1,\tau))d\tau,$$
$$k=\overline{1,N_2} \tag{11}$$

$$J'_{\varepsilon_k} = -\int_{\tau_{\max}}^{\tau_{\max}} \frac{\psi_{M+1}(X_1,\tau)}{(\varepsilon(T)+\varepsilon_h(T_h)-\varepsilon(T)\varepsilon_h(T_h))}\left((1-\varepsilon_h(T_h))k\frac{\partial T}{\partial x}+\sigma\varepsilon_h(T_h^4-T^4)\right)\varphi_k^3(T)d\tau$$
$$k=\overline{1,N_3} \tag{12}$$

$$J'_{\varepsilon_{hk}} = -\int_{\tau_{\max}}^{\tau_{\max}} \frac{\psi_{M+1}(X_1,\tau)}{(\varepsilon(T)+\varepsilon_h(T_h)-\varepsilon(T)\varepsilon_h(T_h))}\left((1-\varepsilon(T))k\frac{\partial T}{\partial x}+\sigma\varepsilon(T)(T_h^4-T^4)\right)d\tau$$
$$k=\overline{1,N_4} \tag{13}$$

where $\psi(x,\tau)$ - solution of a boundary-value problem adjoint to a linearized form of the initial probl em Eqn. (1)-Eqn. (4).

$$-c(T)\frac{\partial\psi_m}{\partial\tau} = k\frac{\partial^2\psi_m}{\partial x^2},$$
$$x\in(x_{m-1},x_m),\ x_0=X_0,\ x_{M+1}=X_1,\ m=\overline{1,M+1},\ \tau\in(\tau_{\min},\tau_{\max}] \tag{14}$$

$$\psi_m(x,\tau_{\max})=0\ ,\ x\in[x_{m-1},x_m],\ m=\overline{1,M+1} \tag{15}$$

$$-\beta_1 k\frac{\partial\psi_1(X_0,\tau)}{\partial x}+\alpha_1\psi_1(X_0,\tau)=0 \tag{16}$$

$$-k\frac{\partial\psi_{M+1}(X_1,\tau)}{\partial x}(\varepsilon+\varepsilon_h-\varepsilon\varepsilon_h)-\frac{d\varepsilon}{dT}\frac{\partial T}{\partial x}(1-\varepsilon_h)\lambda\frac{\partial T}{\partial x}\psi_{M+1}(X_1,\tau)+$$
$$+\frac{d\varepsilon}{dT}\frac{\partial T}{\partial x}\varepsilon_h\sigma(T_h^4-T^4(X_1,\tau))\psi_{M+1}(X_1,\tau)-4\varepsilon\varepsilon_h T^3(X_1,\tau)_2\psi_{m+1}(X_1,\tau)=0, \tag{16}$$

$$\psi_m(x_m,\tau)=\psi_{m+1}(x_m,\tau),\ m=\overline{1,M} \tag{17}$$

$$\lambda(T)\cdot\left(\frac{\partial\psi_m(x_m,\tau)}{\partial x}-\frac{\partial\psi_{m+1}(x_m,\tau)}{\partial x}\right)=2(T(Y_m,\tau)-f_m(\tau)),\ m=\overline{1,M} \tag{20}$$

3. EXPERIMENTAL VERIFICATION

As an example, the results of data processing on experimental study of material specimens are used, which involves the thermo-vacuum facility developed in MAI at the Department of Space System Engineering.

The test facility consists of a horizontally cylindrical set vacuum chamber with a water-cooling system.

A heating element consists of a molybdenum foil with dimensions 80x70x0.05 mm (Fig. 1). The tested specimen is located on a thermo-insulating base made of thermal insulation material so that its heating surface is parallel to heater's plane and at certain distance from it (3-5 mm). A ceramic plate is put over the heater, which reduces the heat withdrawal from it in the direction opposite to the test specimen improving its heating uniformity.

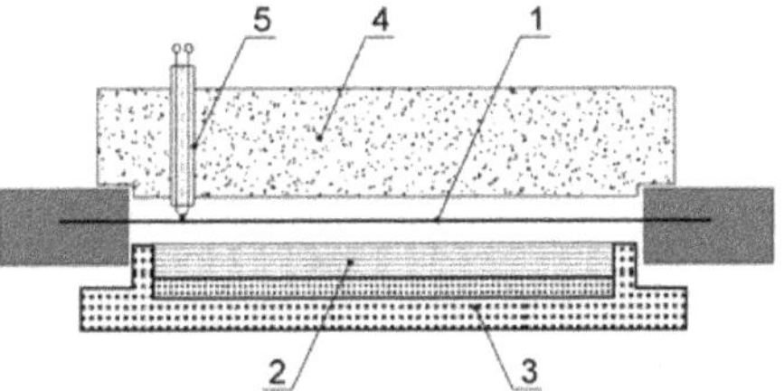

Figure 1: Experimental module: 1 – heater, 2 - test specimen, 3 - insulating basement, 4 - insulating cover, 5 - control thermocouple.

The models of test material are the square plates of 50x50x15 mm (Fig. 2) with four thermocouples installed in the specimen. An installation of thermosensors in specimens was chosen from a solution of the problem of optimal experiment design. The co-ordinates of thermocouple positioning in the first set of specimens, for estimating the material's thermal characteristics, had the following values: $x_0 = 0\ mm$ (for a boundary condition of the first kind sensor readings on the internal surface were used), $x_1 = 7.5\ mm$, $x_2 = 11.8\ mm$, $x_3 = 15\ mm$ (positioned on the exposed surface).

The number of approximation parameters for every characteristic was assumed to be 5. A comparison of experimentally measured and calculated is shown on Fig. 3 (only for one specimen). The results proper of the inverse problem solving - the composite material thermal characteristics and emissivity are given on Fig. 4,5 (the results for two sets by three experiments in vacuum and air conditions).

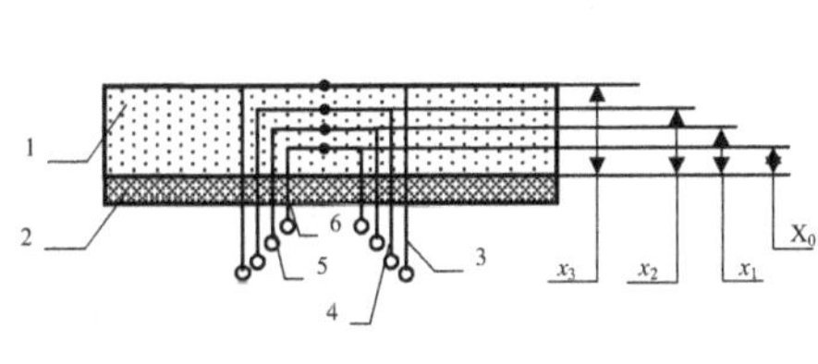

Figure 2: Test specimen: 1 - test materials, 2 - metallic basement, 3-6 - thermocouples.

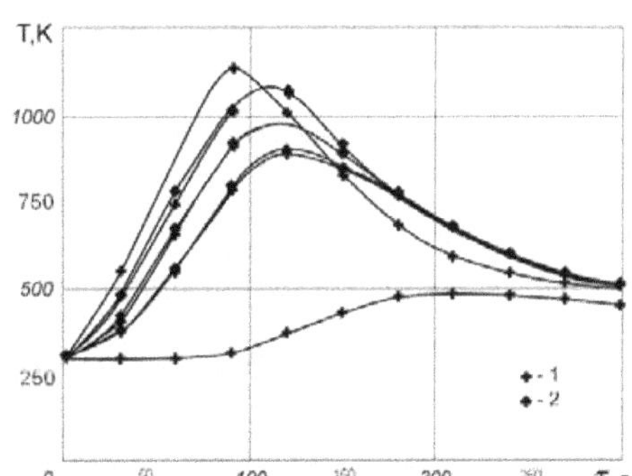

Figure 3: Temperature values in points of thermocouple positioning: 1 - calculated, 2 – experimentally measured

CONCLUSIONS

The paper seeks to describe the algorithm developed to process the data of unsteady-state thermal experiments. The following main factors have an influence on the accuracy of the inverse heat conduction problem (in sequence of significance): the errors in coordinates of thermosensor positions; the errors in values of different characteristics; the errors in estimating the residual level. It was shown that in the cases considered the accuracy of the inverse problems solution is compatible with the errors of the simulated "experimental measurements". Next step in the development of the proposed approach is to consider an estimating interface conductance between periodically contacting surface of specimen and heater foil using the approach similar (Orlande H.R.B. and Ozisik M.N. [11]).

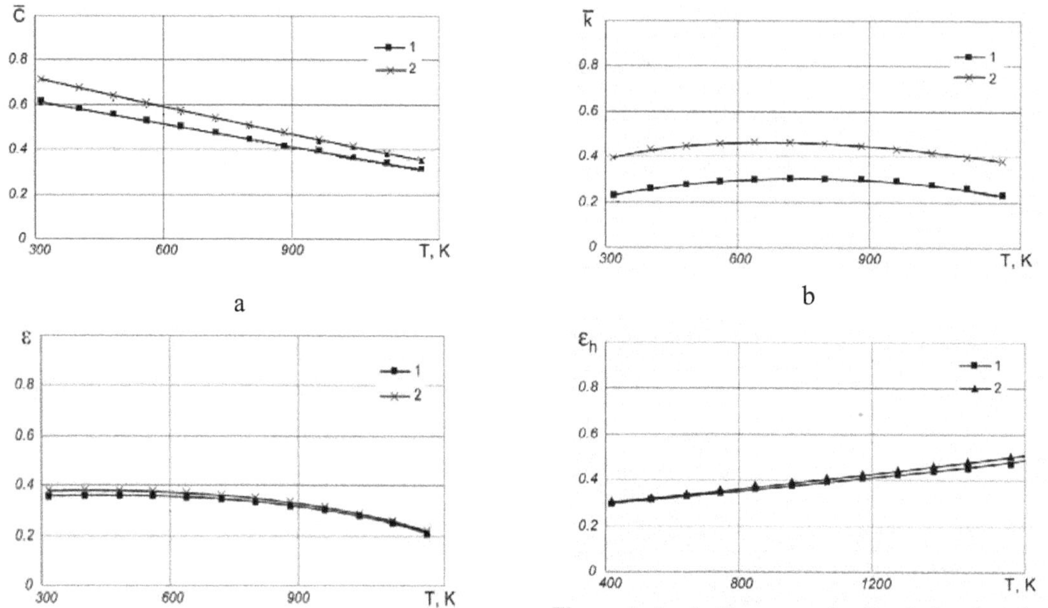

Figure 4: Results of testing: a - heat capacity, b - conductivity, c – emissivity. 1 – vacuum, 2 – air.

Figure 5: A-priori known and estimated emissivity of the heater foil. 1 – vacuum, 2 – air.

REFERENCES

[1] Artyukhin E.A., Ivanov G.A. and Nenarokomov A.V., "Determination of a complex of materials thermophysical properties through data of nonstationary temperature measurements", High Temperature, Vol. 31, pp.199-202, 1993.

[2] Artyukhin E.A. and Nenarokomov A.V., "Coefficient Inverse heat conduction problem", Journal of Engineering Physics, Vol. 53, pp.1085-1090, 1987.

[3] D.Baillis, M.Raynaud and Sacadura J.F., "Spectral radiative properties of open-cell foam insulation", Journal of Thermophysics and Heat Transfer, Vol. 13, pp.292-299, 1999.

[4] Ho C.-H. and Ozisik M.N., "An inverse radiation problem", Int. Journal Heat and Mass Transfer, Vol. 32, pp.335-342, 1989.

[5] Li H.Y., "Estimation of thermal properties in combined conduction and radiation", Int. Journal Heat and Mass Transfer, Vol. 114, pp.1060-1063, 1999.

[6] Li H.Y. and Ozisik M.N. "Inverse radiation problem for simultaneous estimation of temperature profile and surface reflectivity", Journal of Thermophysics and Heat Transfer, Vol.7, pp.88-93, 1993.

[7] Li H.Y. and Yang C.Y., "A genetic algorithm for inverse radiation problems", Int. Journal Heat and Mass Transfer, Vol. 40, pp.1545-1553, 1997.

[8] Nenarokomov A.V. and Alifanov O.M., "Inverse radiative-conductive problems for estimating material properties", Proc. of the Third International Symposium on Radiative Transfer, Begell House, New York\Wallingford (UK), pp.466-473, 2001.

[9] Alifanov O.M., Artyukhin E.A. and Rumyantsev S.V., Extreme Methods for Solving Ill-Posed Problems with Applications to Inverse Problems, Begell House, New York\Wallinford(UK), 1995.

[10] Alifanov O.M., Inverse Heat Transfer Problems, Springer-Verlag, Berlin \ Heidelberg, 1994.

[11] Orlande H.R.B. and Ozisik M.N., "Inverse problem of estimating interface conductance between periodically contacting surfaces", Journal of Thermophysics and Heat Transfer, Vol. 7, pp.319-325, 1993.

[12] E.A. Artyukhin and A.V.Nenarokomov, "Inverse problem of deriving the integral emissivity of a solid",.High Temperature, Vol.24, pp.725-729, 1986.

Destructive Materials Thermal properties determination with application for spacecraft structures testing

A.V. Nenarokomov[1], O.M. Alifanov[1], S. A. Budnik[1], A.V. Netelev[1]
[1]Moscow Aviation Institute (Technical University), Dept. of Space System Engineering
4 Volokolamskoe Sh., Moscow, 125993, Russia
nenar@mai.ru

Abstract

In many practical situations it is impossible to measure directly thermal and thermokinetic properties of analyzed composite materials. The only way that can often be used to overcome these difficulties is indirect measurements. This type of measurements is usually formulated as the solution of inverse heat transfer problems. Such problems are ill-posed in mathematical sense and their main feature shows itself in the solution instabilities. That is why special regularizing methods are needed to solve them. The general method of iterative regularization is concerned with application to the estimation of materials properties. The objective of this paper is to estimate thermal and thermokinetic properties of advanced materials using the approach based on inverse methods. An experimental-computational system is presented for investigating the thermal and kinetics properties of composite materials by methods of inverse heat transfer problems and which is developed at the Thermal Laboratory of Department Space Systems Engineering, of Moscow Aviation Institute (MAI). The system is aimed at investigating the materials in conditions of unsteady contact heating over a wide range of temperature changes and heating rates in a vacuum, air and inert gas medium.

Key Words: iterative regularization, inverse problem, destruction materials, numerical methods.

1. INTRODUCTION

The modern approaches to a thermal control of spacecrafts assume broad application of mathematical and physical simulation methods. But mathematical simulation is impossible if there is no true information available on the external heat fluxes, temperatures, etc. of objects analyzed. In the majority of cases in practice the direct measurement of thermal states of space structures, especially of complex composition, is impossible. There is only one way, which permits to overcome these complexities - the indirect measurement. Mathematically, such an approach is usually formulated as a solution of the inverse problem: through direct measurements of system's state (available temperature, component concentration, etc.) define the total thermal state of a system analyzed, for example, the heat fluxes at the external surface. Violation of cause-and-effect relations in the statement of these problems results in their correctness in mathematical sense (i.e., the absence of existence and/or uniqueness and/or stability of the solution). Hence to solve such problems we develop special methods usually called regularized.

The approaches to estimate thermal states of complex space structure based on methods of ill-posed problem solving were widely analyzed in Russia and in other countries having displayed efficiency in the development and investigation of modern structure in rocket-space, aircraft, automotive industries, metallurgy, power engineering etc. A new metrological system for thermal analysis of space structure being developed is a combination of sufficiently accurate measurements of primary heat values in testing conditions to the maximum approximate to full-scale conditions and ultimately correct mathematical treatment of experimental data based on the theory of inverse problems.

In estimating temperature-dependent properties of modern composite destructive materials the most effective are methods based on solution of the coefficient inverse heat transfer problems. The most promising direction in further development of research methods for destructive composite materials using the procedure of inverse problems is the simultaneous determination of a combination of material's thermal and kinetic properties (thermal conductivity $\lambda(T)$, heat capacity $c(T)$ and some other parameters). Such problems are of great practical importance in the study of properties of composite materials used as destructive surface coating of spacecrafts.

The mathematical model of heat transfer in specimen is

$$C(T(\tau,x))\rho\frac{dT(\tau,x)}{d\tau}=\frac{d}{dx}\left(\lambda(T(\tau,x))\frac{dT(\tau,x)}{dx}\right)+$$
$$+C_g(T(\tau,x))\int_{l_0}^{x}\frac{d\rho(x,\tau)}{d\tau}d\xi\frac{dT(\tau,x)}{dx}+H(T(\tau,x))\frac{d\rho(x,\tau)}{d\tau} \tag{1}$$
$$(x,\tau)\in Q=(0,X_L)\times(0,\tau_m]$$
$$T(0,x)=T_0(x) \tag{2}$$
$$-\alpha_1\lambda_1(T(0,\tau))\frac{dT(0,\tau)}{dx}+\beta_1 T(0,\tau)=q_1 \tag{3}$$
$$-\alpha_2\lambda_L(T(X_L,\tau))\frac{dT(X_L,\tau)}{dx}+\beta_2 T(X_L,\tau)=q_2 \tag{4}$$
$$T(x,\tau_r)=T_r \tag{5}$$
$$\rho(x,\tau_r)=\rho_0 \tag{6}$$
$$\rho(x,\tau_c)=\rho_c \tag{7}$$
$$\frac{d\rho(x,\tau)}{d\tau}=\begin{cases}0, T(x,\tau)<T_r\\ -\rho^{n_l}A(T)\exp\left(\dfrac{-E(T(x,\tau))}{RT(x,\tau)}\right), & \rho(x,\tau)>\rho_c, T(x,\tau)\geq T_r\\ 0, \rho(x,\tau)\leq\rho_c\end{cases} \tag{8}$$
$$T^{exper}(d_i,\tau)=f_i(\tau) \tag{9}$$

In the model Eqn. (1)-(9) the parameters $C(T)$, $\lambda(T)$, $C_g(T)$ and $H(T)$ are unknown. The experimental equipment and the method described below could be applied for estimating of material's seven characteristics; the availability of a few specimens of the material allows us to provide uniqueness of the solution.

The results of temperature measurements inside the specimen are assigned as necessary additional information to solve an inverse problem

$$T^{\exp}(x_m,\tau)=f_m(\tau),\quad \mathrm{m}=\overline{1,\mathrm{M}} \tag{10}$$

In the inverse problem Eqn. (1)-(5) it is necessary first of all to indicate as a temperature range $[T_{\min},T_{\max}]$ of the unknown functions, which is general for all experiments, and for which the inverse problem analysis has a unique solution. For $T_{\min}$ the minimum value of initial temperature is used. Of much greater importance is a correct sampling of value $T_{\max}$. Proceeding from the necessity to provide uniqueness of solution, it seems possible to sample, for $T_{\max}$, a minimum among maximum temperature values gained on the thermocouple positioned on the heated surface at every testing specimen. The same should be done for the heater temperature range $[T_{h,\min},T_{h,\max}]$

2. INVERSE PROBLEM ALGORITHM

Suppose then that the unknown characteristics are given in their parametric form. With this purpose introduce in the interval $[T_{\min},T_{\max}]$, six uniform difference grids with the number of nodes Ni, i=1,..,4.

$$\omega_i=\{T_k=T_{\min}+(k-1)\Delta T,\ \mathrm{k}=\overline{1,\mathrm{N_i}-1}\},\mathrm{i}=\overline{1,4} \tag{11}$$

In case of cubic B-splines, the unknown function is presented as

$$C(T)=\sum_{k=1}^{m_1} C_k \varphi_{1,k}(T) \tag{12}$$

$$\lambda(T)=\sum_{k=1}^{m_2} \lambda_k \varphi_{2,k}(T) \tag{13}$$

$$C_g(T)=\sum_{k=1}^{m_3} C_{g_k} \varphi_{3,k}(T) \tag{14}$$

$$H(T)=\sum_{k=1}^{m_4} H_k \varphi_{4,k}(T) \tag{15}$$

where C_k, k=1,...,N1, λ_k, k=1,...,N2, , c_{gk}, k=1,...,N3, H_k, k=1,...,N4, are parameters.

As a result of approximation, the inverse problem is reduced to the search of a vector of unknown parameters $\overline{p}=\{p_k\}$, k=1,Np, with dimensions Np = N1+...+N4 Writing down a mean-square error of the design and experimental temperature values in points of thermal sensors positioning

$$J(c(T),\lambda(T),c_g(T),H(T))=J(\overline{p})=\sum_{m=1}^{M}\int_{\tau_{\min}}^{\tau_{\max}}\left(T(x_m,\tau)-f_m(\tau)\right)^2 d\tau \tag{16}$$

where $T(x,\tau)$ is determined from a solution of the boundary-value problem Eqn. (1)-(9) for n-th experiment using the approximations of Eqn. (12)-(15). Proceeding from the principle of iterative regularization [1,2], the unknown vector can be determined through minimization of functional Eqn. (16) by gradient methods of the first order prior to a fulfillment of the condition

$$J(\overline{p})\le \delta_f \tag{17}$$

where $\delta_f=\sum_{m=1}^{M}\int_{\tau_{\min}}^{\tau_{\max}}\sigma_m(\tau)d\tau$ - integral error of temperature measurements fm(τ), m=1,M, and $\sigma_m(\tau)$ - measurement variance.

To construct an iterative algorithm of the inverse problem solving a conjugate gradient method was used. In the approach being developed the methods of calculus of variations are used for calculation of the minimized functional gradient. Here an analytic expression for the minimized functional gradient can be obtained

$$J'_{C_k^l}=-\sum_{l=1}^{N}\int_0^{\tau_m}\int_{d_{l-1}}^{d_l}\rho\varphi_k^{C_l}\frac{dT}{d\tau}\psi dx d\tau \tag{18}$$

$$J'_{\lambda_l^k}=\sum_{i=1}^{N}\int_0^{\tau_m}\int_{d_{l-1}}^{d_l}\left\{\varphi_k^{\lambda_l}\frac{d^2T}{dx^2}+\frac{d\varphi_k^{\lambda_l}}{dT}\frac{dT}{dx}\frac{dT}{dx}\right\}\psi dx d\tau+\int_0^{\tau_m}\alpha_1\frac{dT_1(0,\tau)}{dx}\varphi_k^{\lambda_l}d\tau+\int_0^{\tau_m}\alpha_2\frac{dT_L(X_L,\tau)}{dx}\varphi_k^{\lambda_l}d\tau \tag{19}$$

$$J'_{C_{gl}^{k}}=\sum_{l=1}^{N}\int_0^{\tau_m}\int_{d_{l-1}}^{d_l}\varphi_k^{C_g}\int_{l_0}^{x}\frac{d\rho}{d\tau}d\xi\frac{dT}{dx}\psi dx d\tau \tag{20}$$

$$J'_{H_l^k}=\sum_{l=1}^{N}\int_0^{\tau_m}\int_{d_{l-1}}^{d_l}\varphi_k^{H}\frac{d\rho}{d\tau}\psi dx d\tau \tag{21}$$

where $\psi(x,\tau)$ - solution of a boundary-value problem adjoin to a linearized form of the initial problem Eqn. (1)-(9).

$$\frac{d\psi}{d\tau}\rho C+\frac{d}{dx}\left(\lambda\frac{d\psi}{dx}\right)=\left(\frac{d\lambda}{dT}\frac{dT}{dx}+C_g\int_{l_0}^{x}\frac{d\rho}{d\tau}d\xi\right)\frac{d\psi}{dx}++\left(\frac{dH}{dT}\frac{d\rho}{d\tau}+\frac{d\rho}{d\tau}C-C_g\frac{d\rho}{d\tau}\right)\psi+\Phi\frac{\partial F}{\partial T}=0 \quad (22)$$

$$\psi(0,x)=0 \quad (23)$$

$$\Phi(\tau_r)=0 \quad (24)$$

$$\psi_m=\psi_{m+1} \quad (25)$$

$$\frac{d\psi_1}{dx}\lambda_1(T_1(X_1,\tau))-\psi_1C_{g,1}(T_1(X_1,\tau))\int_{l_0}^{x}\frac{d\rho_1(T_1(X_1,\tau))}{d\tau}d\xi+\frac{\psi_1}{\alpha_1}\beta_1=0 \quad (26)$$

$$\frac{d\psi_L}{dx}\lambda_L(T_L(X_L,\tau))-\psi_LC_{g,L}(T_L(X_L,\tau))\int_{l_0}^{x}\frac{d\rho_L(T_L(X_L,\tau))}{d\tau}d\xi+\frac{\psi_L}{\alpha_L}\beta_L=0 \quad (27)$$

$$-\frac{d\Phi}{d\tau}=\Phi\frac{\partial F}{\partial\rho}-H\frac{d\psi}{d\tau}-\left(C\frac{dT}{d\tau}+\frac{dH}{dT}\frac{dT}{d\tau}\right)\psi-\frac{d}{d\tau}\left(\int_{x}^{l}C_g\frac{dT}{dx}\psi d\xi\right) \quad (28)$$

$$\Phi\ (\tau_r)=0 \quad (29)$$

3. EXPERIMENTAL VERIFICATION

By the way of experimental study of the material specimens the results of experiments are used on the automatic thermo-vacuum stand (TVS-1) manufactured in MAI at the Department of Space System Engineering. Two specimens made of the same test material Fig. 1 were located symmetrically about the heating element on a thermoinsulating base made of thermal insulation material so that its heating surface is parallel to the heater and at a certain distance from it (δ = 4-5 mm).

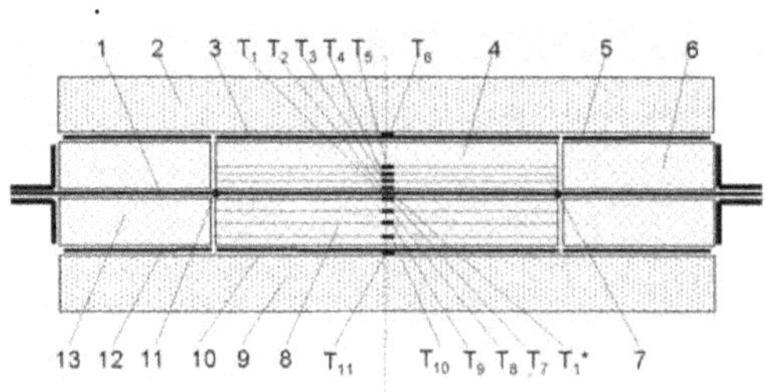

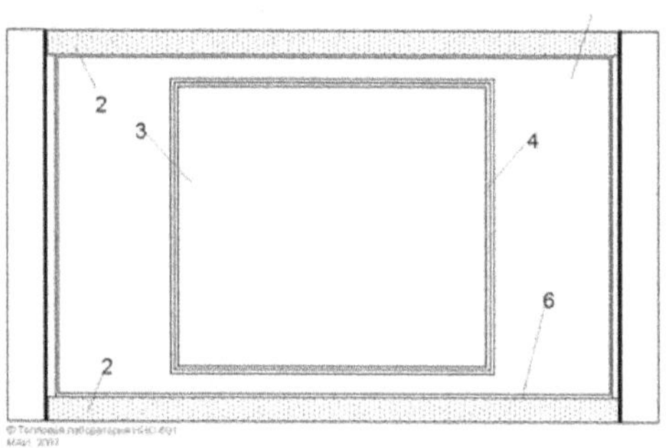

Figure 1: A testing scheme for specimens: 1 – heating element; 2 –thermal insulated slab; 3 – sensitive element (SE) of heat flux on the upper specimen; 4 – upper specimen (1a/2a); 5 –mask of the upper (SE); 6–siding thermoinsulated slab; 7 – voltage measuring point on the heater element; 8 –lower specimen (B); 9-thermo insulate slab; 10 – SE on lower element specimen (B); 11- voltage measuring points on the heating element; 12- mask of the lower SE; 13- siding thermoinsulated slab; T_1, T_1^* - thermocouples on the heater ; $T_2, -T_6$ - thermocouples on the upper specimen (A). $T_2, -T_6$ - thermocouples on the lower specimen (B).

Comparisons of the calculated and measured temperatures on the specimens surfaces for testing material presented in Figure 2. The result of estimating the unknown coefficients for material are presented in Figures 3-5. Table 1 includes the obtained values of the least squares and the maximum deviation of the calculated temperatures from that measured in the experiments.

experiment	Least-squares temperature deviation (K)	Maximum temperature deviation (K)
1	5.64	27.5
2	6.46	33.5
3	8.41	41.2
4	4.98	20/8

Table 1: The deviation of the calculated temperatures

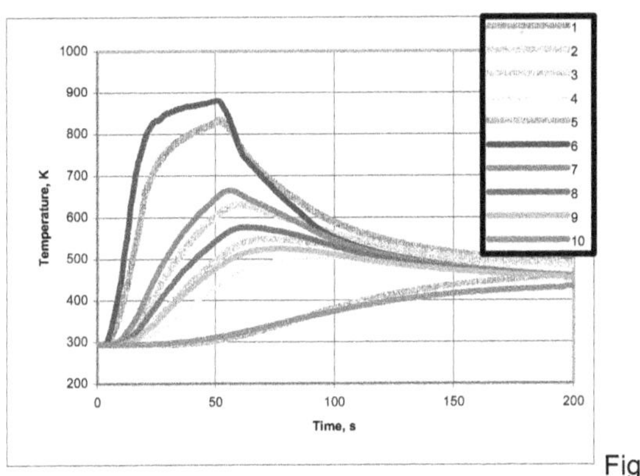

Figure 2: Comparing the calculated (T_1 and T_2) and measured (T_3 and T_4) temperatures for material ETTI-CF-ERG

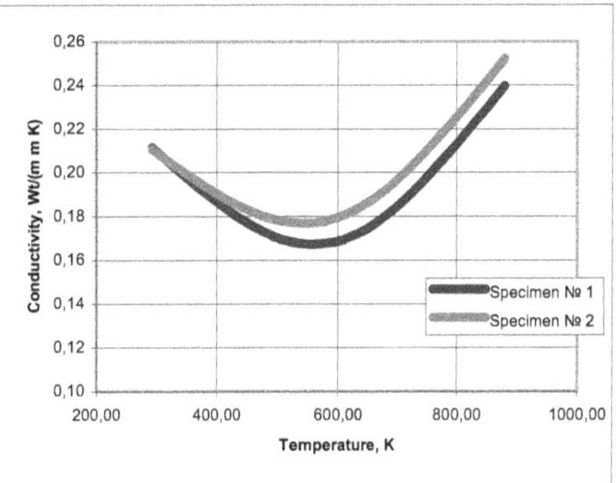

Figure 3: Estimated value of the thermal conductivity (specimens №1 and №2)

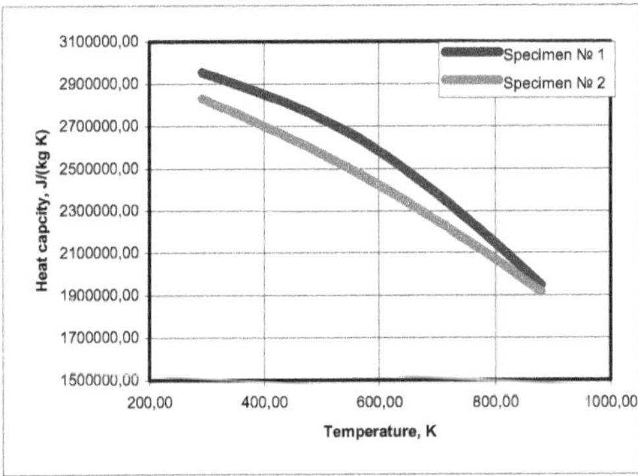

Figure 4: Estimated value of the thermal conductivity (specimens №1 and №2)

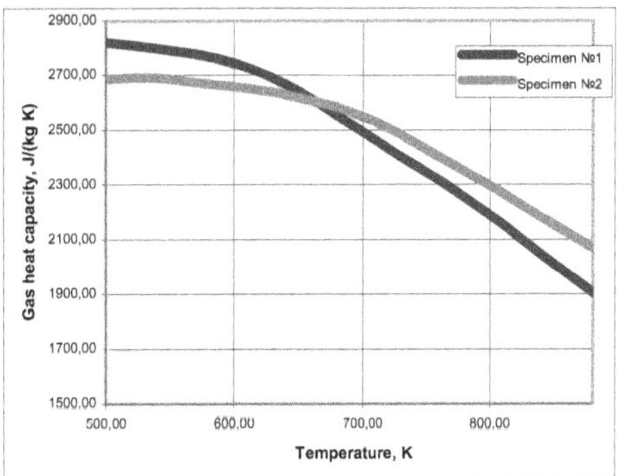

Figure 5: Estimated value of the gas heat capacity for (specimens №1 and №2)

CONCLUSIONS

Proceeding from calculations performed the following conclusions can be made; the following main factors have an influence on the accuracy of the inverse heat conduction problem (in sequence of significance): the errors in coordinates of thermosensor positions; the errors in values of different characteristics; the errors in estimating the residual level. It was shown that in the cases considered the accuracy of the inverse problems solution is compatible with the errors of the simulated "experimental measurements".

For partially decomposed materials the model of heat conduction with temperature-dependent thermal characteristics is approximate, and characteristics are effective, since the heat transfer in such material is provided not only by heat conduction but also by different transformation processes depended on conditions of heating. A deviation of of calculated and experimental temperature values in the experiments did not exceed 8 K, that confirms the possibility of using, for the given material, a model of heat conduction with the effective thermal characteristics. But the presented method can be used used only for determining the effective thermal characteristics of composite materials for partial heating conditions

ACKNOWLEDGMENTS

This work was done with the financial support of Russian Government (grant # NSh-65566.2010.8).

REFERENCES

1. A.V.Nenarokomov and O.M.Alifanov, Inverse radiative-conductive problems for estimating material properties, M.P.Menguc, N.Selcuk ed. Proc. of the Third International Symposium on Radiative Transfer, Begell House, New York\Wallingford (UK), 2001, p. 8.

2. O.M.Alifanov, Inverse Heat Transfer Problems, Springer-Verlag, Berlin \ Heidelberg, 1994, p. 348.

3. O.M.Alifanov, E.A.Artyukhin and S.V.Rumyantsev, Extreme Methods for Solving Ill-Posed Problems with Applications to Inverse Problems, Begell House, New York\Wallinford(UK), 1995, p. 306.

4. E.A.Artyukhin, G.A.Ivanov and A.V.Nenarokomov, Determination of a complex of materials thermophysical properties through data of nonstationary temperature measurements, High Temperature, 31, 235 (1993).

5. E.A.Artyukhin and A.V. Nenarokomov, Coefficient Inverse heat conduction problem, Journal of Engineering Physics, 53, 1085 (1987).

Sequential estimation of temperature-dependent thermal diffusivity in cherry pomace during nonisothermal heating

Ibrahim Greiby[1], *Dharmendra K. Mishra*[1,2], *Kirk Dolan*[1,3]

[1]*Department of Biosystems & Agricultural Engineering, Michigan State University, East Lansing, MI 48824*
[2]*Nestle Nutrition, PTC Fremont, MI 49413*
[3]*Department of Food Science & Human Nutrition, Michigan State University, East Lansing, MI 48824*
Phone: 517-355-8474 ext 178
Fax: (517) 353-8963

ABSTRACT

Fruit and vegetables are a rich source of many bio-active compounds from which value-added nutraceuticals can be produced. To design processes for these solids over a large temperature range, temperature-dependent thermal properties should be used. The objective of this work was to estimate temperature and moisture-dependent thermal diffusivity of cherry pomace. The sequential method of estimation was used to determine how the addition of each datum affected the thermal property estimates. Tart cherry pomace(70%) was equilibrated to 25% and 41% moisture content (MC) and heated in sealed 202×214 cans (0.027 m dia, 0.073 m height) at 126 °C in a steam retort for 50 minutes. Can center temperature was measured to obtain time-temperature data. The sum of squares of errors = (center temperature observed – center temperature predicted)^2 was minimized by nonlinear regression in Matlab with Comsol. Thermal diffusivity of 41% MC pomace was estimated as 2.49 $\pm$ 0.002 x 10^{-7} m^2/s and 1.63$\pm$ 0.001 x 10^{-7} m^2/s at 25°C and 126°C, respectively, changing linearly with temperature. Sequential parameter estimates came to a constant within 20 minutes.This study gives the first step for process design by estimating thermal diffusivity at elevated temperatures common in extrusion, spray-drying and other food drying methods.

1. Introduction

Thermal diffusivity (α) is an essential property necessary to design thermal processes. Because estimation of parameters is very sensitive to measured quantities, the determination of thermal diffusivity is very challenging. (Ukrainczyk 2009) used an iterative procedure based on minimizing a sum of squares function with the Levenberg-Marquardt method to solve the inverse problem. Another method based on least square optimization of a finite difference solution of Fourier's equation, has been established for thermal diffusivity estimation for different food products (Betta, Rinaldi et al. 2009). A study suggested that the transient plane-source method was an accurate method for simultaneously measuring thermal conductivity and thermal diffusivity of foods (Lihan and Lin-Shu 2009). Thermal diffusivity was estimated for a food subjected to constant heat flux and insulated on the other side using a computer program involving nonlinear sequential parameter estimation method based on Gauss minimization (Mohamed 2009). Many methods for determining thermal diffusivity have been developed. These include the ac method (Calzona, Cimberle et al. 1993), thermal-wave cavity (Balderas-Lopez and Mandelis 2001), thermal lens technique (Bernal-Alvarado, Mansanares et al. 2003) Thermal Lens Spectrometry (Jimenez-Perez, Cruz-Orea et al. 2009), and the sequential parameter estimation technique (Mohamed 2009). A review on simultaneous determination of thermal diffusivity and heat transfer coefficient has been published by (Erdogdu 2008). However, none of these methods estimated thermal diffusivity over a large temperature range and at temperatures above 100°C. Because many foods experience elevated temperatures where nutritional compounds may be degraded rapidly, and the thermal diffusivity may

increase significantly, methods are needed to determine the thermal properties at these more severe conditions.
Therefore, for this work a least-squares method to estimate of thermal diffusivity of cherry pomace during non-isothermal heating at elevated temperatures at different moistures was established. Sequential estimation of diffusivity was presented as well.

2. Materials and methods

2.1 Thermal parameters estimation

The heat transfer model for this experiment was for heat conduction in cylindrical coordinates:

$$\frac{1}{r}\frac{\partial}{\partial r}\left(k(T)r\frac{\partial T}{\partial r}\right)\frac{\partial}{\partial z}\left(k(T)\frac{\partial T}{\partial z}\right)=C(T)\frac{\partial T}{\partial t} \quad (1)$$

The governing boundary conditions were:

$$\frac{\partial T}{\partial t}(0,z,t)=0 \quad (2)$$

$$T(r,z,0)=T_i \quad (3)$$

$$\frac{\partial T}{\partial z}(r,0,t)=0 \quad (4)$$

$$-k(T)\frac{\partial T}{\partial r}(R,z,t)=h(T(R,z,t)-T_\infty) \quad (5)$$

$$-k(T)\frac{\partial T}{\partial z}(r,H,t)=h(T(r,H,t)-T_\infty) \quad (6)$$

The term *C(T)* was considered constant with temperature in Eq.(1) We are assuming that the *k*(*T*) relation varies much more with *T* than does *C*(*T*). In this case(MISHRA 2010), Eq. (1) can be written as:

$$\frac{1}{r}\frac{\partial}{\partial r}\left(\alpha(T)r\frac{\partial T}{\partial r}\right)\frac{\partial}{\partial z}\left(\alpha(T)\frac{\partial T}{\partial z}\right)=\frac{\partial T}{\partial t} \quad (7)$$

This transient conduction heat transfer problem was solved with the finite element model using Comsol (Comsol Inc., Burlington, MA) and Matlab (The MathWorks Inc., Natick, MA). Measured retort temperature was supplied to Comsol as the time-varying (during come-up) boundary condition. Thermal diffusivity as a function of temperature was estimated via an inverse technique by minimizing the sum of squares of measured temperature and the predicted temperature of the center of the can and □(*T*) at two different temperatures was estimated using Eq. (8).

$$\alpha(T)=\left(\frac{(T_2-T)}{(T_2-T_1)}\times\alpha_1+\frac{(T-T_1)}{(T_2-T_1)}\times\alpha_2\right) \quad (8)$$

2.2 Sample Preparation.

Three different samples of cherry pomace (fresh 70%, 41% and 25 % MC, wb) obtained from Michigan Cherry Growers were filled in 202 × 214 steel cans (radius 0.027m, and height 0.073m) in duplicate. Density of each sample was measured. The retort was brought up to temperature in approximately 5-8 min, and was then held at 126°C.

3. Results and discussion

3.1 Temperature vs. time results

A. An example of different times of thermal processing (retort processing at 126°C) of cherry pomace, is shown in the following figure:

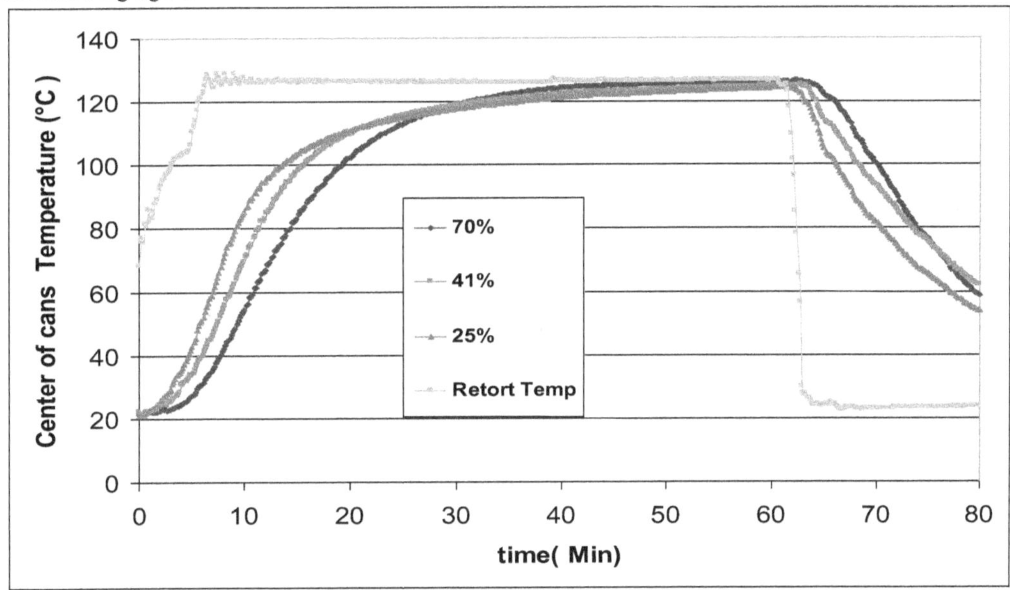

Figure 1: Thermal processing of cherry pomace (three different moisture contents) at 126 °C for 50 minutes.

3.2 Parameter Estimates

Using eq. (8) thermal diffusivity was estimated at two different temperatures, at 25°C and 126°C for α1 and α2 ,respectively. Results are summarized in Table 1 for all heating time (50 min):

Table1: Estimation of thermal diffusivity parameters for cherry pomace by ordinary least squares:

Moisture content %(wb)	Parameter	Parameter estimates m^2/s	Standard error	Correlation coefficient	Sequential estimation parameters	95% asymptotic confidence interval	RMSE (°C)
25	α1 α2	3.7015×10^{-7} 1.4696×10^{-7}	0.0251×10^{-7} 0.0029×10^{-7}	-0.2026	3.68×10^{-7} 1.48×10^{-7}	$(3.6522\text{-}3.7509)\times 10^{-7}$ $(1.4640\text{-}1.4753)\times 10^{-7}$	2.7958
41	α1 α2	2.4914×10^{-7} 1.6385×10^{-7}	0.0020×10^{-7} 0.0013×10^{-7}	-0.9973	2.55×10^{-7} 1.56×10^{-7}	$(2.4875\text{-}2.4953)\times 10^{-7}$ $(1.6359\text{-}1.6411)\times 10^{-7}$	2.688
70	α1 α2	1.9713×10^{-7} 1.4227×10^{-7}	0.000122×10^{-7} 2.44×10^{-7}	0.7611	1.96×10^{-7} 1.46×10^{-7}	$(1.9710\text{-}1.9715)\times 10^{-7}$ $(1.4227\text{-}1.4227)\times 10^{-7}$	2.2135

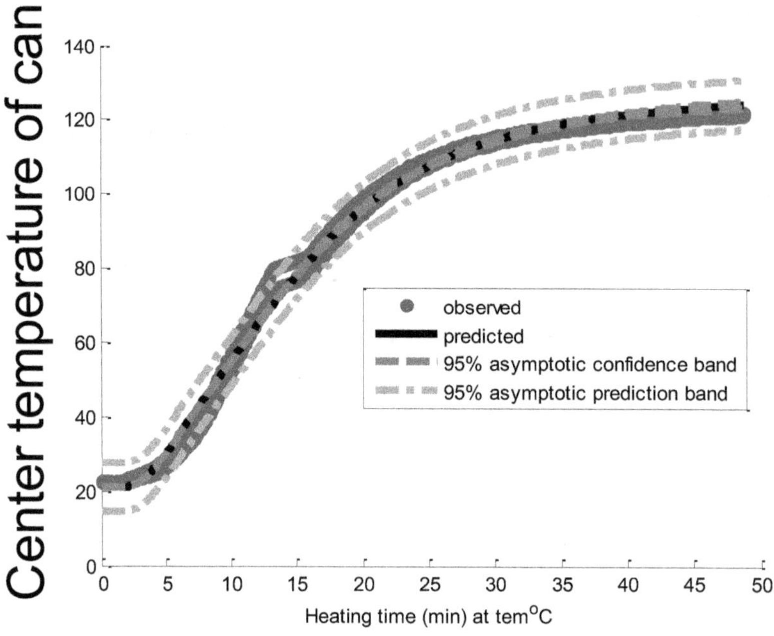

Figure 2: Example time-temperature plot with 95% confidence and predicted intervals of center temperature of the can of cherry pomace (41%MC) heated for 50 minutes

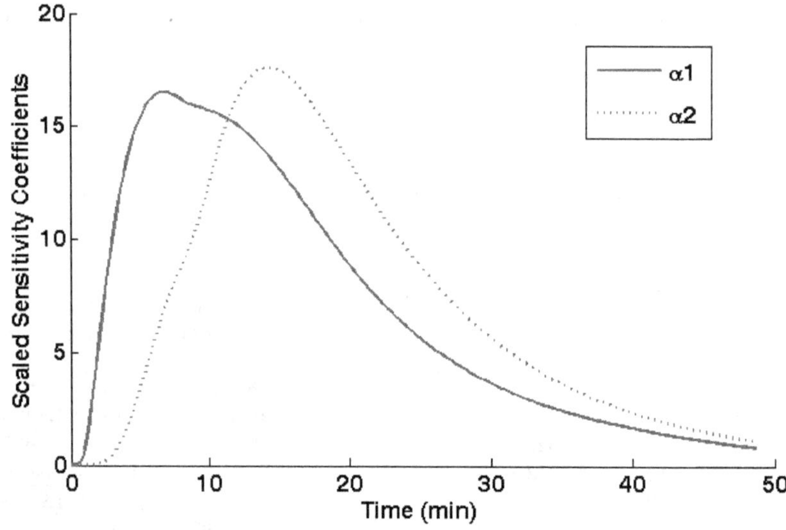

Figure 3: Example of Scaled sensitivity coefficients of thermal diffusivity of cherry pomace (41% MC) at 50 minute retorting

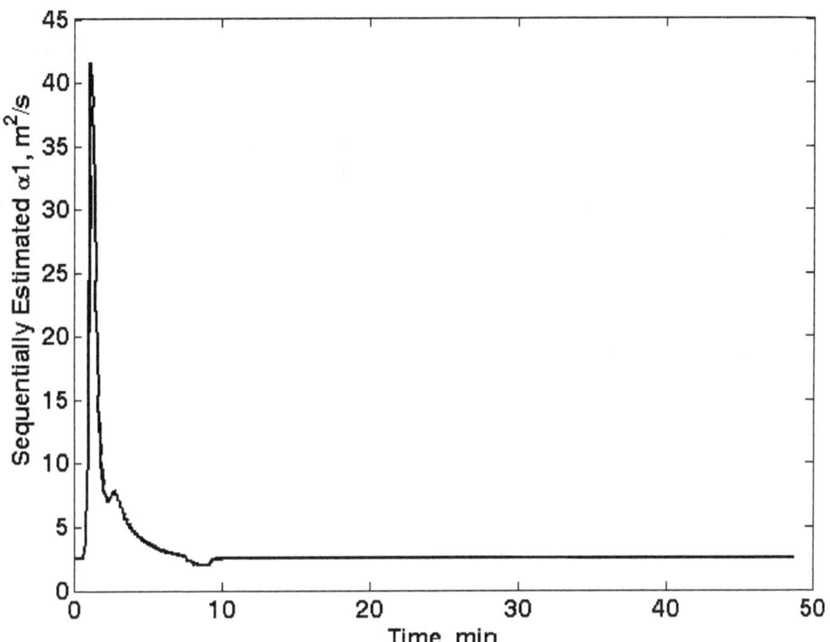

Figure 4: Sequential estimation of thermal diffusivity (α1) of Cherry pomace (41% MC) at 50 minutes of retorting

4. Conclusions

Thermal diffusivity values for cherry pomace at three different moisture contents were successfully estimated using both ordinary least squares estimation and sequential estimation. Thermal diffusivity values decreased with temperature and decreased with moisture content. Sequential estimation showed that the estimated thermal diffusivity attained nearly a constant within 20 min, so the experiment could be ended at that time. Advantages of this experimental procedure are that thermal diffusivity can be estimated as a function of temperature by conducting only one experiment at actual nonisothermal conditions, rather than numerous isothermal experiments at unrealistic laboratory conditions.

Nomenclature:

H	Height m
k	thermal conductivity($Wm^{-1}k^{-1}$)
r	Radius m
t	Time(s)
T	Temperature(°C)
Cp	specific heat capacity ($J\ kg^{-1}\ K^{-1}$)
ρ	density ($kg\ m^{-3}$)
α	Thermal diffusivity(m^2s^{-1})

5. References

Balderas-Lopez, J. A. and A. Mandelis (2001). "Simple, accurate, and precise measurements of thermal diffusivity in liquids using a thermal-wave cavity." Review of Scientific Instruments **72**(6): 2649-2652.

Bernal-Alvarado, J., A. M. Mansanares, et al. (2003). "Thermal diffusivity measurements in vegetable oils with thermal lens technique." Review of Scientific Instruments **74**(1): 697-699.

Betta, G., M. Rinaldi, et al. (2009). "A quick method for thermal diffusivity estimation: Application to several foods." Journal of Food Engineering **91**(1): 34-41.

Calzona, V., M. R. Cimberle, et al. (1993). "A New Technique to Obtain a Fast Thermocouple Sensor for Thermal-Diffusivity Measurements in an Extended Temperature-Range." Review of Scientific Instruments **64**(12): 3612-3616.

Erdogdu, F. (2008). "A review on simultaneous determination of thermal diffusivity and heat transfer coefficient." Journal of Food Engineering **86**(3): 453-459.

Jimenez-Perez, J. L., A. Cruz-Orea, et al. (2009). "Monitoring the Thermal Parameters of Different Edible Oils by Using Thermal Lens Spectrometry." International Journal of Thermophysics **30**(4): 1396-1399.

Lihan, H. and L. Lin-Shu (2009). Simultaneous determination of thermal conductivity and thermal diffusivity of food and agricultural materials using a transient plane-source method. Journal of Food Engineering. **95:** 179-185.

MISHRA , K. D. D. a. L. Y. (2010). "Bootstrap Confidence Intervals for the Kinetic Partameters of Degradation of Anthocyanins in Grape Pomace." Journal of Food Process Engineering.

Mohamed, I. O. (2009). "Simultaneous estimation of thermal conductivity and volumetric heat capacity for solid foods using sequential parameter estimation technique." Food Research International **42**(2): 231-236.

Ukrainczyk, N. (2009). "Thermal diffusivity estimation using numerical inverse solution for 1D heat conduction." International Journal of Heat and Mass Transfer **52**(25-26): 5675-5681.

Estimation Of Roughness And Dispersion Coefficients With The Particle Collision Algorithm In Open Channels

Y Martínez González[1], J B Martínez Rodríguez[1], P P G Watts Rodrigues [2] and A J Silva Neto[2]

[1] Facultad de Ingeniería Civil, Instituto Superior Politécnico José Antonio Echeverría, Calle 114 No. 11901 e/ 119 y 127, Marianao, Ciudad de La Habana, Cuba.

[2] Instituto Politécnico, IPRJ, Universidade do Estado do Rio de Janeiro, UERJ, P.O. Box 97282, 28601-970, Nova Friburgo, RJ, Brazil.

E-mails: ymarq@cih.cujae.edu.cu, bienvenido@cih.cujae.edu.cu,pwatts@iprj.uerj.br, ajsneto@iprj.uerj.br

Abstract.

In this contribution, the one-dimensional flow and transport equations for open channels were numerically solved. Based on data collected by Rodríguez et al (1996) in Albear channel, the longitudinal dispersion coefficient and the roughness parameter were estimated by a global search optimization algorithm (Particle Collision Algorithm-PCA) coupled to the models, in a way that strategies were proposed to yield stable and convergent solutions. PCA is inspired on the scattering and absorption of a given incident nuclear particle by an objective nucleus (Sacco et al, 2005). In this model, if the particle-core reaches a low value of the objective function, it is absorbed, otherwise, it is scattered. This allows that the searched space of problem solution is widely explored, in a way that the most promising regions are searched through scattering and absorption successive events (Silva Neto and Becceneri, 2009). The results obtained in this contribution were supported by the high correlations that were achieved.

1. Introduction

Open channel hydraulics has always been a very interesting domain of scientific and engineering activity because of the great importance of water for human living. According to Szymkiewicz [19], the source of difficulties is the proper description of flow processes and their mathematical description, as well as the solution of derived equations and identification of parameters involved. For this reason the great progress in open channel flow that took place during last years, paralleled the progress in computer technique, informatics and numerical methods.

2. Governing equations

The equation that governs the one-dimensional steady and spatially varied flow in the presence of input and/or lateral discharge in a open-channel with arbitrary shape and alignment can be expressed formally as (Martínez, [10]):

$$\frac{dh}{dx} = \frac{S_o - \dfrac{Q^2 n^2}{R^{2m+1} A^2} + (\varphi + 1 - k_e)\dfrac{Q}{A^2 g}\hat{q}}{1 - (1 - k_e)\dfrac{Q^2 T}{g A^3}} \tag{1}$$

where: g is the gravity acceleration; h = flow depth; x = space variable; A= wetted area; Q = flow; T = surface width; S_o = bottom slope of channel, R = hydraulic radius; n = roughness coefficient; m = exponent (eg. Manning equation m = 1/6); k_e = longitudinal coefficient of expansion or contraction; q = input or lateral discharge and the parameter φ = 0 for a bulk discharge, φ = 0.5 in the presence of discharge by filtration and φ = 1 for lateral contributions to the channel.

According to (Fischer et al, [4]), the one-dimensional transport of a conservative substance in a open channel with arbitrary geometry and alignment, subjected to the processes of advection-dispersion in open channels can be expressed as:

$$\frac{\partial}{\partial t}(AC) + \frac{\partial}{\partial x}(AUC) - \frac{\partial}{\partial x}\left(AD\frac{\partial C}{\partial x}\right) = 0 \tag{2}$$

where C = substance concentration , D = longitudinal dispersion coefficient, $U = Q/A$, and represents the mean velocity. If the flow can be assumed as steady and uniform, for a prismatic channel and assuming D constant, Eq.2 reduces to:

$$\frac{\partial C}{\partial t}+U\frac{\partial C}{\partial x}-D\frac{\partial^2 C}{\partial x^2}=0 \tag{3}$$

3. Numerical approaches

Given the flow regime in channel, if the observed remains constant over a period of time (eg 6, 8, 10 hours, etc) , Eq.1 is valid and can be solved by Runge-Kutta method (Carnahan et al, [2]; Martínez [10]) . This method is implemented in MatLab by *ode45* function, which solves initial value problems in ordinary differential equations (The MathWorks, Inc, [20]).

Eq.3 has been solved analytically for different initial and boundary conditions (Fischer et al, [4]). The development of these solutions has been closely associated with the introduction of procedures with a reasonable accuracy order to estimate the longitudinal dispersion coefficient. Such is the case of the method known as *routing procedure*, developed by Fischer [3], which, from a reasonable estimative of the coefficient D, predicts the concentration distribution of a substance at a given distance from the injection point x_2, based on a known distribution at point x_1, where $x_2 > x_1$.

$$C(x_2,t)=\int_{-\infty}^{+\infty}\frac{U C(x_1,\tau)}{\sqrt{4\pi D(\bar{t}_2-\bar{t}_1)}}\exp\left\{-\frac{\left[U(\bar{t}_2-\bar{t}_1-t+\tau)\right]^2}{4D(\bar{t}_2-\bar{t}_1)}\right\}d\tau \tag{4}$$

where $\bar{t}_1$ and $\bar{t}_2$ = mean travelling times respectively at at x_1 and x_2; τ = integration variable. The use of equation (4) involves an iterative process for estimate the parameter *D*. Usually, as a first approach, is introduced the *change of moment* method (Fischer et al, [4]). The Eq.4 can be solved by trapezoidal numerical integration, implemented in MatLab by *trapz* function (The MathWorks, Inc, [20]).

4. Inverse problem formulation

4.1 Objective function

According to Mesa [11]-[12], the inverse problem is generally ill-posed, since it can lead to non-uniqueness and instability of the identified parameters; beside that e several errors can be associated with these problems. It is a usual practice in parameters estimation; make use of the minimization of the root mean square, as a criterion for determining the optimal set of parameters, given by

$$\min_{\Theta} F(\Theta) \quad F(\Theta)=\sqrt{\frac{\sum_{j=1}^{N_T}\sum_{i=1}^{N_o}\left[\Psi_{ij}^{o}-\Psi_{ij}(\Theta)\right]^2}{N_T N_o}} \tag{5}$$

where $\Theta_i \leq \Theta \leq \Theta_s$, where Θ = vector of parameters, here with N components, Θ_i Θ_s are, respectively the vector of lower and upper bounds of parameters , N_T = number of times, N_o = number of nodes with comments, Ψ_{ij}^{o} = measured values and Ψ_{ij} = computed values of the simulated variable . To solve the inverse problem it was adopted a stochastic method, recently developed by Sacco et al [14], the **P**article **C**ollision **A**lgorithm (PCA), which up to now had no applications in hydraulic problems.

4.2 Global search algorithm based on particle collision (PCA)

According to Silva Neto and Becceneri [18], the PCA is inspired on the scattering and absorption of a given incident nuclear particle by an objective nucleus. In this model, if the particle-core reaches a low value of the objective function, it is absorbed, otherwise, it is scattered. This allows that the searched space of problem solution is widely explored, in a way that the most promising regions are searched through scattering and absorption successive events. This version of PCA, despite its simplicity, has been successful in some engineering applications, especially in radioactive transfer problems (Knupp et al, [7]), and optimal design of nuclear reactors (Sacco and Oliveira, [14]; Sacco et al, [15], Sacco et al [16], Sacco et al, [18]).

5. Study case: the Albear channel

The particular study case in this contribution is the Albear channel, the oldest of the aqueducts of Havana (Lenzano, [8]). It´s considered a masterpiece of hydraulics in Cuba, recently declared a National Monument and is one of the seven engineering wonders of Cuba (Garcia et al, [5] and Garcia, [6]). The channel has presented serious problems due to a continuous use without a proper maintenance, as the dome structure cracking, roots penetration and seepage losses (Fig. 1). In 2001, the channel was submitted to an important restoration, in order to fix some observed damages. However, even after that, there are still unsolved problems (Aguas de la Habana, [1]), the most important being the absence of post-rehabilitation studies to assess the hydraulic behavior of flow in the aqueduct.

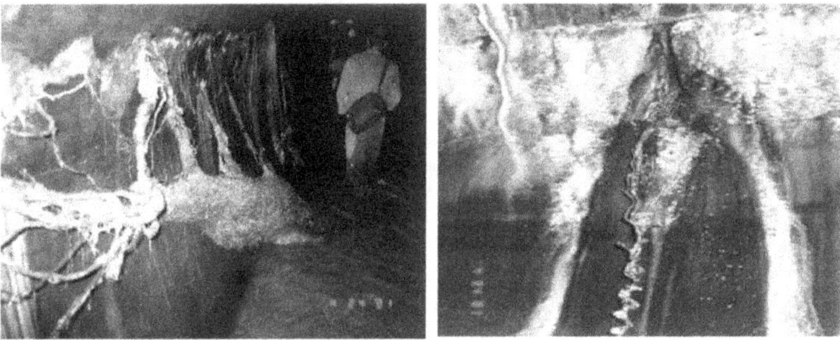

Figure 1: Roots penetration and infiltrations.(Courtesy: Aguas de La Habana)

6. Sensitivity analysis

The sensitivity analysis can be done through the evaluation of variable fluctuations with respect to parameters *n* and *D,* once known seepage losses *q* in Albear channel. Here will be adopted the modified sensitivity coefficient *X*, which can be expressed formally as (Lugon et al,[9] and Knupp et al,[7]):

$$X_{\Theta_j}(x,t) = \Theta_j \frac{\partial \psi(x,t)}{\partial \Theta_j} \tag{6}$$

where ψ = simulated variable (which can be measured), Θ = particular parameter of the problem and j = 1,2, ..., M, where M = total number of unknowns. According to Lugon et al [9], when two or more parameters are estimated simultaneously, their effects on the variable must be independent, otherwise they may affect the variable in the same way, bringing difficulties to distinguish their influences separately and leading to poor estimative as a consequence.

In order to tackle the problem of a possible simultaneous estimation for *n* and *q* in equation (1), it is carried out a sensitivity analysis of *h* based on observed data (Rodríguez et al, [13]). In each monitoring stations (Torre Norte, Torre Cuadrada, etc), the variable is remarkably sensitive to *n*, which is a measure of energy losses due to the friction. The sensitivity increases as a result of increasing the depth *h*, wherever flow conditions remain constant. However, for the range of *q* and for all stations, the sensitivity is almost constant, decreasing in flow direction (Fig. 2). This trend does not favor the estimation of this parameter simultaneously with roughness coefficient.

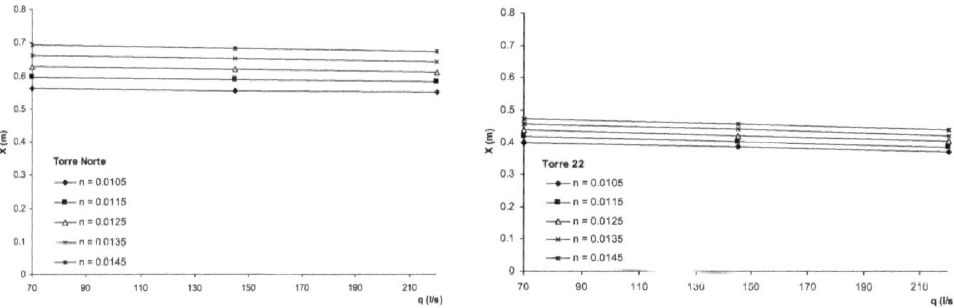

Figure 2: Variation of sensitivity with respect to roughness and total seepage losses.

These results motivated an assessment of *C* sensitivity regarding *q* and *D* in the transport Eq. 4. For this analysis, the calculated *D* varied between 0.05 and 0.35 m^2/s, confirming that advection dominates the phenomenon of dispersion in Albear channel. The sensitivity in this case varies along the space and time, as shown in Fig. 3.

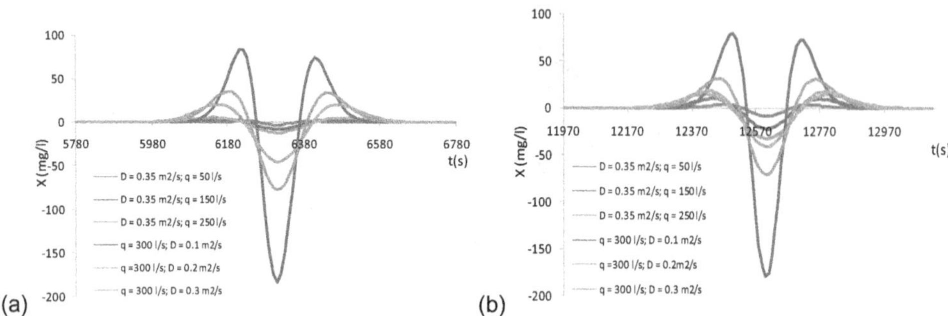

Figure 3: Sensitivity with respect to the dispersion coefficient and total seepage losses in (a) Torre 12 (b) entrance of Palatino reservoirs

The sensibility analysis showed that q and D differently influenced on tracer concentration, keeping the boundary condition at the beginning of the analyzed section (Torre Cuadrada) unchanged in each case. For each stations, C is more sensitive to variation of D coefficient during the whole time interval between the beginning and the end of tracer concentration cloud. The tracer sensitivity showed to be directly proportional to the total seepage losses q, being markedly higher around the beginning and the end of tracer cloud. The sensitivity was higher close to Palatino reservoirs, as a result of cumulative losses throughout the channel, and lower at Torre Norte. The influence of parameters D and q on the distribution of tracer concentration are correlated because $D= D(q)$. This indicates that it is inappropriate to estimate simultaneously.

7. Results

A numerical experiment was designed for each reach, in which the number of iterations N of PCA and the number of iterations M to carry out a local exploration are 10, 50 and 100. To compute seepage losses, was focus attention on the procedure to determine the average velocity U. Table 1 shows the results of numerical experiment for all reaches (the best estimative of D, mean μ_Θ, standard deviation, σ_Θ, and coefficient of variation C_v). PCA was able to make good estimative in all channel reaches, although a slight increase in the value of the longitudinal dispersion coefficient between Torre 22 and Palatino could be observed. This increase is a consequence of increase in velocity near the entrance of Palatino reservoirs.

Table 1. Identification parameters in Eq. 4

Reach	q (m^3/s/m)	D (m^2/s)		F(Θ) – eq.(5)-(mg/l)	Elapsed time (s)
Torre Cuadrada-Torre 12	1.9594E-05	0.1948		7.9261	48.9976
		μ_Θ	0.1950	N = 100 M = 50	
		σ_Θ	0.0013		
		C_v	0.0066		
Torre 12 -Torre 22	2.5646E-05	0.2210		5.5761	24.7021
		μ_Θ	0.2219	N = 50 M = 10	
		σ_Θ	0.0014		
		C_v	0.0064		
Torre 22 - Palatino reservoirs	3.6562E-05	0.6092		7.4947	90.4511
		μ_Θ	0.5829	N = 100 M = 10	
		σ_Θ	0.0394		
		C_v	0.0067		

The searched range for the parameter n was [0.01, 0.015], values related to the type of material covering the channel. The channel was divided into 306 subintervals, with a length of 29.7 m. The measurements of depth are made in Torre Norte, Torre 12, Torre 18 and Torre 22 allowed the identification of roughness coefficient. Table 2 indicates that PCA was able to make a good estimative. In addition, small values for C_v pointed out that it is possible to obtain good estimates with a small number of iterations of PCA.

Table 2. Identification parameter in Eq. 1

Reach	n 50×100	μ_Θ	σ_Θ	C_v
Torre Norte-Torre Cuadrada	0.0130	0.0127	3.43E-04	0.0267
Torre Cuadrada-Torre 12	0.0129	0.0129	1.20E-04	0.0093
Torre 12- Torre 18	0.0141	0.0140	2.12E-04	0.0151
Torre 18-Torre 22	0.0133	0.0134	1.12E-04	0.0083
Torre 22- Palatino reservoirs	0.0113	0.0113	1.41E-04	0.0125
F(Θ) – eq.(5)-(m)	0.0026			
Elapsed time (s)	890.88			

7. Conclusions

A formulation for the characterization of flow and advection-dispersion in Albear channel was presented. The consequent optimization problem was solved using stochastic method PCA. Also the sensitivity analysis has been proved to be an important tool to verify which parameters could be more easily identified, developing strategies for the parameters search, subjected to constraints imposed by the space search algorithm.

8. References

[1] Aguas de la Habana: Informe Técnico de Visita al Canal. Empresa Aguas de la Habana. 2002.

[2] Carnahan, B.; Luther, H.A.; Wilkes, J.O.: Applied Numerical Methods. John Wiley and Sons. 1969

[3] Fischer, H.B: Dispersion Predictions in Natural Stream. J. Sanit. Eng. Div. ASCE. 94, pp 927-944. 1968.

[4] Fischer, H.B.;List, E. J.;Koh, R. C. Y; Imberger. J; Brooks, N. H.: Mixing in Inland and Coastal Waters. Academic Press. ISBN-13: 978-0-12-258150-2; ISBN-10: 0-12-258150-4. 1979

[5] García Blanco,R; Pérez Monteagudo, F.; Aruca Alonso, L.J.; Álvarez Hernández, A.: Una Obra Maestra: El Acueducto Albear de La Habana. Editorial Científico Técnica. La Habana. ISBN: 959-05-0286-5. 2002.

[6] García Blanco, R.: Acueducto de Albear: Una Obra que Nació Verdaderamente Monumental. Revista Voluntad Hidráulica. INRH. Cuba. pp 64-67. ISSN: 0505-9461. 2009.

[7] Knupp, D.C, A.J. Silva Neto e W. F. Sacco: Radiative Properties Estimation With the Particle Collision Algorithm Based on a Sensitivity Analysis, High Temperatures - High Pressures. 2009

[8] Lenzano Paneque, G.: Declaración Oficial del Acueducto de Albear como Monumento Nacional. Revista Voluntad Hidráulica. INRH. Cuba. pp 60-61. ISSN: 0505-9461. 2009.

[9] Lugon Junior,J.; Silva Neto, A.J; Watts Rodrigues, P. P G.: Assessment of Dispersion Mechanisms in Rivers by Means of An Inverse Problem Approach. Inverse Problems, Design and Optimization Symposium. Miami, Florida,U.S.A., April 16-18. 2008.

[10] Martínez González, Y: Análisis del Comportamiento Hidráulico de Canales en Régimen Impermanente Mediante la Modelación Matemática. Tesis de Doctorado en Ciencias Técnicas. Facultad de Ingeniería Civil. Centro de Investigaciones Hidráulicas, CIH. ISPJAE.Ciudad de la Habana. Cuba. ISBN: 2409-2007. 2006.

[11] Mesa, H. R.: Solución del Problema Inverso en Modelos de Flujo del Agua Subterránea Mediante un Algoritmo de Convergencia Global. Tesis de Doctorado en Ciencias Técnicas. ISPJAE. CIH. Ciudad de La Habana. Cuba. 2000.

[12] Mesa, H. R.: Calibración Automatizada de Parámetros Hidrogeológicos para Acuíferos en Régimen Impermanente. XX Congreso Latinoamericano de Hidráulica. IAHR. Ciudad de La Habana. Cuba. 2002.

[13] Rodríguez, T.; Alfonso Mon, E.; Alfonso, J.L.: Estudio sobre los Ciclos del Agua en La Habana. Informe. Metrópolis-Unión Europea. 1996.

[14] Sacco, W.; Oliveira, C.R.E: A New Stochastic Optimization Algorithm based on a Particle Collision Metaheuristic. 6th World Congresses of Structural and Multidisciplinary Optimization. Rio de Janeiro, 30 May - 03 June, Brasil. 2005.

[15] Sacco, W. F.; Oliveira, C. R. E.; C. Pereira, M. N. A.: Two stochastic optimization algorithms applied to nuclear reactor core design. Progress in Nuclear Energy, 48. pp 525-539. 2006.

[16] Sacco, W. F.; Filho, H. A.; Pereira, C. M. N.A.: Cost-Based Optimization of a Nuclear Reactor Core Design: A Preliminary Model. 2007 International Nuclear Atlantic Conference - INAC 2007.

Santos, SP, Brazil, September 30 to October 5, 2007. Associação Brasileira de Energia Nuclear – ABEN. ISBN: 978-85-99141-02-1. 2007.

[17] Sacco, W. F.; C. Lapa, M. F.; Pereira, C. M. N. A.; Filho, H. A.: A Metropolis Algorithm applied to a Nuclear Power Plant Auxiliary Feedwater System Surveilance Tests Policy Optimization. Progress in Nuclear Energy, 50. pp 15-21. 2008.

[18] Silva Neto, A.J.; Becceneri, J.C.: Técnicas de Inteligência Computacional Inspiradas na Natureza: Aplicação em Problemas Inversos em Transferência Radiativa. 2009.

[19] Szymkiewicz, R (2010): Numerical Modeling in Open Channel Hydraulics. Springer. Water Science And Technology Library. ISBN 978-90-481-3673-5.

[20] The MathWorks, Inc: MatLab ® Mathematics. 2006

Estimation of leakage flow rates inside soil dikes through temperature measurements

Stéven KERZALÉ[1], Denis MAILLET[1], Alexandre GIRARD[2]

[1]LEMTA, Nancy-Université & CNRS
2 Avenue de la Forêt de Haye – 54504 Vandœuvre Lès Nancy Cedex
Email: steven-externe.kerzale@edf.fr
Email: denis.maillet@ensgsi.inpl-nancy.fr
[2]EDF R&D, département STEP
6 Quai Watier – 78401 Chatou Cedex
Email: alexandre.girard@edf.fr

Abstract

An estimation of the velocity field in a 2D section of an anisotropic homogenous soil dike, using internal temperature measurements, is proposed here. Non-linear-Kalman-filter-based inversion algorithms are applied on the advection-diffusion equation coupled with Darcy's equation in this dike, which is considered as a porous medium with variable saturation.

A sensitivity study shows that the thermal behavior of the dike is purely advective in the horizontal direction, because of the anisotropy of the dike, and the horizontal hydraulic conductivity is the main physical parameter that governs the velocity field in this porous medium. As a consequence, a reduced one dimensional model is used for inversion.

Key Words: inverse problem, non-linear Kalman's filter, heat equation, Darcy's law, finite volume method

Nomenclature

T	temperature, K^{1}
$\vec{v}$	Darcy's velocity of water, $m.s^{-1}$
C	heat capacity, $J.K^{-1}.kg^{-1}$
K_h, K_v	horizontal and vertical hydraulic conductivity, $m.s^{-1}$
K	relative permeability of water
S_o	storage compressibility, m^{-1}
S	water saturation
Greeks symbols	
Λ	thermal dispersion tensor, $W.m^{-1}.K^{1}$
ψ	hydric potential, m
ε	soil porosity
ρ	density, $kg.m^{-3}$
α_L, α_T	longitudinal and transverse thermo-dispersivity, m
Superscripts	
^{t}	transpose
^{e}	soil
^{w}	water
Subscripts	
$_r$	residual
$_s$	saturation
$_h$	horizontal
$_v$	vertical
$_L$	longitudinal
$_T$	transverse

1. Introduction

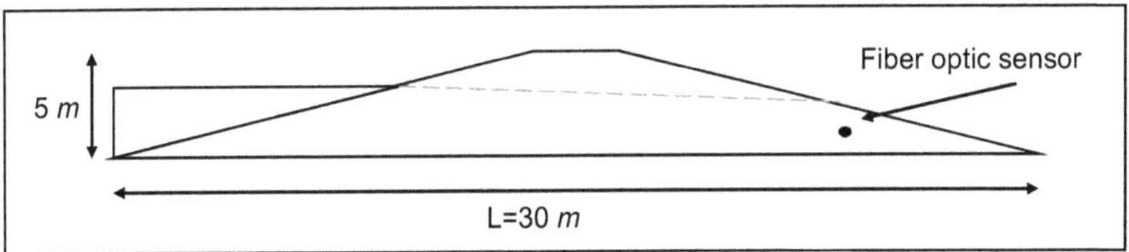

Figure 1: Cross section of canal and dike

Detection of possible leakages of a canal through water seepage through the soil dike, as well as their quantitative evaluation in real time, is the cause of major concerns for a good canal management. Temperature measurements by fiber optic sensors over very large distances (several kilometers) are now possible: they provide information on the local state of the dike. In any cross section, see Figure 1, air temperature T_{air}, water temperature T_{water} inside the canal, and local soil temperature T_{dike} inside the dike can be monitored this way during several months. Knowledge of intrinsic thermo-physical and hydro-geological properties of the constitutive soil material(s) of the dike is also available.

2. Physical state-space model

The temperature field in the dike is given by the one-temperature heat dispersion equation in a porous medium [1]:

$$(\varepsilon S(\psi)(\rho C)^w + (1-\varepsilon)(\rho C)^e)\partial_t T = -(\rho C)^w \vec{v}\cdot\vec{\nabla}T + \vec{\nabla}\cdot(\Lambda\vec{\nabla}T), \tag{1}$$

with $\vec{v}$ the velocity given by Darcy's law

$$\vec{v} = -K\big(S(\psi)\big)(K_h, K_v)^t \cdot \vec{\nabla}(\psi + y), \tag{2}$$

and the hydric potential ψ given by the law of mass conservation [2]:

$$\big(S_0 S(\psi) + \varepsilon\partial_\psi S(\psi)\big)\,\partial_t\psi + \vec{\nabla}\cdot\vec{v} = 0, \tag{3}$$

Van Genuchten and Mualem represent the hydrodynamics properties S, the water saturation, and K, the relative permeability of the water phase, by

$$\begin{cases} S(\psi) = H(\psi) + \big(1 - H(\psi)\big)\left(S_r + \dfrac{S_s - S_r}{(1+(\alpha\psi)^n)^m}\right) \\ K(S) = \sqrt{\dfrac{S - S_r}{S_s - S_r}}\left(1 - \left(1 - \left(\dfrac{S - S_r}{S_s - S_r}\right)^{1/m}\right)^m\right)^2, \end{cases} \tag{4}$$

This is an empirical model where H represents the Heaviside's operator. The thermal dispersion tensor Λ is given by the following expression

$$\Lambda_{ij} = (\varepsilon S(\psi)\lambda^w + (1-\varepsilon)\lambda^e)\delta_{ij} + (\rho C)^w\left(\alpha_T ||\vec{v}||\delta_{ij} + (\alpha_L - \alpha_T)\frac{v_i v_j}{||\vec{v}||}\right). \tag{5}$$

3. Results of simulations and sensitivity study

With appropriate boundary conditions corresponding to air and water temperatures measurements during two years, see Figure 2, the previous equations have been solved numerically by the finite volume method in space and by BDF (*backward differentiation formulae*) discretization in time, implemented in a MATLAB® code. We compare those results with the output of COMSOL® and FEFLOW®, see Diersch [2], commercial softwares. The corresponding simulation of water flow and heat transfer in a soil with different hydraulic conductivities is shown in Figure 3, in order to study the influence of hydraulic conductivity on temperature.

We note that the temperature simulated at the sensor location in the dike and temperature of water, see Figure 3, show similar behavior: the amplitudes of their variations are very close, for hydraulic conductivities between 10^{-2} $m.s^{-1}$ and 10^{-3} $m.s^{-1}$, but their phases are a bit different. The dike can be considered as a low pass filter for water. A 10^{-4} $m.s^{-1}$ hydraulic conductivity yields a similar thermal behavior with a larger phase shift (lower pass filter).

Analyzing the results for a 10^{-5} $m.s^{-1}$ hydraulic conductivity is more involved: the phase difference between the dike and the air temperatures becomes very small with a larger attenuation for the variations of dike temperature. This damping may stem from the influence of water on the dike sensor.

Of course, a lower conductivity (10^{-6} $m.s^{-1}$ or less) reduces the influence of water on the dike: sensor temperatures become almost the same as the air ones.

Table 1 gives the maximum velocity and the Péclet number ($P_e = (\rho C)^w v_x L/\Lambda_{xx}$) for different hydraulic conductivities. This number allows an evaluation of the relative influences of diffusion and advection in the transfer of heat: it is the ratio between the advection term (caused here by water seepage through the dike) and the diffusion term in the heat equation: it is thus related to the influence of either water or air on the dike temperature.

We see that the higher the hydraulic conductivity of the dike is, the better the quality of velocity estimation becomes: this can be numerically observed in our results.

Table 1 : Characteristic values for different hydraulic conductivities

K_h (m.s^{-1})	0	10^{-6}	10^{-5}	10^{-4}	10^{-3}	10^{-2}
v_{max} (m.s^{-1})	0	1.84^{-7}	1.84^{-6}	1.84^{-5}	1.84^{-4}	1.84^{-3}
P_e	0	0.4987	2.6279	4.5860	4.9553	4.9955

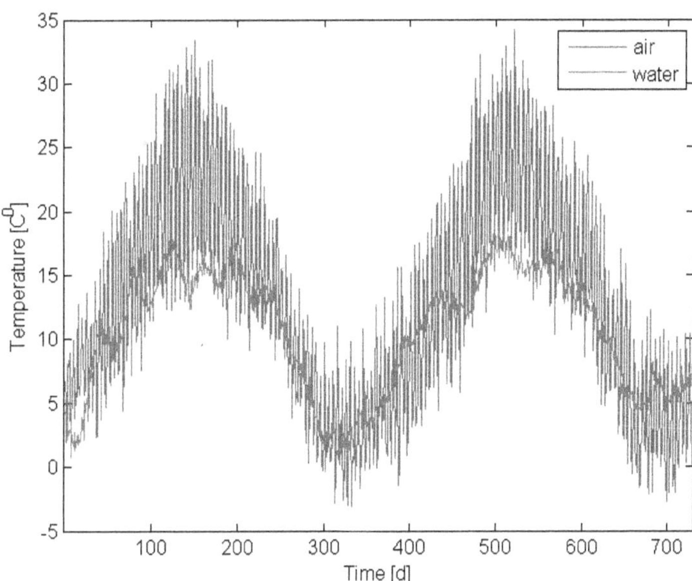

Figure 2: Recording of air and water temperatures

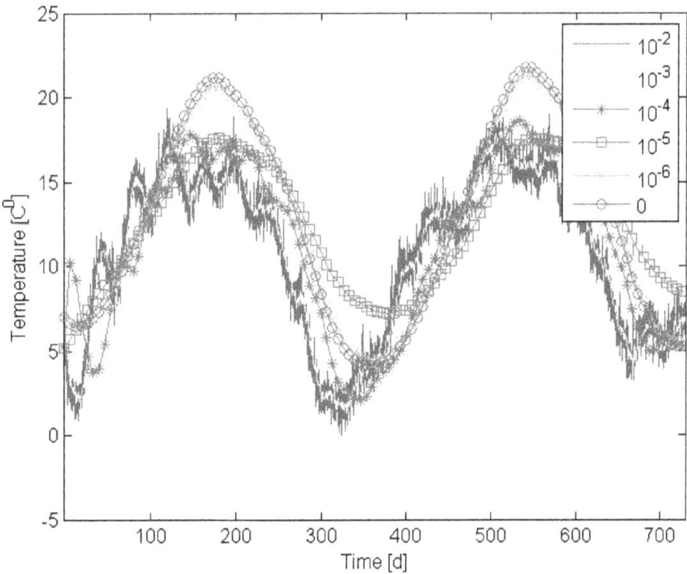

Figure 3 : Sensitivity of the temperature to hydraulic conductivity

4. Inverse problem

The inverse problem considered here consists of an estimation of physical parameters (hydraulic and thermal) of a porous medium by observation of partial data: water, air and sensor-in-the-dike temperatures.

Inverse problem can be resolved in many different ways. Variational methods consist in minimizing the L_2 distance between prediction (provided by the model) and observation (provided by the sensor) during a predefined temporal window, whereas sequential methods, or observing methods, update the estimation with each new available partial observation. For our problem, we choose a sequential method, because a real time estimation is required.

In this section, a series of Kalman's filters of different types [3] is considered. The Kalman filtering is made in two steps. First, prediction provides a deterministic model, and then thanks to the difference between model and observation, a correction can be computed. Implicitly, the Kalman's filter minimizes the cost function:

$$\mathcal{J}(X_n) = \mathrm{E}[(y_n - HX_n)^t R^{-1}(y_n - HX_n)] + \mathrm{E}[X_n^t Q^{-1} X_n], \tag{6}$$

where y_n represents the observation at time t_n, X_n the state of the system, that is looked for, and H is the relation between X_n.

In order to avoid an *inverse crime*, see Kaipio [4] , our algorithm is tested, for a homogenous medium case with data computed with FEFLOW.

4.1. Hypothesis and estimation parameter

Before testing the inversion algorithm, we must define the parameter to be estimated. The velocity field in the dike is looked for, but one can wonder under which form it can be parameterized. Richards equations, Eqn. (2) and (3), provide information about the field, but their implementation requires an *a priori* knowledge of the hydraulic and thermal parameters. The assumption of a stationary flow is made here, because the time required for the development of a leakage is much smaller than the thermal diffusion time.

We observed in section 3 that temperatures of water and at the sensor are similar for high hydraulic conductivities. In that context, the diffusive term can be ignored and the heat equation turns purely advective, see Péclet number values in Table 1. Thus, precise knowledge of $(\rho C)^w$, $(\rho C)^e$, Λ^w, Λ^e, ε, α_T and α_L is not necessary, since these parameters have no big influence on the results.

The parameter with larger effect on the ground temperature is the hydraulic conductivity K_h (K_v/K_h~1/100), so it has been decided to estimate it. This parameter is an intrinsic characteristic of soil and its estimation can only be done in a leakage context. Without a leakage, we wish our algorithm to return the value 0 (a dry ground implies a null velocity field according to Darcy's law). Physically speaking, this is wrong, as hydraulic conductivity is not equal to zero, otherwise, no leakage could be possible. Consequently, we prefer to speak of an *apparent hydraulic conductivity*. In a context of a leakage, hydraulic conductivity will be estimated first, using Kalman's estimation algorithms, and velocities will be computed afterwards, as an output of the model.

4.2. Kalman's filter on a 2D model

4.2.1. Extended Kalman's filter

Principle – Extended Kalman's filter principle is the same as linear Kalman's filter, but the system needs to be linearized in the neighborhood of the current prediction. This filter is theoretically under-optimal and does not warrant convergence and stability.

Results – After fifty days or more, a good estimation of hydraulic conductivity is reached, see Figure 5. If one looks at the early time steps (days), after 3 or 4 days, the system matrix becomes well-conditioned and the algorithm begins to converge. Moreover, constructions of matrices of model noise and measurement noise are quite involved.

4.2.2. SEEK's Filter: Singular Evolutive Extended Kalman's Filter

Principle – Many versions of the filter are available for reducing the computation cost. Pham et al. [6] study a new Kalman's filter introduced by Moireau et al. [5] for data assimilation in oceanography problem, called singular evolutive extended Kalman's filter. The singularity of this filter comes from the "correction direction" evaluated with the dynamical system.

Results – Results are less precise than the ones of the extended Kalman's filter, but adjusting noise matrices is easier and a more stable algorithm is obtained. The convergence needs few hours to be achieved.

4.3. Extended Kalman's filter on a 1D model

It has been shown in section 3 that the thermal behavior becomes mainly advective, as soon as a high enough level for the apparent hydraulic is reached. Moreover anisotropy of the dike makes the horizontal velocity dominant. So, for inversion purposed, a physically reduce model, where the heat equation is written in a one dimensional space frame, will be considered. It brings the following advantages: first the time of the calculus is reduced (about one minute to simulate 2 years of data, for the inverse problem), effect of numeric noise is attenuated, and the numerical scheme is more stable.

The reduced heat equation model is

$$(\varepsilon S(\psi)(\rho C)^e + (1-\varepsilon)(\rho C)^s)\partial_t T = -(\rho C)^e v_x \partial_x T + \partial_x(\Lambda_{xx}\partial_x T). \tag{7}$$

A numerical test, see Figure 4, confirms that the thermal behaviors at the sensor location are the very closed for the 1D and 2D models.

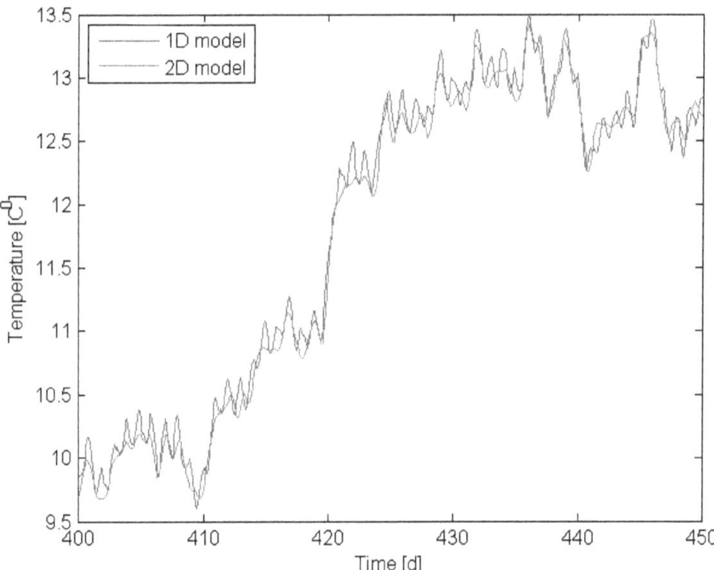

Figure 4: Comparison of temperature outputs of 1D and 2D models.

Results – Results are very promising for the continuation of this study. We have no problem of convergence and the convergence is very fast. We do not test the SEEK's filter because we have only 30 points of spatial discretization.

5. Conclusion

We designed three methods for real time leakage's velocity estimation in a soil dike, starting from transient temperature measurements at one single internal point. Observation of data and their inversion using a pertinent reduced model make the estimation of the hydraulic conductivity possible. This parameter is the dominant parameter in the model that governs water leakage. The three methods, that use estimation algorithms based on Kalman's filter, yield unbiased estimations, but the last one (with the one dimension model), is more performant in terms of robustness, computing time and precision.

If application filed of this method is now considered, estimation of a single parameter starting from recorded values of three temperatures (soil, air and water), is probably not robust enough, since many other parameters or assumptions of our state model are not precisely known (porosity, thermal dispersion coefficients, non-homogeneous character of the dike material, ...). So two different approaches deserve to be considered, in order to get more robustness to the leakage flow rate estimation. One approach would consist in adding a second parameter (together with K_h) in the estimation, such as a parameter present in the definition of Λ_{xx} or in studying the bias on K_h caused by departure of this second parameter from its nominal (non-estimated) value. An alternative approach would be to use an experimental identification of the transfer function of the system, the air temperature being the input and the water and the soil temperatures being the two outputs: this approach would be pertinent for leakage detection but can also provide interesting informations if it is associated to the state approach presently used.

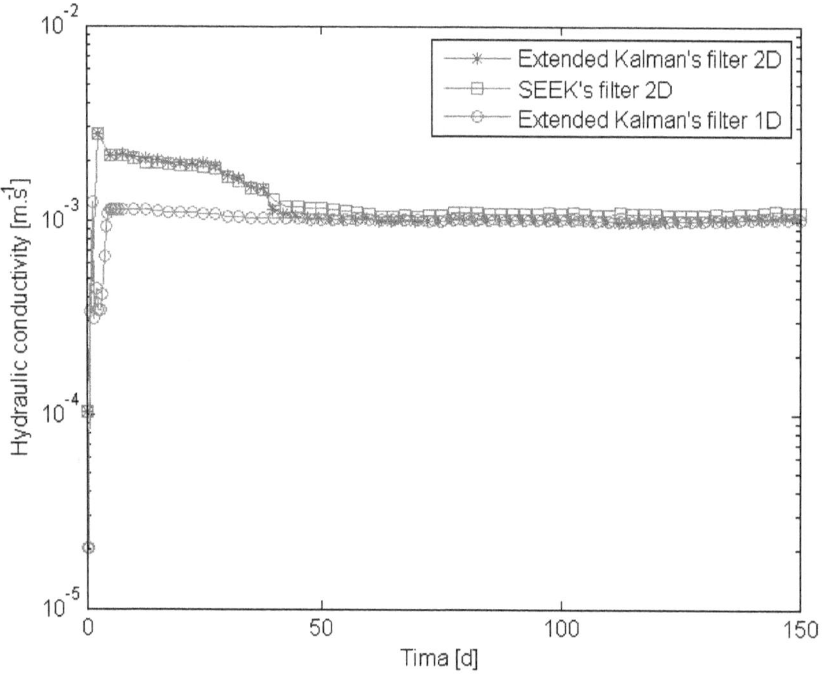

Figure 5 : Estimation of hydraulic conductivity equal to 10^{-3} m.s^{-1}

References

[1] C. Moyne, S. Didierjean, H.P. Amaral Souto, O.T. Da Silveira, *Thermal dispersion in porous media: One-equation model*, Int. J. Heat Mass Transfer 43, 3853–3867, 2000.

[2] J.J.G. Diersch, *FEFLOW 53 Reference Manual*, Wasy GmbH, 2005.

[3] C. Gerkens, C. Ullrich, M. Mateus, G. Heyenan, *Comparaison de techniques de validation dynamique de données*, SISMO 2006, http://www.lassc.ulg.ac.be/bibli/gerkens-2006-1.pdf

[4] J. Kaipio, E.Somersalo, *Statistical inverse problems: Discretization, model reduction and inverse crimes*, Journal of Computational and Applied Mathematics 198, 2, 493-504, 2007.

[5] Ph. Moireau, D. Chapelle & P. Le Tallec, *Joint state an estimation parameter for distributed mechanical systems,* Comput. Methods Appl. Mech. Engrg. 197 (2008) 659-677.

[6] D.T Pham, J. Verron & M.C Roubaud, *A singular evolutive extended Kalman filter for data assimilation in oceanography*, Journal of Marine Systems 16 (1998) 323-340.

Retrieving of the initial moisture field in porous material based on temperature measurements during drying

Z.P. Buliński[1], K. Kasza[2], Ł. Matysiak[2], and A.J. Nowak[1]

[1] Institute of Thermal Technology, Silesian University of Technology
Konarskiego 22, 44-100 Gliwice, Poland
Emails: zbigniew.bulinski@polsl.pl, andrzej.j.nowak@polsl.pl
[2] ABB Research Centre, Starowiślna 13a, 31-038 Kraków, Poland
Emails: krzysztof.kasza@pl.abb.com, lukasz.matysiak @pl.abb.com

Abstract

In this paper some computational aspects of a mathematical model developed for the vacuum paper drying process are discussed. Paper in this model is treated as a porous medium, which is in thermodynamic equilibrium with a humid air. Water distribution within the paper is resolved by solving a so-called lumped diffusion equation. Developed model utilises numerous experimental data and parameters having unknown or at least uncertain values. Their values are found by application of the inverse thermal analysis in which temperature measurements are processed. The objective function contains a sum of squares of the deviations between measured and calculated temperatures at all measurement points and one regularisation term. Minimization of the objective function is carried out using the Levenberg-Marquardt (L-M) method.

Key Words: inverse thermal problems, vacuum drying, sorption curves, diffusion coefficient

1. Introduction

One of the best insulation material commonly used to produce such elements like bushings, transformer windings and many others is paper of special type (crepe paper, pressboard, *etc.*). Unfortunately, its insulating properties depend strongly on the moisture content. It occurs, for instance, that when the moisture content within the paper exceeds the value of 3% the dielectric breakdown strength of the paper weakens drastically. Similarly, the dissipation factor increases rapidly when the moisture content within the paper exceeds the value of 0.5%. In addition, the presence of water in the insulator has negative influence on the manufacturing process of the product. The exemplary phenomena related to this kind of problems are: uncontrolled water evaporation during manufacturing process and presence of the vapour, decrease of the dielectric strength of transformer oil, overheating of the product due to high values of dielectric losses and many others. Hence, it is strongly recommended to decrease the water content in the insulation material below 1% or even lower if it is possible. That is why a special attention should be paid to the proper planning and carrying out the drying stage of the production process.

The vacuum drying process of the bushing starts from the heating up a conducting core (usually aluminium pipe) wrapped with the crepe paper in a vacuum chamber to the elevated temperature. Then air is evacuated and pressure drops inside the chamber causing a rapid water evaporation from the paper. Such a heating-vacuum cycle is repeated couple of times. The knowledge of the moisture distribution within the paper at the beginning of each vacuum stage is absolutely essential for rational controlling of the drying process. Because direct measurement of the moisture field is practically not possible, its determination can be a subject of mathematical modelling and inverse analysis.

Numerical modelling of the industrial drying processes of the paper in most cases treats it as a porous medium, e.g. [1]. Preliminary results of such modelling [2], addressed to drying of the electric bushings, utilised temperature measurements in a fairly classical way. Namely, the analysis consisted of the minimisation of the objective function being a sum of squares of the temperature deviations at all measurement points. In this work the formulation is extended by including to the objective function additional regularisation term accounting for an excursion of one unknown *i.e.* the evaporation constant) from its expected value.

2. Governing equations of the model and experiments

The most popular way of deriving conservation equations of mass, momentum and heat transfer for fluid entrapped in pores and solid matrix entails spatial averaging over volume containing many pores. Comprehensive coverage of this topic can be found for instance in [3]. As it was already mentioned, it is a common practice to treat paper as a porous medium. Properties of such porous media together with fundamentals of convection within the porous media can be found in [4].

The mathematical model describing the considered drying process has been built (reader interested in details should look at reference [2]) utilising a commercial Computational Fluid Dynamic (CFD) software *ANSYS Fluent* [5]. Model equations are solved for the parameters of the humid air. The crepe paper is in thermodynamic equilibrium with a humid air and as a consequence only one energy equation, common for both phases, is solved. Water is treated as a one phase regardless its state, *i.e.* no matter if it is free or bounded water. Its distribution within the paper is resolved by considering so-called 'lumped diffusion equation' containing the effective diffusion coefficient of water within porous material. This coefficient, dependent on the water content and temperature, is calculated based on the sorption/desorption curves. Such curves have been determined experimentally [6].

The water diffusion equation is added to the remaining governing equations using User Defined Scalar (UDS) capability. For modelling of water evaporation and condensation process the modified Hertz-Knudsen equation [6] was adopted.

$$\dot{m}_{H_2O} = -2C_0(1-\gamma)\sqrt{\frac{M_{H_2O}}{2\pi R}}\,\frac{p_{eq}(T)-p_v}{\sqrt{T}} \tag{1}$$

Where $\dot{m}_{H_2O}$ is the water mass flow rate, γ stands for porosity, R represents gas constant and M_{H2O} is the water molar mass. Pressure p_v is the vapour pressure (in Pa) and T is the temperature of the porous material (in K) (temperature of water – dry paper composition), p_{eq} is the vapour equilibrium pressure at the porous material surface (in Pa) and C_0 stands for the evaporation constant (dimensionless). It is a function of the temperature T and the local water content X:

$$p_{eq}(T) = p_{sat}(T)\varphi(X,T) \tag{2}$$

where $p_{sat}(T)$ is the saturation pressure (in Pa) of the water for a given temperature T while $\varphi(X, T)$ is the equilibrium air humidity (dimensionless) at the porous material surface. It is determined by the sorption isotherm and in this work parameters of the modified Henderson model were determined by fitting the measurements [6]:

$$\varphi(X,T) = 1-\exp\left[-(1.3792T+1555.6)X^{1.9126}\right] \tag{3}$$

The process dependent constant C_0 occurring in this equation has a fairly strong influence on the amount of evaporating water and because of this its value together with the initial moisture content has to be estimated through inverse analysis. At the same time this constant will be involved into the regularisation term of the objective function.

Because of the deep vacuum and high humidity of the air inside the vacuum chamber, the thermal radiation is an important heat transfer mode and was also considered in the model. Both phases, the moist air and the paper, were treated as media participating in radiation. Radiation transport equation was integrated employing the Discrete Ordinate method.

To close developed mathematical model of the vacuum drying process but also to validate it a special experimental stand was built and selected experiments were carried out. The most important element of this stand was a vacuum chamber with a model bushing (paper coil) inside. They are shown in Fig. 1. The stand allows one to carry out one heating-vacuum cycle of the process. The heating period lasts about 6 hours after which the vacuum is applied for about half an hour. Temperatures are recorded at the walls of the vacuum chamber as well as inside the paper coil by the set of thermocouples. The location of thermocouples as well as initial temperature field within a half of the coil vertical cross-section are shown in *cf.* Fig. 2.

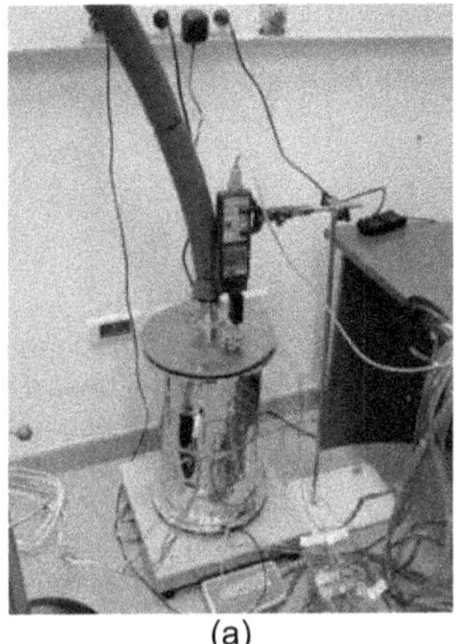

(a)

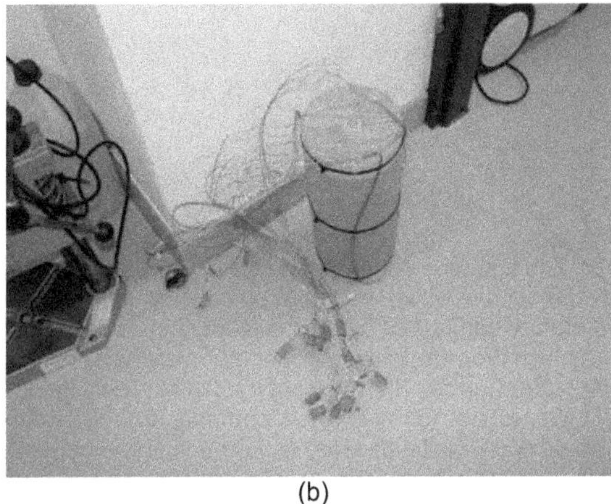

(b)

Figure 1: Parts of the paper vacuum drying experimental stand: (a) vacuum chamber, (b) paper coil with thermocouples.

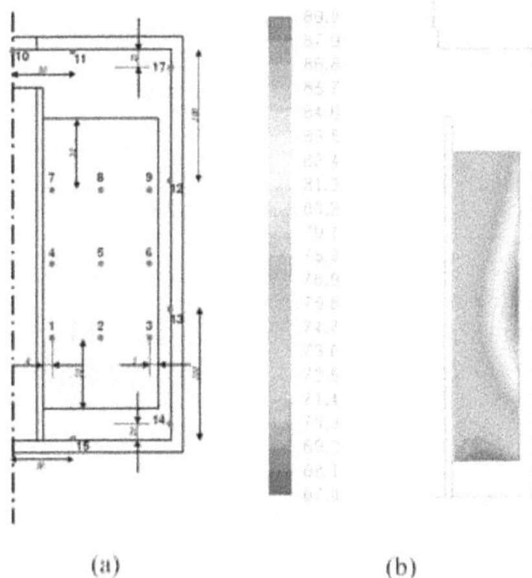

(a) (b)

Figure 2: Thermocouples arrangement within the paper coil (a) and initial temperature field (b). Dimensions are in millimetres and temperatures are in Celsius degrees.

3. Approximation of the initial temperature and moisture fields

The initial temperature field $T_0(\boldsymbol{x})$ at an arbitrary spatial point $\boldsymbol{x}$ was approximated with a linear combination of nine (this is the number of measuring points within a half of the coil vertical cross-section) two-dimensional Multiquadrics Radial Basis Functions (MQ RBF):

$$T_0(\boldsymbol{x}) = \sum_{i=1}^{9} \alpha_i \, \phi\left(\left\| \boldsymbol{x} - \boldsymbol{x}^i \right\| \right) = \sum_{i=1}^{9} \alpha_i \sqrt{\left(x_1 - x_1^i\right)^2 + \left(x_2 - x_2^i\right)^2 + c^2} \tag{4}$$

where α_i is the approximation coefficient, $\boldsymbol{x}^i$ is the RBFs collocation vector and ϕ is multiquadrics radial basis function. RBFs are commonly used approximation functions which arguments are distances from function base point, defined as an Euclidean norm. Quantity x_j is the *j*-th component of the position vector $\boldsymbol{x}$ while x_j^i is the *j*-th component of the *i*-th RBFs collocation point and *c* is the function shape parameter.

Coefficients of the approximation function (4) were found as a solution of the following set of equations

$$T_0(\boldsymbol{x}_j)=\sum_{i=1}^{9}\alpha_i\,\phi\left(\left\|\boldsymbol{x}_j-\boldsymbol{x}_j^i\right\|\right)=T_{0,j}^{*}(\boldsymbol{x}_j)\qquad \text{for } j=1,\ldots,9 \tag{5}$$

$$\frac{\partial}{\partial x_2}T_0(\boldsymbol{x}_j)=\sum_{i=1}^{9}\alpha_i\,\frac{\partial}{\partial x_2}\phi\left(\left\|\boldsymbol{x}_j-\boldsymbol{x}_j^i\right\|\right)=0\qquad \text{for } j=18,\ldots,32 \tag{6}$$

The first nine equations (5) reflect condition that the estimated temperature (4) at the measurement points must be equal to the measured value. Following five equations (6) impose constraints that temperature derivatives within the aluminium rod with respect to the vertical coordinate are negligible, due to very high thermal conductivity of that material. Those equations add extra five collocation points located along aluminium rod (they are not shown explicitly in Fig. 2.).

Set of Eq. (5) and (6) is certainly overdetermined – contains fourteen equations and only nine unknown coefficients. Its least square solution produced resulting temperature field which is shown in the Fig. 2.

The two-dimensional distribution of the water content inside the paper roll was approximated with a second-order polynomial, *i.e.*

$$X_0(x)=\beta_1+\beta_2\,x_1+\beta_3\,x_2+\beta_4\,x_1\,x_2+\beta_5\,x_1^2+\beta_6\,x_2^2 \tag{7}$$

Higher order polynomial can not be used since there were only nine temperature sampling points during the experiment. Six unknown coefficients $\beta_1,\ldots,\beta_6$ are found as the least square solution of nine equations formulated for nine collocation points. The water content at the collocation point were treated as unknowns parameters and underwent estimation with the use of the inverse analysis.

4. Bases of inverse analysis

The measurements collected by thermocouples 1 – 9 are divided into two sets: "teaching data" and "testing data". It is obvious that the longer a sampling time of the "teaching data" set is, the more accurate results of the inverse analysis are. On the other hand, the long sampling time lengthens the computation time. Therefore the length of sampling time needs to be a compromise between accuracy and CPU time. In this work the first 50 seconds of measurements were chosen as a teaching data set. Remaining measurement data (i.e. after starting 50 seconds of the experiment) were used to validate the final model.

As already mentioned the aim of the inverse analysis is to retrieve value of selected quantities – the evaporation constant C_0 and the moisture distribution $\boldsymbol{X}_0$ within the paper coil at the beginning of the sucking out stage of the process. It should also be noted that vector $\boldsymbol{X}_0$ contains initial water content at the collocation points. Hence, all together ten unknowns are searched utilizing temperature measurements at above nine points located inside the paper coil.

This research should be seen as the extension of the work [2] in which the objective function consists of not only a sum of squares of deviations between temperature measured and that predicted by the mathematical model. Here an extra regularization term was added to that function, *i.e.*

$$\Delta=\left[\boldsymbol{T}(\boldsymbol{P})-\tilde{\boldsymbol{T}}\right]^{T}\boldsymbol{W}^{-1}\left[\boldsymbol{T}(\boldsymbol{P})-\tilde{\boldsymbol{T}}\right]+\eta\left(C_0-\tilde{C}_0\right)^2 \tag{7}$$

where $\boldsymbol{T}(\boldsymbol{P})$ and $\tilde{\boldsymbol{T}}(\boldsymbol{P})$ are vectors of predicted and measured temperatures, respectively. The vector $\boldsymbol{P}$ contains the unknown model parameters which are to be computed through the inverse procedure. As it was already mentioned it consists of the evaporation constant C_0 (see Eq. (1)) and the values of initial of water content field at the nine collocation points. Since all sensors used for temperature measurements are of the same type, the covariance matrix can be $\boldsymbol{W}$ neglected.

Coefficient η stands for the regularization parameter while $\tilde{C}_0$ is the expected value of the evaporation constant. In current calculations it was assumed $\tilde{C}_0=0.3$ (based on literature). Value of regularization parameter η varied within relatively wide range and results for η = 10 and η = 100 are presented below.

The complexity of the featured inverse problem arises from the fact that simultaneously model parameters and initial condition are being estimated. To solve this problem the Levenberg-Marquardt (L-M) method [7] is used to update unknown parameters values (collected in the vector $\boldsymbol{P}$) in an iterative loop.

Although L-M method is well known to be very efficient for solving the least square problems, its performance significantly depends on the initial guess. In analyzed problem this initial guess is even more important than usually, because to reduce CPU time, the sensitivity matrix for L-M algorithm will be determined only once (just for guess solution). Hence due to the fact that the resulting algorithm may fail to converge, the initial guess cannot be a completely arbitrary. Proper determination of the initial guess requires some experience.

5. Selected results and discussion

The key results of the inverse analysis are presented in Table 1. It contains values of the evaporation constant C_0 as well as the total moisture content at the beginning of the sucking out stage.

Table 1: Results of inverse analysis depending on the regularization parameter η.

	Regularization parameter η = 10	Regularization parameter η = 100
Evaporation constant C_0	0.029997	0.030010
Total moisture content [kg]	0.0201	0.0197

The initial moisture distribution must fulfil constrain that the total moisture content at the beginning of the sucking out stage has to be lower than its value before experiment. Total moisture content within the bushing before experiment was measured to be equal to 0.0334 kg. It can be seen that both results satisfy that constrain.

The initial moisture distribution obtained taking into account results included in Table 1 are shown in Fig. 3. These results are technically justified and differences between them are practically negligible.

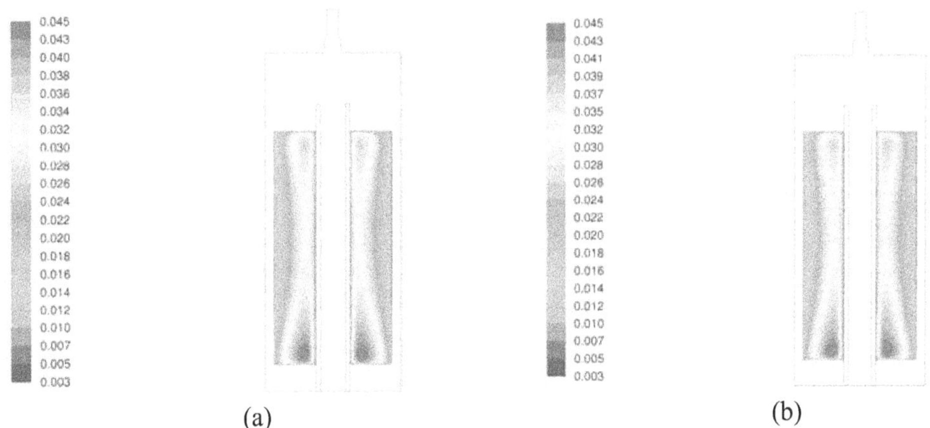

Figure 3: Initial moisture distribution obtained for regularization parameter η = 10 (a), and for regularization parameter η = 100 (b).

Exemplary comparison of temperature histories with the measurements at two selected points is shown in Fig. 4. Results have been obtained for regularization parameter η = 100. Temperature history for point no 2 can be characterized as the best fitting while temperature history for point no 3 seems to be described as the worst fitting. For regularization parameter η = 10 results are very similar.

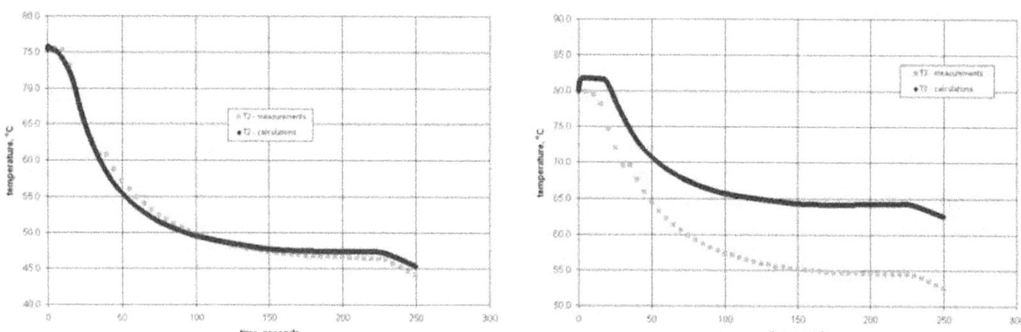

Figure 4: Comparison of temperature histories with the measurements at points no 2 (a) – the best fitting and 3 (b) – the worst fitting.

Since obtained results are very similar it is difficult to compare them directly. They should rather be perceived through the moisture field that they produce in the course of the experiment. This however requires more time consuming simulations.

6. Conclusion

Although all inverse problems are generally thought to be ill-posed, estimation of the unknown or uncertain model parameters as well as some fields through inverse analysis was successful. In this way the evaporation constant and the moisture filed at the beginning of the sucking out period of experiment were computed. All cases carried out up-to-day produced quite similar results independently on regularization parameter η. They also satisfied condition imposed on the total moisture content within the bushing and predicted satisfactorily a course of the process after 50 seconds of measurements. However, it obvious that more computer simulations should bring more knowledge what is really the optimal value of the regularization parameter.

Although comparison of the calculations carried out with estimated values of the evaporation constant and the initial water distribution revealed satisfactory agreement with the testing data set some improvements can still be suggested. First of all, sampling (*i.e.* collocation) points should be positioned closer to the external surface of the bushing domain. Their current locations introduce noticeable errors on the external surface. Secondly, experimental information on the total amount of water at the beginning of the sucking out period incorporated into the objective function could considerably help to stabilize the inverse problem solution. Finally, instead of polynomial approximation of the initial water content the Radial Basis Functions could also be considered. All these possibilities will be subject of further research.

7. References

[1] W. J. Coumans and W. M. A. Kruf. Mechanistic and lump approach of internal transport phenomena during drying of paper sheet. Drying Technology, Vol. **13**, 985--998, 1995.

[2] Nowak A.J., Buliński Z., Kasza K. and Matysiak Ł., Thermal inverse problems in computational modelling of the paper vacuum drying process. Inverse Problems in Engineering and Science, **19** (2011), pp. 59-73.

[3] Bear J., Bachmat Y., Introduction to modeling of transport phenomena in porous media. Kluwer Academic. Dordrecht, 1990.

[4] Nield}Nield D.A., Bejan A., Convection in Porous Media. Springer-Verlag. New York, 1999.

[5] FLUENT 6.3 User's guide, Fluent Inc., Lebanon, USA, 2006.

[6] Nowak A.J., Buliński Z., Kasza K. and Matysiak Ł., Experimntal analysis of vacuum drying of the high voltage bushing. *Proc. of the World Conference on Experimental Heat Transfer, Fluid Flow and Thermodynamics, ExHFT-7*, Cracow, 2009, pp. 1175 - 1182, invited paper.

[7] M. N. Ozisik and H. R. B. Orlande. Inverse heat transfer: fundamentals and application, 1st Edition, Taylor & Francis, 2000.

Mathematical Modeling of Heat Transfer in High-Porous Materials Based on Inverse Problems Results

Oleg M. Alifanov[1], Valery V. Cherepanov[2], Sergey A. Budnik[1] and Alexey V. Nenarokomov[1]

[1]Department of Space Systems Engineering,
[2]Department of Physics,
Moscow Aviation Institute, Moscow, Russia
Email: bold2010@live.ru

Abstract

The convenient and effective complex method intended for theoretical prediction and experimental-computational determining of some properties of modern constructional materials (Fibrous Materials (FM), Reticulated Foamwith Open Porosity (RFOP) *etc.*) is being developed. The theoretical basis of the offered approach is formed by the methods of Direct Mathematical Simulation (DMS) for global structure of the complex irregular systems *with a local regularity*, and by the Inverse Heat Transfer Problem (IHTP) methods. Combining these methods with the thermal non-stationary experiment it is possible to define and to predict the properties of a material. In particular the radiating and contact heat conductivity components, accommodation of energy coefficient, the complex refractive index, scattering indicatrix, scattering and absorption coefficients, etc.

Key Words: high-porous materials, physical properties, definition and prediction, thermal experiment, mathematical model, inverse problems, methodology.

1. Introduction

The general tendency in the development of space engineering is connected with an increase of the number of thermally loaded engineering objects, with the toughening of conditions of their thermal loading by a simultaneous increase of reliability and safe life and a reduction of consumption of materials. For space vehicles and reusable transportation systems the support of thermal conditions is one of the most important aspects of design, which determine the main design solutions. The distinctive features of modern heat-loaded structures in space engineering are non-stationarity, non-linearity, multidimensionality and conjugate nature of heat-and-mass transfer processes. These distinctions confine a possibility of using many traditional design-and-theory and experiment methods for the materials and structures development. The main direction of new system development is the creation of advanced thermal protective and insulating materials. The development of such materials and solving the mentioned above problems of thermal design of systems requires wide use of mathematical and physical simulation methods. But mathematical simulation is impossible without reliable information about the characteristics (properties) of objects under consideration. In practice, the direct measurement of thermal and radiation properties of materials, especially of complex composition, is impossible. There is only one way, which permits to overcome these complexities: the combination of DMS for global structure of the complex irregular systems, which have the property of local regularity, and indirect measurements, based on IHTP, when the required properties of a system are analyzing through direct measurements of another values (temperature, component concentration, etc.) [1]. A diagram of approach stated above, is presented in Figure 1.

Methods of IHTP were developed to increase the information value of thermal experiments and tests, to improve the accuracy and reliability of experimental data processing and interpretation, and also to estimate the unknown parameters of the thermal systems. In the majority of cases, this methodology is used as optimal, but in a number of practical situations, it is the sole technique available, as, for example, in measuring the transient heat fluxes and determining thermal propertiesof materials witha memory

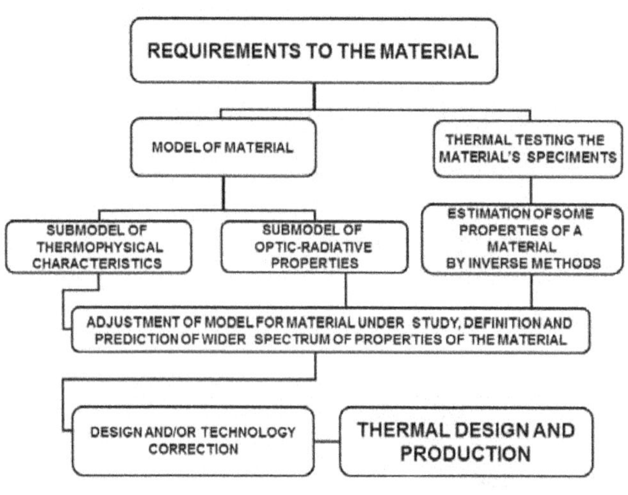

Figure 1: Analysis and prediction of material's properties.

of heating/cooling conditions. Violation of cause-and-effect relations in the statement of these problems results in their correctness in mathematical sense (i.e., the absence of existence and/or uniqueness and/or stability of the solution). Hence to solve such problems we develop special methods usually called regularized. This methodology is based on the mathematical theory of ill-posed problems of mathematical physics. So, a successful implementation of IHTP methodology is possible only by a rational combination of physical simulation of heat transfer processes in the specimen, exact measurements of initial thermal parameters and correct mathematical processing of experimental data based on the solution of the IHTP. The results of IHTP solving can be used for the creation of the detailed model of heat transfer, which can be used for the thermal design of real space structure or design of advanced thermal protective and insulating materials.

Light highly-porous heat-shielding materials, such as FM, RFOP etc., are of very interest to many aerospace applications. Usually, the study of their properties is conducted through suitablephysical experiments with using the material's specimens. Similar an approach is possible only after actual creation of materials. In doing so is the potentiality to determine and predict of many properties of a material under study is critically limited or even eliminated without invoking means of mathematical simulation. Such parameters include the contact and radiation components of overall thermal conductivity, effective energy accommodation coefficients, complex refractive index, radiation coefficients of scattering and absorption, scattering indicatrix, and so on, because the values of these parameters are defined by the processes of particularly local or spectral nature.

The materials studied consist of randomly oriented fibers or ligaments-nodal point's skeleton (Figure 2). Its elements may be made of one or several materials. The space between the fibers is filled with gas. The global irregularity, inhomogeneity, and anisotropy of high-porosity materials significantly narrow down the spectrum of approaches to mathematical simulation of their structure and properties. At the same time, such systems are locally regular [2].

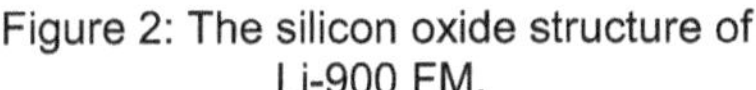

Figure 2: The silicon oxide structure of Li-900 FM.

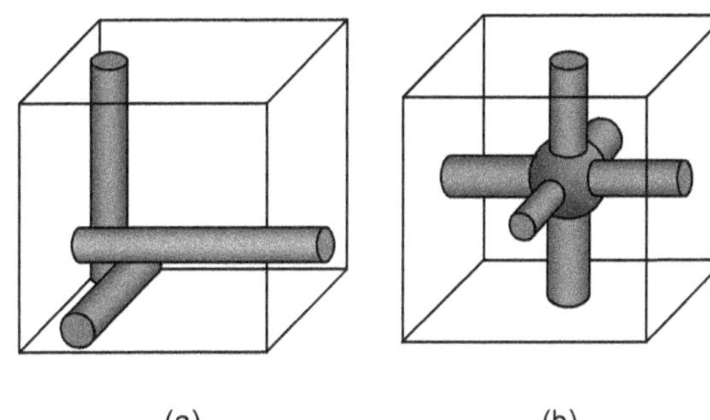

(a) (b)

Figure3: Orthogonal representative elements for FM (a) and RFOP (b).

This enables one to identify regular representative elements in a system, describe their properties and, on this basis, perform calculations for both individual aspects of the problem and the properties of material as a whole. Note that it is in simulating a material as a whole that one can form and calculate the entire complex of its determining properties. Note further that the approach to the simulation of high-porosity materials such as VALOX, KR, TZM, ETTI, CFOAM etc. including the effects of microstructure by using a regular arbitrary inhomogeneous *orthogonal*fibrous and spheres model system [2,3] still remains without alternative [4,5]. A system of random successively generated elementary volumes is considered, a typical representatives of which are given in Figure 3a, 3b.

Unfortunately, the volume of theses does not allow stopping in more details on this question. The detailed description of mathematical model of a material can be found in works [2,3].More schematically this material is stated in [6,7].In summary we will notice, that at construction of mathematical model we started with following requirements:

- The stochastic system of illuminated in a random wayorthogonal representative elements is considered.
- The account of real structural properties of a material and forming substances.
- The account of anisotropy of a material and requirements for effective density.
- Isothermality and adiabaticity within one representative element are supposed.
- Each new representative element to consider shipped in substance, which properties aredefining by all earlier generated elements.
- Neglect heat convective flux, percolation, globules and foreign inclusionsin first level of the model.
- The description of radiation transfer over in scattering system at level of theMie theory, taking into account a cooperative effects.
- Conditions of illumination of each of representative volumes should be co-ordinated with properties of radiation in material as a whole.

It is the objective of the further study to describe to use this model for determining the most important thermal and physical properties of high-porosity materials based on results of the corresponding IHTP solving.

2. The material's properties estimation

The mathematical model of heat transfer process in the material's specimen (infinite plate of known thickness) can be presented as

$$C(T)\frac{\partial T}{\partial \tau}=\frac{\partial}{\partial x}\left(\lambda(T)\frac{\partial T}{\partial x}\right),\ x\in(X_0,\ X_1),\ \tau\in(\tau_{\min},\ \tau_{\max}]\ (1)$$

$$-\lambda(T)\frac{\partial T(X_m,\ \tau)}{\partial x}=q_m(\tau),\ m=\overline{0,1},\ \tau\in(\tau_{\min},\ \tau_{\max}],\ T(x,\ \tau_{\min})=T_0(x),\ x\in[X_0,X_1]\ (2)$$

In models (1) – (2) the coefficients $C(T)$ and $\lambda(T)$ are unknown. The complimentary information needed for solving the inverse problem prescribed are the results of the temperature measurements

$$T^{\exp}(X_m,\ \tau) = f_m(\tau),\ m = \overline{0,1} \quad (3)$$

Let us introduce in the interval $[T_{min}, T_{max}]$ two uniform difference grids with the number of nodes N_i, i=1,2, namely

$$\omega_i = \left\{T_k = T_{\min} + (k-1)\Delta T,\ k = \overline{1,N_i}\right\},\quad i = \overline{1,2} \qquad (4)$$

We now approximate the unknown functions on grids (6) using cubic B-splines as follows

$$C(T) = \sum_{k=1}^{N_1} C_k \varphi_k^1(T),\ \lambda(T) = \sum_{k=1}^{N_2} \lambda_k \varphi_k^2(T), \qquad (5)$$

where $C_k, k=1\text{-}N_1$, λ_k, $k=1 \div N_2$, , are parameters. As a result of such approximation the functional inverse

problem is replaced it is essential more simple parametrical in which it is required to define a vector of parameters $\mathbf{p} = \left\{p_k\right\}_{k=1}^{N_p}$, $N_p = N_1 + N_2$.

The detailed description of a method of the decision of a problem (1) - (5) can be found in [8].

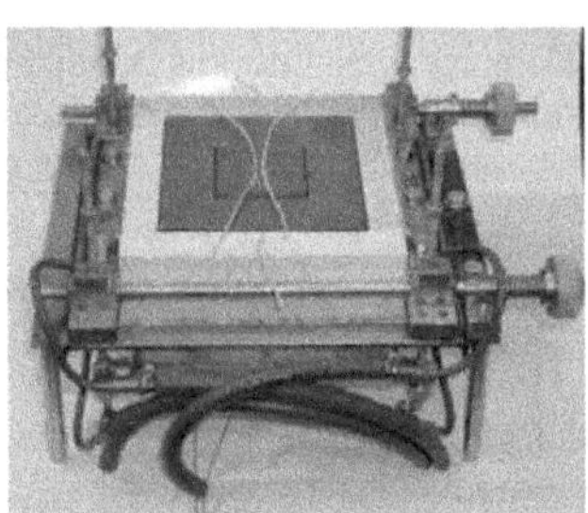

Figure 4: The experiment module with installed specimens and removed upper plate of the heat-insulating holder.

A picture of the experimental module with installed specimens and removed upper plate of the heat-insulating holder is shown in Figure 4. The assembled experiment module was installed in the vacuum chamber of stand TVS-1M.

The result of estimating the functions $\lambda(T)$ and $C(T)$ for material ETTI-CF-ULT and ETTI-CF-ERG are presented in Figures 5 and 6.

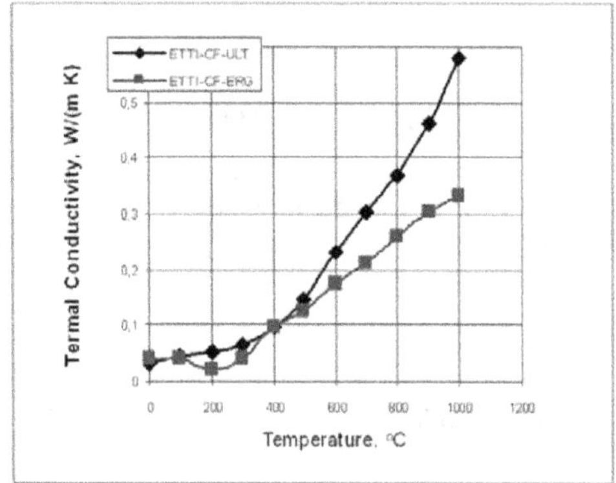

Figure 5: Estimated value of the thermal conductivity for material ETTI-CF-ERG and ETTI-CF-ULT.

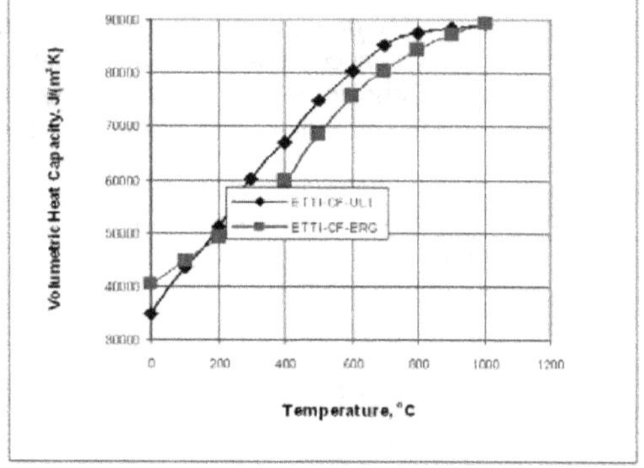

Figure 6: Estimated value of the volumetric heat capacity for material ETTI-CF-ERG and ETTI-CF-ULT.

3. Modeling results

The model [2,3,6,7] was used for developing computer codes which enable one to adequately describe the materials of the types under consideration. Some results for FM TZMK-10 are presented in

Figures 7-8. This material is executed from fibers of high-purity fused quartz with $n \approx 1.45$ and $\rho_f = 2180$ kg/m^3. The samples of the material at our disposal had the effective density of 144 kg/m^3. In the calculations we used also data on physical properties of the quartz and air, available in [8,9].

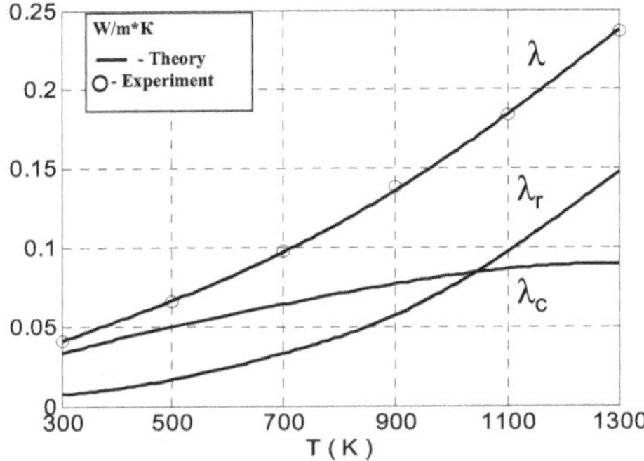

Figure7: Calculation and experimental results for thermal conductivity ofTZMK-10.

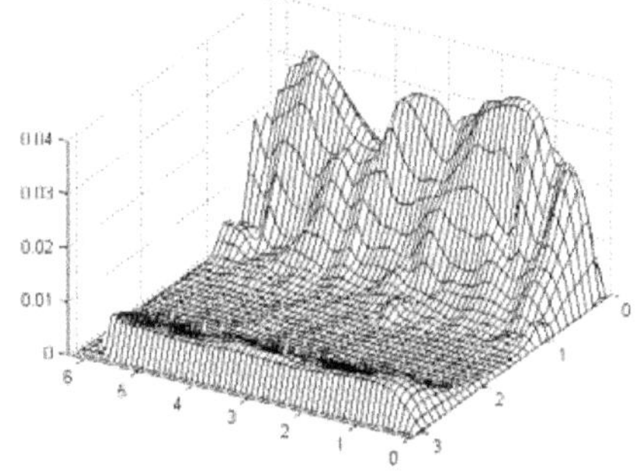

Figure 8: Spectral scattering indicatrix of the Mie for TZMK-10 and wavelength λ_w=3.39 μm.

Figure 7 gives the calculated curves of temperature dependences for both the overall thermal conductivity $\lambda$and for its radiation λ_r and conduction λ_c components at normal pressure. Figure 8 gives the spectral scattering indicatrix of the material for wavelength λ_w=3.39 μm and T=573K, as a function of polar angle θ and azimuth φ. The polar axis was orientedin a direction of an external heat flux. The result was obtained for statistics of 30000 representative elements.

Similar models and results are available and for RFOP materials. So, some results of prediction of the overall thermal conductivity of *ETTI-CF* - type materials as a function of density are shown in Figure 9. The results of corresponded inverse problem solution (part 1) were used for this forecast.

Spectra of absorption and scattering for one of representative elements of ETTI-CF-ULT material are shown in Figure 10.

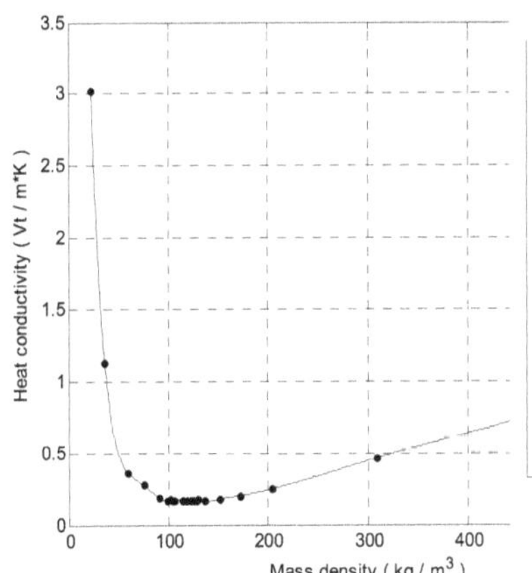

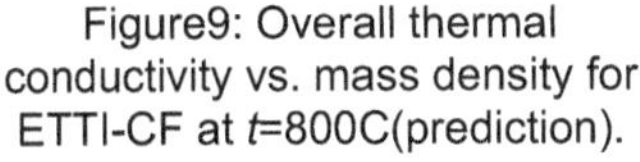

Figure9: Overall thermal conductivity vs. mass density for ETTI-CF at t=800C(prediction).

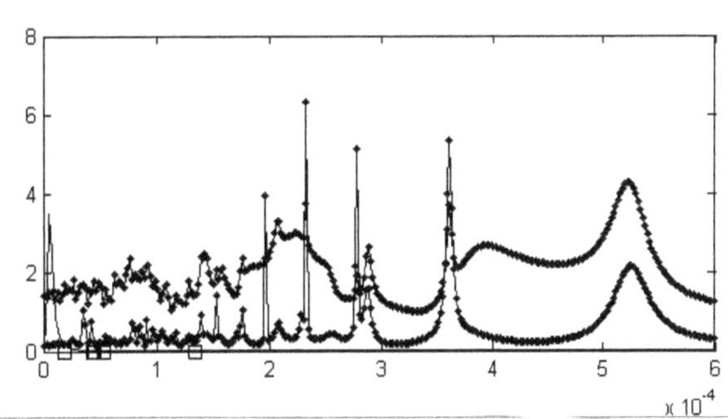

Fugure 10: Absorption α (more low) and scattering $\beta 10^{-3}$ (above) for a representative element of ETTI-CF-ULT as afunction of wavelength Λ_w (m). Diameters of knot andstruts are noted by small squares on a horizontal axis.

4. Conclusion

The software complex and experimental means, allowing with the high accuracy to determine and predict thermal and optical properties of the advanced structural and thermal engineering materials is developed. The deviations of the calculated temperatures (using thermal property estimations) from the temperatures measured in the experiments are insignificant showing sufficient accuracy in the estimations of thermal properties of the analyzed materials.

5. References

[1] Alifanov O.M., "Mathematical and experimental simulation in aerospace system verification", Acta Astronautica, vol. 41, p.43,1997.

[2] Alifanov O.M. and Cherepanov V.V. "Mathematical Simulation of High-Porosity Fibrous Materials and Determination of Their Physical Properties", High Temperature, vol. 47, № 3, pp. 438-447,2009.

[3] Alifanov O.M.,Budnik S.A.,Cherepanov V.V.,Nenarokomov A.V. "Experimentally-Theoretical Research of Heat Transfer in High-Porous Materials". Thermal Processes in Engineering, Vol.3, №2,pp.53-65, 2011 (in Russian).

[4] Zhao C.Y., Lu T.J., Hodson H.P. "Thermal radiation in ultralight metal foams with open cells",Int. J. Heat and Mass Transfer, Vol.47, pp.2927-2939,2004.

[5] LoretzM., CoquardR., BaillisD., MaireE. "Metallic foams:Radiative properties/comparison between different models", J.of Quant.Spect.& Radiative Transfer, 109, pp.16-27, 2008.

[6] Alifanov O.M.,Budnik S.A.,Cherepanov V.V.,Nenarokomov A.V. "High Porous Aerospace Materials: Identification and Prediction of Phisical Properties",*IAC Proceeding ISSN 1995-6258, 60th International Astronautical Congress*, October 2009, Daejeon, Republic of Korea.

[7] Alifanov O.M., Budnik S.A., Cherepanov V.V., Nenarokomov A.V. "Mathematical Model of Heat Transferin Highly Porous Materials", *ASME paper IHTC14-22792, Proceedings of 14th International Heat Transfer Conference*,August 8-13, 2010,Washington DC, USA.

[8] Alifanov O.M., Artyukhin E.A. and Rumyantsev S.V., Extreme Methods for Solving Ill-Posed Problems with Applications to Inverse Heat Transfer Problems, Begell House, New York, 1995.

[9] Vargaftik N.B. Tables of the Thermophysical Properties of Liquids and Gases, New York, Halsted Press, 1975.

[10] Banner D., Klarsfeld S., Langlais C. "Temperature Dependence of the Optical Characteristics of Semitransparent Porous Media", High Temp.-High Pressures, Vol. 21,pp.347-354, 1989.

ANALYSIS OF THE STRUCTURAL INTEGRITY OF SPECTRUM THROUGH INVERSE RESPONSE RESIDUAL FUNCTIONS

Marcus Alexandre Noronha de Brito[1,2], *Luciano Mendes Bezerra*[1], *Ramon Saleno Yure Costa Silva*[1]

[1]*Graduate Program and Structures and Construction*
University of Brasília- UnB,Brasília, Federal District, +55 61 3321-1000
Email: marcusanb@yahoo.com.br
lmbz@unb.br
ramon@unb.br;

[2]*Federal Institute of Education, Science and Technology of Brasília – IFB*
Email: marcus.brito@ifb.edu.br

Abstract

The development of numerical algorithms with computational inverse formulation in this paper allows the identification and quantification of internal damage to structures of steel trusses. The application of appropriate residue function enables the capture of the variation in static and dynamic responses of the structure intact and the same with the prediction of damage, being capable of generating spectra that bring structural information such as the magnitude of this waste and damage evolution the structural element. With this analysis is fully possible to determine the damage that is found inside the structure, indicating their position and their size. This paper presents a numerical and mathematical principles of this algorithm, and the residue function used in analysis of the examples.

Keywords: Inverse problems, structural integrity, residual functions.

1. Introduction

The program using the finite element method (FEM), developed for plane trusses and named TRUSS, is given below. The program was developed in FORTRAN 90, capable of analyzing planar lattice structures providing static and dynamic information, such as axial forces, support reactions, nodal displacements, vibration modes and frequencies of the structure.

TRUSS is divided into subroutines as shown in the flowchart in Figure 1. Contains: (a) sub-routine data entry, (b) assembling the stiffness matrix and mass, (c) calculation of eigenvalues and eigenvectors, (d) application of boundary conditions of the structure, (e) calculation of displacements Nodal, (f) axial forces (g) sub-routine application of structural signatures, with their functions as waste to be defined subsequently in this work and (h) a sub output, output, or print data .

In data entry are also included the nodal coordinates, the connectivity between the elements and nodes with their loading and shipments. With these data can form the geometry of the structure to be analyzed and the corresponding finite element mesh.

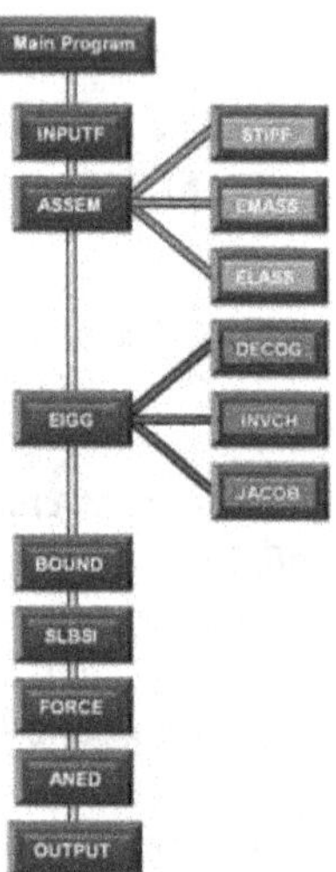

Figure 1: Flowchart of the program subroutines TRUSS

In the subroutine ASSEM, subdivided into three other sub-routines, STIFF, putty and ELASS, as shown in Figure 1, we make the assembly of the stiffness matrix and mass of the truss in the global system set for the structure.

The Subroutine STIFF initiates the creation of the stiffness matrix via the subroutine call ASSEM. This reads the data provided in the previous stanza INPUTF such as connectivity, area of cross sections of each bar, modulus of elasticity and length of each element. Therefore, they are calculated with these data, the cosine directors to mount stiffness matrix of each bar element of the structure; EMASS rides the mass matrix elements of the plane truss structure and how it eliminates the nodal rotational inertia, the mass matrix; ELASS assembles the stiffness matrix and global mass of the structure analyzed.

The program TRUSS, the static calculation, the system uses to solve the equations of Gauss Elimination Method, SLBSI subroutine, modified for solving systems with symmetric and with a band.

The subroutine Eigg, subdivided into three other subroutines: DECOG, JACOB and INVCH calculates the natural frequencies and modes of vibration of the lattice. The subroutines DECOG, JACOB and INVCH are described later. It begins the process of extracting the eigenvalues and eigenvalues from the equation of free vibration of the structure described by the displacement vector, $X = \Phi\, cos\left(\omega t\right)$

$$KX + M\ddot{X} = 0 \;\therefore\; K\Phi = \omega^2 M\Phi = 0 \qquad (1)$$

Both the mass matrix and the stiffness matrix is symmetric. A solution of symmetric systems can be the Choleski method, sometimes also called Banachiewicz Method, Brebbia [1]. This method makes use of the fact that a symmetric matrix can be written by the product of two triangular matrices.

2. Search Strategy for Damage

Define a function residue on a proper metric space, Kolmogorov and Fomin [2].

The function $F_{res}^{(j)}(z)$ compares the signature structure chosen for the assessment of damages with one of the structural features of signature $F_j(z_i)$ between the situation for actual loss $F_j(z_r)$ and damage is predicted $F_j(z_i)$. The vector $\{z_i\} = \{b_i, \alpha\}$ indicates a bar with a fraction b_i of predicted and the cross section α, and $z_r = \{b_r, \alpha_r\}$ being b_r the location bar for actual loss and α_r the fraction of the cross section that quantifies the actual damage to the bar b_r. The function $F_{res}^{(j)}(z_i)$ can be mathematically defined as:

$$F_{res}^{(j)}(z_i) = \left\{ \left[F_j(z_r) - F_j(z_i) \right]^q \right\}^{1/q} \quad (2)$$

At Eqn. (2) the constant q, second Gersztenkorn and Scales [3] defines a metric that is often used as a Euclidean norm (i.e. $q = 2$) very commonly adopted in minimizing regression problems. Note that the value $q = 2$ is appropriate for a hypothetical Gaussian distribution of errors in reference $F_j(z_r)$ data, Bezerra [4,5].

The location of the damage corresponds to the minimum of Eqn. (2), ie:

$$Min\left[F_{res}^{(j)}(z_i) \right] \rightarrow 0 \quad (3)$$

The closer to zero the value of Eqn. (3), the prognosis $\{z_i\} = \{b_i, \alpha\}$ will be closer to the actual measured values $z_r = \{b_r, \alpha_r\}$, of the actual damage. The farther from zero the value of Eqn (3), the worse the prognosis, ie, the reality $\{z_i\} = \{b_i, \alpha\}$ is far from $z_r = \{b_r, \alpha_r\}$.

The location of the damage by the program is done using the algorithm of minimization of residual function, Eqn. (3), making it possible combinations $\{z_i\} = \{b_i, \alpha\}$ with $i \in N^*$ and $\alpha \in (0, 1)$ defined in the subroutine ANED in program TRUSS.

Figure 2 below shows a schematic representation of the use of Eqn (3) in the combination between the magnitudes of the residual functions used, in order to locate the bar in the structure damaged by the convergence between the accepted values and the values read.

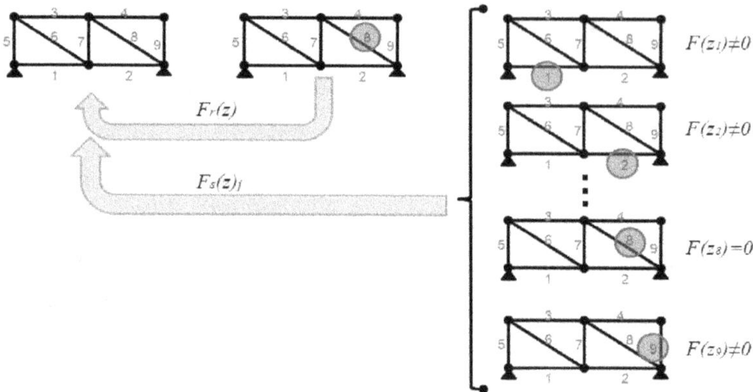

Figure 2: Schematic representation of the convergence between the values read F_r *(z)* and simulated values or accepted F_s *(z)*.

3. Validation

The structure used, Figure 3 for the validation of the program TRUSS, it is a triangular structure with three bars, with 5000 lb load applied horizontally arranged on node 2. Submits to the two lattice damage. The first real damage is located at 3 bar and corresponds to 40% of the cross section, ie the vector $\{z_1\} = \{b_1 = 3, \alpha_1 = 0,40\}$. The second damage is admitted to the bar 2 with 20% of quantification, the vector $\{z_2\} = \{b_2 = 2, \alpha_2 = 0,20\}$.

So, can easily observe the program's efficiency in minimizing the role TRUSS residue composed by signing $F_5(z) = \frac{\sum_{j=1}^{N} \Delta u_j}{\Delta \omega_j^2}$ used, Paz [6].

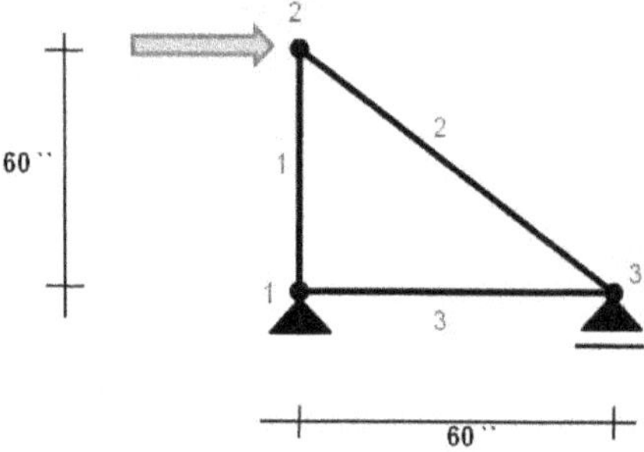

Figure 3: Validation structure - configuration of elements and loading.

The residue function in this case can be seen in Figure 4 represents the behavior of individual lattice bars when in different situations of damage, the function minimum residual is at 3 bar when the curve touches the horizontal axis, with exactly 40% reduction in area, thus signifying the location and quantification damage proposed.

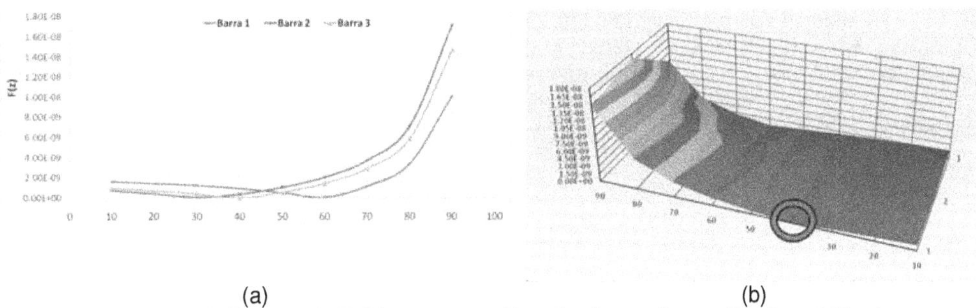

(a) (b)
Figure 4: Analysis of 2D (a) and 3D (b) spectrum of localization and quantification of frist damage validation.

From observing the graphs of Figure 4 can be deduced that the residue function seems to generate a smooth surface. The position of the minimum is not easily detected given that in both Figure 4, the minimum is at a nearly flat region of the 3D surface generated by the residue function.

Described as the second corresponds to the damage actual damage cross section reduction of 20% in area and located on barr 2, shown in Figure 5.

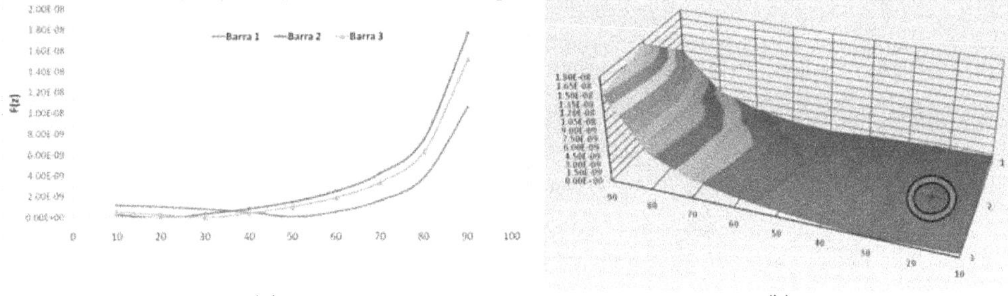

(a) (b)
Figure 5: Analysis of 2D (a) and 3D (b) spectrum of localization and quantification of second damage validation.

4. Conclusions

Examination of these two damage in this small structure can be concluded that damage to small, less than 50% of the cross section, the residue function is not very visible, however, for damage with large percentage reduction in cross-section of the bar's visibility residue function is high. However, in both cases the combinatorial minimization identifies the damage it has been able to find the function of zero waste.

Determine damage in truss structures appears to be an easy task at first, because the bars of the lattice, roughly speaking, are under the effect of axial forces only. So, a defect or multiple defects in a bar under axial load bearing capacity of this diminishes the bar, regardless of position along it. So finding fault lattice boils down to identifying the number of the bar $i \in N^*$ damaged b_i and the fraction $\alpha \in (0,1)$ damage cross section which decreases the load bearing capacity of the bar - in short $b_i = b_r$, and $\alpha \approx \alpha_r$ the index r is indicative of the actual damage. The major problem encountered in the determination of lattice damage is the difficulty of obtaining sensitive and structural signatures that generate residual functions well behaved.

The method of structural damage identification through the use of solutions for inverse problems was very efficient with respect to the reduced processing time of the structure, the small cost incurred in this type of operation when large amount of information extracted by the structure numerical algorithm developed computational and accuracy of results, ensuring the reliability of the answers.

5. References

[1] Brebbia, C. A.Ferrante A. J. "Computational Methods for the Solution of Engineering Problems" 3rd Ed. 1986.

[2] Kolmogorov A. N ., and S. V. Fomin. Introdution to real analysis. New York: Dover.1970.

[3] Scales J. A., and A. Gersztenkorn. (1988). Robust methods in inverse problems. Inverse Problems. 4: 1071-1091.

[4] Bezerra L. M. "Inverse Elastostatics Solutions with Boundary Elements", Ph.D. thesis, Carnegie Mellon University, Pittsburgh, PA, 1993.

[5] Bezerra, L. M.; Saigal, S. "A Boundary Element Formulation for the Inverse Elastostatics Problem (IESP) of Flaw Detection". International Journal for Numerical Methods in Engineering, USA, v. 36, n. 13, p. 2189-2202, 1993

[6] Paz, Mário, Leigh, W. E. "Structural Dynamics: theory and computation". Boston: Kluwer Academic Publishers,2004.

DETECTING BEAM DAMAGES WITH STATIC AND MODAL ANALYSES USING WAVELET TRANSFORMS

Ramon S. Y. C. Silva[1], Luciano M. Bezerra[1], Marcus A. N. Brito[1]

[1] Department of Civil Engineering, University of Brasilia,70910-900- Brasília-DF, Brazil.

Emails: ramon@unb.br,lmbz@unb.br

Abstract.

There are several techniques of non-destructive damage detection in structures. However, these techniques are expensive and require an accurate analysis of large extension of the structure. The numerical techniques can help in non-destructive tests of structures, showing the possible location of damage and thus decreasing the area of analysis and becoming the non-destructive tests less expensive. Among the numerical methods most used to detect damage stand out the finite element method and the boundary element method. This paper presents theoretical applications of the Wavelet Transform for damage detection in a cantilever beam subjected to static and modal analyses. The effects of errors noises in the response signals are also investigated. The modeling of damage is done in finite elements using ANSYS program and the damage simulated deleting some finite elements from the mesh. The static response (in displacement) of the structure with simulated damage is used in the analysis to detect the location of damage using the Wavelet Transform. Two types of mother-wavelet, the Daubechies and Biorthogonal are used. The results of analyses are presented and discussed in this paper.

Key Words: wavelet, damage, displacements, mode shapes.

1. Introduction

In recent years, there is a large interest in the scientific community in the researches associated on damage detection in structures using numerical methods with the goal to help in the non-destructive techniques applied to performance monitoring, pathology and structural health of civil structures [6].

Generally, the numerical methods make the comparison of signatures obtained before and after the onset of damage, and these signatures are defined in terms of displacements, frequencies, mode shapes, stress, among others. It should be important that indication of damage position could be showed without using the comparison of signatures but just based on the response of damaged of the actual structure.

Although the literature on damage detection has been so far dominated by studies based on methods that utilize frequency or stiffness variation information, methods based on wavelet transform, a recently developed mathematical theory rooted in signal analysis [1,2,4], are emerging.

These methods are based on discrete measurements in some points of structure. Such measurements could be subjected to equipment errors or human errors and generally follow a normal distribution.

2. Wavelets Theory

Considering a signal of interest in the time or space domain and ψ(t) the values of wavelet function in the time and frequency domains. The wavelets are generated from the mother wavelet ψ(t) by translation and dilation, as follow below:

$$\psi_{a,b}(t) = \frac{1}{\sqrt{a}}\psi\left(\frac{t-b}{a}\right) \tag{1}$$

Where a and b are integer numbers which represents the dilation and translation parameters respectively. The wavelet transform of a signal f(t) is defined by:

$$C_{a,b} = C_{a,b}(t_o) = \int_{-\infty}^{\infty} f(t)\Psi_{a,b}(t_o)\, dt \tag{2}$$

The results of this transformation are called wavelet coefficients and show how well the function correlates with the signal. These wavelet coefficients are very sensitive to discontinuities and singularities present in the analyzed signal. Considering this property, it was found that damage due to a sudden loss of stiffness can be detected through mode shapes with wavelet coefficients which achieve large amplitudes like a spike or an impulse in the damage location. This perturbation of wavelet coefficients due to this damage is clearer in the finest scales of the wavelet transform. This procedure is the basis of the wavelet transform damage detection [3].

During computation, the analysing wavelet is shifted smoothly over the full domain of the analysing function. Then the wavelet is scaled and the above procedure is repeated [5].

3. Numerical Study

This section presents the finite element model of cantilever beams cracked used in this research subjected to static and modal analyses and also errors noises in the response signals. The beams were modeled using the element PLANE42 of ANSYS11.0 program.

The element PLANE42 has four nodes and three degrees of freedom by node: translations in *x*,*y* and *z* directions. The signals of static response obtained in ANSYS were analysed in MATLAB program to compute the wavelet coefficients.

3.1 Static analysis

The cantilever beam was submitted to a load F=500kN in the free end and to a transverse crack a' = 0,025m positioned at d=0,125m from the clamped end, see Figure 1. The material and geometric properties of the beam analyzed are shown in Table1.

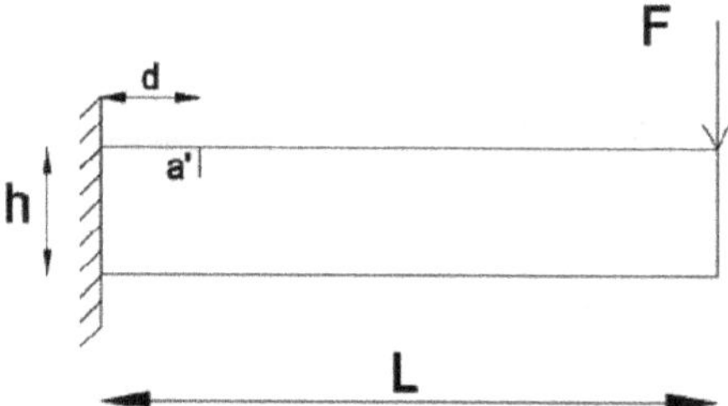

Figure 1: Schematic model of cantilever beam.

Table 1: Geometric and materials properties of cantilever beam

Properties	Symbol	Value	Unity
Base da viga	B	0,10	m
Beam height	H	0,10	m
Area	S	0,01	m^2
Moment of inertia	I	$8{,}333 \times 10^{-6}$	m^4
Beam lenght	L	0,50	m
Modulus of elasticity	E	200,00	GPa
Density	ρ	7850,00	Kg/m^3
Poisson coefficient	ν	0,30	-

The finite element model of cantilever beam was discretized in 4000 elements and 4509 nodes, the crack was simulated deleting some elements from the mesh and the boundary conditions were applied in the all nodes of left end restricting the degrees of freedom *x*, *y* and *z*, see Figure 2.

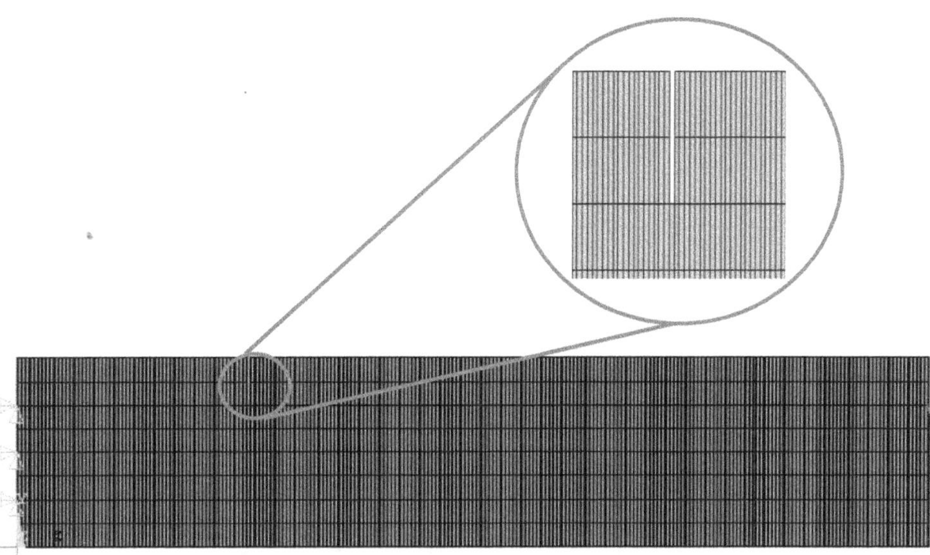

Figure 2: Finite element model of cantilever beam.

The nodal displacements of beam with crack and without crack are presented in the Figure 3.

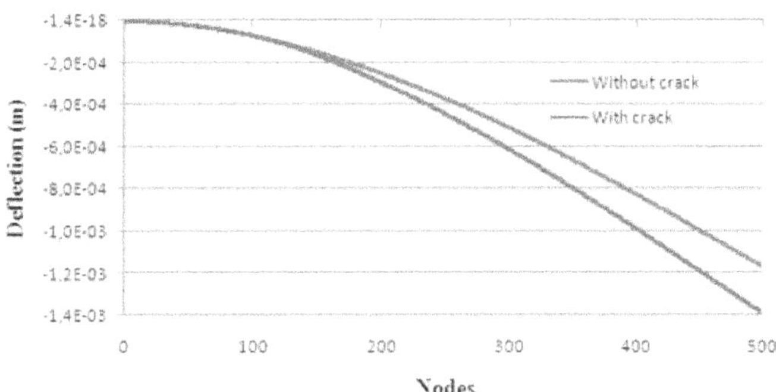

Figure 3: Deflection of cantilever beam.

The wavelet transform was applied in the signal of displacements obtained in the nodes of bottom line of beam using the wavetoolbox of MATLAB to compute the wavelet coefficients for two differents mother-wavelet: biorthogonal6.8(bior6.8) and daubechies2(db2). The Figures 4 and 5 shows the results of these wavelet transforms.

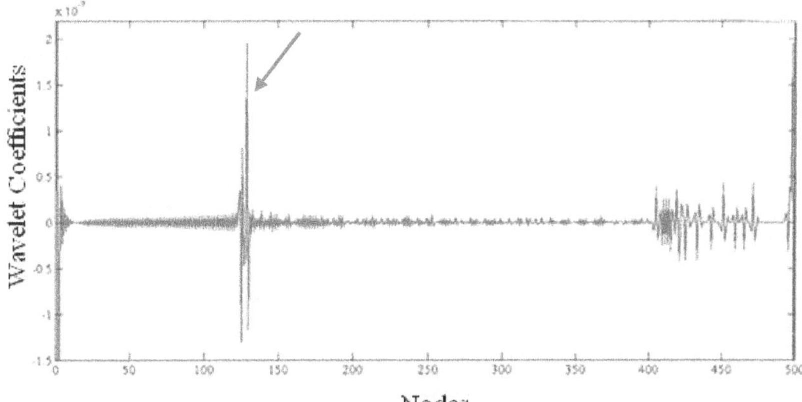

Figure 4: Wavelet coefficients using db2.

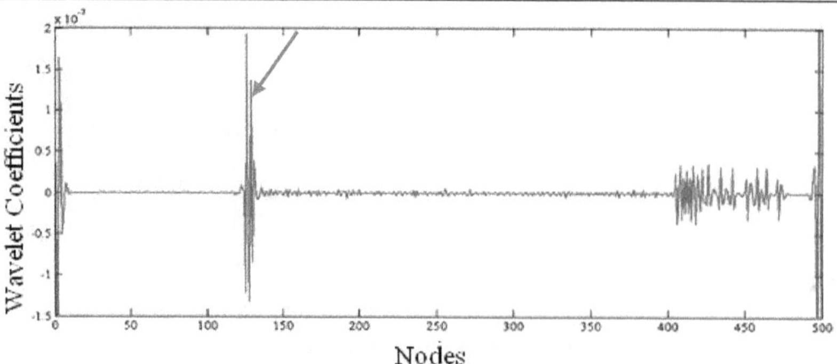

Figure 5: Wavelet coefficients using bior6.8.

The two mother-wavelet were able to detect the exact position of damage (node 125), moreover, the graphics presented little perturbations in the ends due to geometric discontinuities.

3.2 Static analysis with noise

For evaluate the effect of noise in the damage detection process, 1% of noise was added in the static signals and the results are presented in Figures 6 and 7.

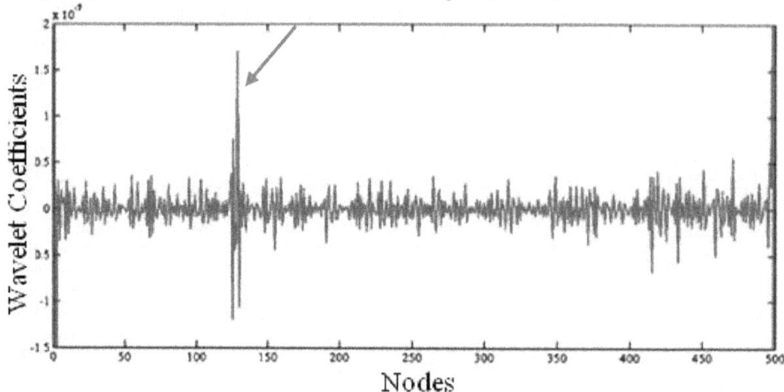

Figure 6:Wavelet coefficients with 1% of noise (db2).

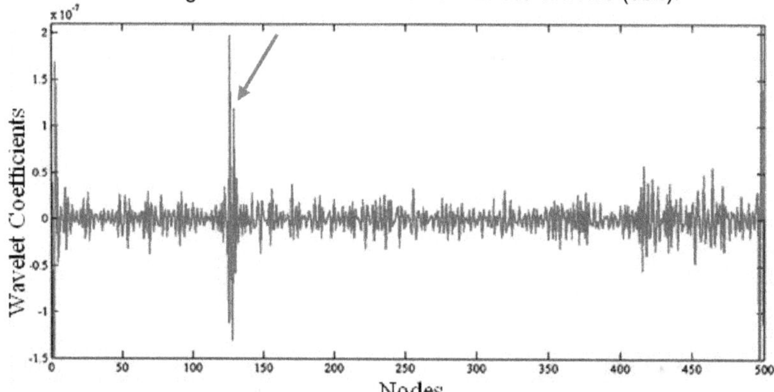

Figure 7:Wavelet coefficients with 1% of noise (bior6.8).

Comparing the Figures 4 and 5 with the Figures 6 and 7, one observes that the introduction of noise caused an increase in the disturbances along of signal, but nevertheless, the crack was detected.

3.3 Modal analysis

In the modal analysis, the three mode shapes were computed and the first mode shape was used in the application of wavelet transforms, as can be seen in Figures 8,9 and 10.

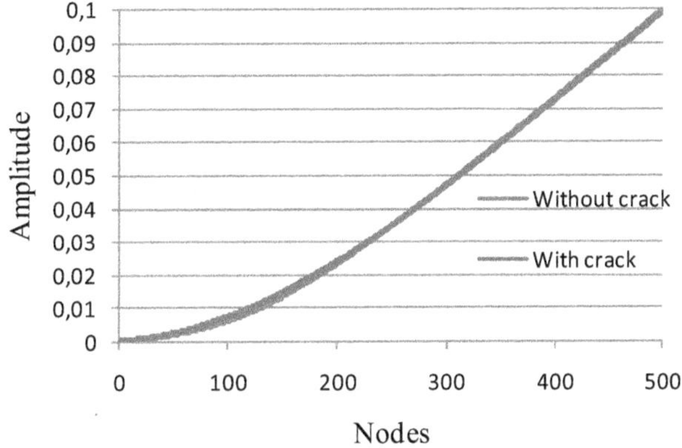

Figure 8:First mode shape of cantilever beam.

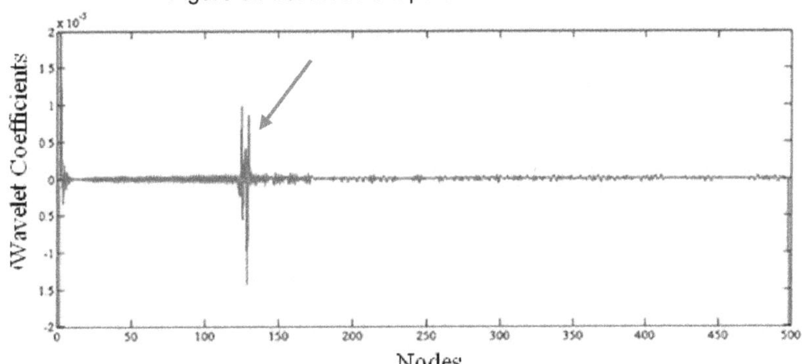

Figure 9:Wavelet coefficients using db2.

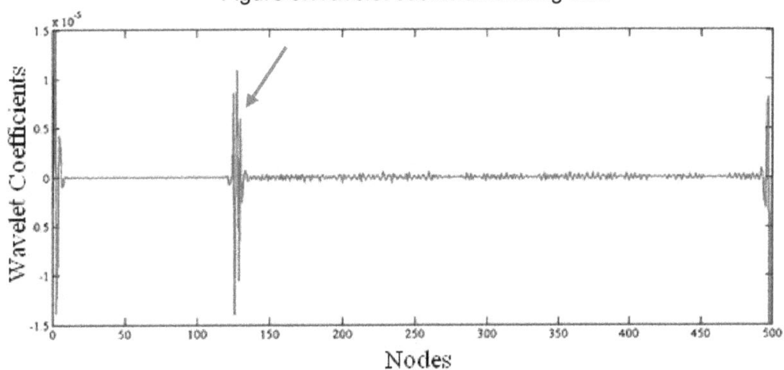

Figure 10:Wavelet coefficients using bior6.8.

The results of the modal analysis show that the application of wavelet transforms in the modal signals, like mode shapes, could be used in the damage detection process, furthermore, the modal signals transformed presented less disturbances than static ones.

4. Conclusions

This paper presented a recent methodology applied to the inverse problem of damage detection using static and modal signals.

The wavelet transform applied in the static and modal signals could detect the exact position of damages and also, singularities due to geometric discontinuities.

The main advantage of the the methods based on wavelets, is that it can detect the position of damage just from the response of damaged structure.

5. References

[1] Daubechies, I.,"Orthonormal bases of compactly supported wavelets". Comm. Pure and Applied Maths", XLI, 909-996, 1988.

[2] Daubechies, I., "The wavelet transform, time-frequency localization and signal analysis". IEEE Trans. On Information Theory 36,961-1005,1990.

[3] Estrada, E. S., "Damage detection methods in bridges trough vibration monitoring:evaluation and application". PhD. Thesis, University of Minho, 298p, 2008

[4] Mallat, S. A., "Theory for multiresolution signal decomposition: the wavelet representation". IEEE Trans. Pattern Anal and Machine Intell. 11, 674-693,1989

[5] Ovanesova, A. V. "Applications of wavelets to crack detection in frame structures". PhD. Thesis. University of Porto Rico, 235p, 2000.

[6] Silva, R. S. Y. C. "Determinação de patologias estruturais utilizando modelagem numérica e transformadas de wavelet".MSc. Dissertation. University of Brasilia. Department of Civil Engineering, 117p, 2011.

Solving Cauchy problems for non linear hyperelasticity

Stéphane Andrieux[1] and Thouraya N. Baranger[2]

[1]LaMSID, UMR CNRS-EDF 2832,
1, avenue du Général de Gaulle, 92141 Clamart, France
Email :stephane.andrieux@edf.fr

[2]LaMCoS, INSA-Lyon, CNRS UMR5259, F69621, France ;
Université de Lyon, Lyon, F-69003, France;
Université Lyon 1, Villeurbanne, F-69622, France
Email :Thouraya.Baranger@univ-lyon1.fr.

Abstract

The problem of expanding measured fields at the surface of a solid within the solid and up to inaccessible parts of its boundary is addressed for a non linear hyperelastic medium. The problem is formulated as a non linear Cauchy problem and is solved by a technique consisting in splitting the unknown field into two solutions of well posed problems and minimizing a specially designed energy-like error gap between the two fields. The minimization involves as unknowns the boundary conditions fields in the inaccessible part of the boundary of the solid. A simple numerical example illustrates the method (hyperelastic thick hollow sphere).

Key Words: Cauchy problem, boundary condition identification, hyperelasticity, inverse problem

1. Introduction

The problem of exploiting overspecified (measured) boundary data on a part of a solid in order to extend the mechanical fields up to unknown parts of the boundary, or more concisely to identify missing or unknown boundary conditions (BC), arises in a large class of applications in science and industry. Indeed, identification of BC or of physical parameters involved in the formulation of the BC are the main practical issues.

At the same time, hyperelastic potentials are used for the modeling of materials such as polymers, metals and geomaterials as concrete or rocks where the nonlinearity arises from the asymmetric behavior of these cracked media in the traction or compression range. There is then a great interest in solving such *data completion problems* in the framework of hyperelastic media.

2. Statement of the Cauchy problem for data completion in non linear hyperelasticity

Let Ω be a domain in IR^n, n=2 or 3, with piecewise Lipschitz boundary $\partial\Omega = \overline{\Gamma}_m \cup \overline{\Gamma}_u$, where Γ_m and Γ_u are open disjoint parts of $\partial\Omega$. The family of Cauchy problems addressed here can be stated as follows. Provided the Cauchy data on Γ_m are given, find an n components vector field $\boldsymbol{u}$ in IR^n such that:

$$\begin{cases} div\,\sigma = 0 & in\ \Omega \\ \sigma = \dfrac{\partial \varphi}{\partial \varepsilon}(\varepsilon(\boldsymbol{u})) & in\ \Omega \\ \varepsilon(\boldsymbol{u}) = \dfrac{1}{2}\left(\nabla \boldsymbol{u} + {}^t\nabla \boldsymbol{u}\right) & in\ \Omega \\ \sigma.\boldsymbol{n} = \boldsymbol{T}^m,\ \boldsymbol{u} = \boldsymbol{U}^m & on\ \Gamma_m \end{cases} \tag{1}$$

The potential φ is a convex, lower semi-continuous differentiable real function, σ is the (second order) Cauchy stress tensor and ε is the (second order) linearized strain tensor, both being symmetric. If the potential φ is not quadratic, then the material is no longer linearly elastic, and the second order operator $div(\partial_\varepsilon \varphi(\varepsilon(\boldsymbol{u}))$ turns out to be nonlinear. The Cauchy problem is then a nonlinear elliptic Cauchy problem. Theoretical results for existence and data compatibility conditions for nonlinear elliptic Cauchy problems have been addressed by Leitao [1].

Numerous numerical approaches are available in the literature for linear elliptic Cauchy problems, although the complexity of the algorithms and the large amount of computation needed restrains the applications in the quasi totality of the papers to 2-dimensional problems. For non linear scalar elliptic problems, a fixed-point method has been proposed by Kügler and Leitão [2]. Energy-error based methods associated with field-splitting techniques have been designed in particular by the authors (Andrieux *et al.* [3], Baranger and Andrieux [4] for linear vectorial elliptic problems, and evolution ones, Andrieux and Baranger [5]). This paper extends to the above class of nonlinear problems this energy-error approach.

3. Decomposition into two usual hyperelastic problems

To derive a general solution method, the main idea is to seek for the boundary conditions on the part Γ_u of the boundary where no data are available. We proceed in two steps as for the derivation of the energy error method for linear elliptic Cauchy problems (Andrieux *et al.* [3],[4]). The first step is exactly similar: two well-posed usual direct problems are defined, the solutions of which are denoted by $\boldsymbol{u}_1$ and $\boldsymbol{u}_2$. These problems are parameterized by two fields: respectively □$\boldsymbol{\tau}$ (a natural boundary condition quantity, that is a surface traction field) and $\boldsymbol{\upsilon}$ (a Dirichlet boundary condition, that is a prescribed displacement field) defined on the boundary Γ_u ,namely:

$$\begin{cases} div\,\sigma_1 = 0 & in\ \Omega \\ \sigma_1 = \dfrac{\partial \varphi}{\partial \varepsilon}(\varepsilon(\boldsymbol{u}_1)) & in\ \Omega \\ \boldsymbol{u}_1 = \boldsymbol{U}^m & on\ \Gamma_m \\ \sigma_1.\boldsymbol{n} = \boldsymbol{\tau} & on\ \Gamma_u \end{cases} \qquad \begin{cases} div\,\sigma_2 = 0 & in\ \Omega \\ \sigma_2 = \dfrac{\partial \varphi}{\partial \varepsilon}(\varepsilon(\boldsymbol{u}_2)) & in\ \Omega \\ \sigma_2.\boldsymbol{u} = \boldsymbol{T}^m & on\ \Gamma_m \\ \boldsymbol{u}_2 = \boldsymbol{\upsilon} & on\ \Gamma_u \end{cases} \tag{2}$$

Because if the solutions of these two problems are equal, then the common solution field $\boldsymbol{u}_1 = \boldsymbol{u}_2$ is obviously a solution $\boldsymbol{u}$ of the Cauchy problem (1), the second step of the method is to derive a functional "measuring the gap" between the two vector fields $\boldsymbol{u}_1$ and $\boldsymbol{u}_2$ as a function of the boundary unknowns $\boldsymbol{\tau}$and $\boldsymbol{\upsilon}$. Then the solution method will be to minimize this functional.

For linear symmetric elliptic operators such as the Lamé operator arising in linear elasticity problems, a magnitude of the error $\boldsymbol{u}_1 = \boldsymbol{u}_2$ has been proposed by the authors as the "energy" of the difference field. By energy, it must be understood the functional appearing as a pseudo potential of the system of equations, which are usually and for such symmetric operators characterized by a weak formulation:

$$a(\boldsymbol{u},\boldsymbol{v}) = l(\boldsymbol{v}) \quad \forall \boldsymbol{v} \in \mathrm{V} \qquad \text{or} \qquad \boldsymbol{u} = \underset{\mathrm{V}}{\mathrm{ArgMin}}\ \frac{1}{2}a(\boldsymbol{v},\boldsymbol{v}) - l(\boldsymbol{v}) \tag{3}$$

In this case, the "error in energy" function is simply:

$$E(\tau,\upsilon)=\frac{1}{2}a(\boldsymbol{u}_1-\boldsymbol{u}_2,\boldsymbol{u}_1-\boldsymbol{u}_2) \qquad (4)$$

This functional is positive, quadratic and achieves its minimum when $u_1 = u_2$. The solution of the Cauchy problem is then obtained by the minimization of the E functional. When combined with *ad hoc* optimization algorithms, it turns out to be very efficient and allowed to deal with three-dimensional applications (Andrieux and Baranger, [6]). This choice has also been used for the resolution of Cauchy problems for aquifers by Escriva *et al.* [7]).

4. Building of the energy gap error functional for hyperelasticity

To derive error functionals with similar desirable properties for non linear elliptic Cauchy problems (1), a different way must be followed and advantage must be taken of the convexity of the potential φ. The error will no more be an "error in energy" but rather an "error in constitutive equation". The Fenchel inequality or conjugacy formula (Ekeland and Témam, [8]) is the key-point of the building of the new error functional.

Let ψ be a lower semi-continuous (lsc), convex function from IR^n into $\overline{IR}$, then for any pair (x,y) of IR^n:

$$\begin{aligned}&\psi(x)+\psi^*(y)-x.y\geq 0\\&\psi(x)+\psi^*(y)-x.y=0 \iff y\in\partial\psi(x) \iff x\in\partial\psi^*(y)\end{aligned} \qquad (5)$$

$\partial\psi(x)$ is the subdifferential set of ψ of at x :
$\partial\psi(x)=\left\{g\in IR^n \text{ s.t.} \psi(x)-\psi(z)\geq g.(z-x)\,\forall z\in IR^n\right\}$

ψ^* is the Legendre-Fenchel conjugate function of ψ: $\psi^*(y)=\underset{x}{Sup}\left[x.y-\psi(y)\right]$

The error in constitutive equation for the hyperelastic material with potential φ is then defined for a pair (σ, ε) by the scalar function e_φ:

$$e_\varphi(\boldsymbol{\sigma},\boldsymbol{\varepsilon})\equiv\varphi(\boldsymbol{\varepsilon})+\varphi^*(\boldsymbol{\sigma})-\boldsymbol{\sigma}:\boldsymbol{\varepsilon} \qquad (6)$$

and enjoys then the following properties :

$$e_\varphi(\boldsymbol{\sigma},\boldsymbol{\varepsilon})\geq 0 \quad \forall(\boldsymbol{\sigma},\boldsymbol{\varepsilon}) \ , \quad e_\varphi(\boldsymbol{\sigma},\boldsymbol{\varepsilon})=0 \iff \boldsymbol{\sigma}\in\partial\varphi(\boldsymbol{\varepsilon})\iff\boldsymbol{\varepsilon}\in\partial\varphi^*(\boldsymbol{\sigma}) \qquad (7)$$

We define the (positive) *Error in Constitutive Equation* for the fields $\boldsymbol{u}_i$ solutions of (2) as:

$$ECE(\boldsymbol{u}_1,\boldsymbol{u}_2)=\int_\Omega\left[e_\varphi(\boldsymbol{\sigma_1},\varepsilon(\boldsymbol{u}_2))+e_\varphi(\boldsymbol{\sigma_2},\varepsilon(\boldsymbol{u}_1))\right]d\Omega \qquad (8)$$

This functional is positive and its minimum is zero. Furthermore, if the potential φ is lsc, differentiable and strictly convex and if the data are compatible, then the following equivalence holds:

$$ECE(\boldsymbol{u}_1,\boldsymbol{u}_2)=0 \text{ for } (\boldsymbol{u}_1,\boldsymbol{u}_2) \text{ satisfying}(2)\iff\begin{cases}\boldsymbol{u}_2(x)=\boldsymbol{u}_1(x)+\boldsymbol{u}+\boldsymbol{\omega}\wedge\boldsymbol{u}\\ \boldsymbol{u}=\boldsymbol{u}_1 \text{ solves the Cauchy problem (1)}\end{cases} \qquad (9)$$

The proposed method for solving the non linear Cauchy problem reduces then to the identification of lacking boundary conditions ($\boldsymbol{T}^u$, $\boldsymbol{U}^u$) enjoying the variational property:

$$\left(T^u, U^u\right) = ArgMin\ E(\tau, v) \equiv ECE(u_1(\tau), u_2(v)) \quad , \quad (u_1(\tau), u_2(v))\, given\, by\, (2) \tag{10}$$

The energy error in constitutive equation can be calculated without the knowledge or computation of the conjugate potential φ^*, it can be expressed simply as:

$$ECE(u_1, u_2) = \int_\Omega (\sigma_1 - \sigma_2) : (\varepsilon(u_1) - \varepsilon(u_2))\, d\Omega \tag{11}$$

Furthermore, this expression leads in turn to another alternative equivalent form of the E functional involving only boundary terms. This expression is more computationally efficient and simplifies also the expression of the adjoint problem when computing the gradient of the functional E.

$$E(\tau, v) = \int_{\Gamma_m} \left(\sigma_1(\tau).n - T^m\right).\left(U^m - u_2(v)\right) d\Gamma + \int_{\Gamma_u} (\tau - \sigma_2(v).n).(u_1(\tau) - v)\, d\Gamma \tag{12}$$

Note that the Fenchel inequality has also been used in the context of non linear mechanics by Ladevèze [9] and for identification problems by Barbu and Kunish [10].

5. Numerical illustration

To illustrate the method, a simple example is tackled with. The solid is a hyperelastic isotropic thick hollow sphere (ratio of internal radius to external radius equal to 0.3).

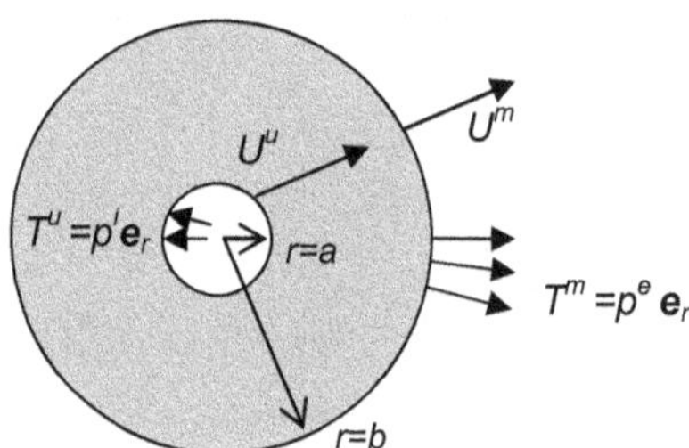

Figure 1 : Geometry and boundary quantities

The potential is taken as a polynomial function of the first two invariants of the strain tensor ε:

$$\varphi(\varepsilon) = G(I_1(\varepsilon), I_2(\tilde{\varepsilon})) \quad , \quad I_1(\varepsilon) = Tr\,\varepsilon \,, \quad I_2(\varepsilon) = I_2(\tilde{\varepsilon}) = \frac{1}{2}\tilde{\varepsilon} : \tilde{\varepsilon}$$

$$G(X, Y) = \frac{1}{2}KX^2 + \frac{1}{4}\alpha K X^4 + 2\mu Y \quad K, \mu, \alpha > 0$$

where $\tilde{\varepsilon}$ is the deviatoric part of the strain tensor ε. This lead to the following stress-strain relation:

$$\sigma = K\,Tr\,\varepsilon\,\mathbf{I} + \alpha K (Tr\,\varepsilon)^3 \mathbf{I} + 2\mu\tilde{\varepsilon}$$

For this material, only the compressibility behaviour is non linear. K and μ are the Lamé coefficients, and α is a dimensionless parameter controlling the non linear part of the compressibility (linear isotropic elasticity is recovered when α =0). Due to the symmetry, the displacement field is only r dependant and radial, and the principal strains are: $e_1=\varepsilon_{rr}$, $e_2=\varepsilon_{qq}=\varepsilon_{\varphi\varphi}$ in the spherical coordinate system. To illustrate the non linearity, the level lines of the potential in the (e_1, e_2) plane are displayed on figure *2a*, and the stress–strain curve (ε_1, σ_1) is depicted on figure *2b* for two values of the orthogonal strain ε_2. The α parameter is 5.

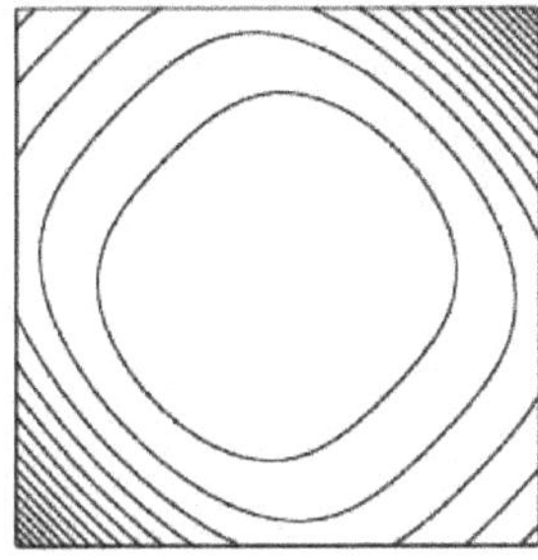
Figure 2a : Level lines of the potential φ in the (e_1, e_2) plane

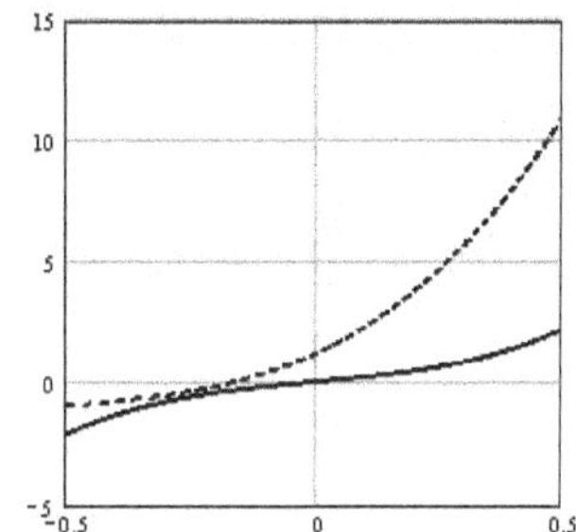

Figure 2b : Strain-stress curve (ε_1, σ_1) (continuous ε_2=0, dotted ε_2=1)

The overspecified data are the external radial displacement U^m and the external pressure p^e, while the unknowns are the internal pressure p^i and the internal radial displacement U^i. The energy error gap reduces to a two-variables function and has a very simple expression (using *eq.* 12):

$$E(U^i,p^i)=4\pi\left[a^2(p^i+\sigma_{rr}^2(a;U))(u_1(a;p^i)-U^i)+b^2(\sigma_{rr}^1(b;p^i)+p^e)(U^m-u_2(b;U))\right] \qquad (13)$$

Consider the following material data: *E=1*, ν*=0.4*, α*=1*. The external and internal pressure are p^e*=-1* and p^i*=1*, and the corresponding displacements *are* U^i*=-1.928* 10^{-2} and U^e*=-6.427* 10^{-2}, which are obtained by solving the forward problem.

The minimization of the functional $E(U^i,p^i)$ is performed by using the Optimization Toolbox of Matlab [11]. The optimization process is initialized with the pair (U^i_o=0,p^i_o=0). Figures 3a, 3b and 3c show the evolution during the optimization process of *E*, U^i and p^i, respectively. Figure 4 shows the surface illustrating the functional *E*. The functional reached its minimal value after 6 iterations: *E= 1.168e-15*. The identified data are U^i= *-1.928* 10^{-2} and p^i= *0.999*.

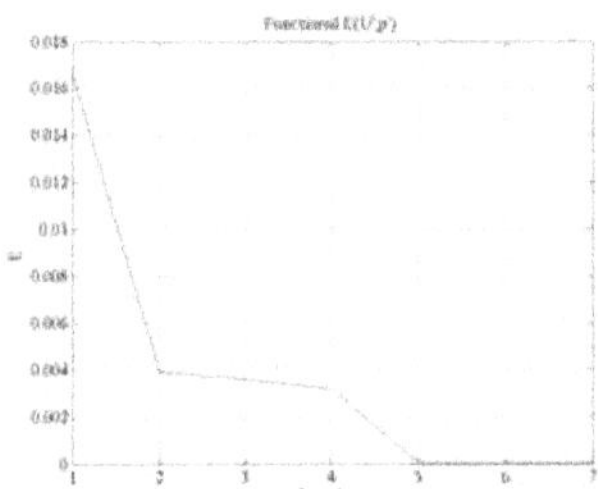
Figure 3a: Functional evolution during optimization process

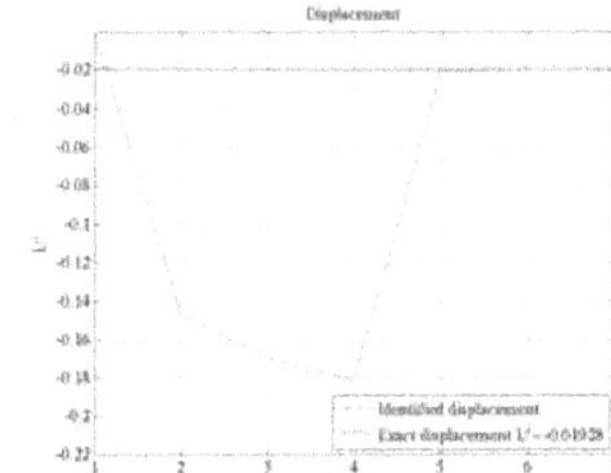
Figure 3b: Identified displacement evolution during optimization process

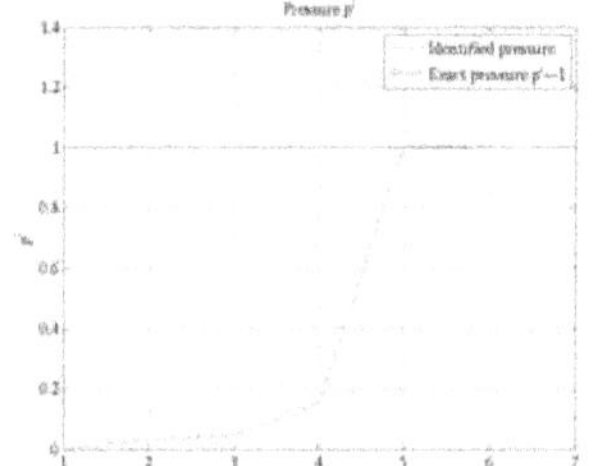
Figure 3c: Identified displacement evolution during optimization process

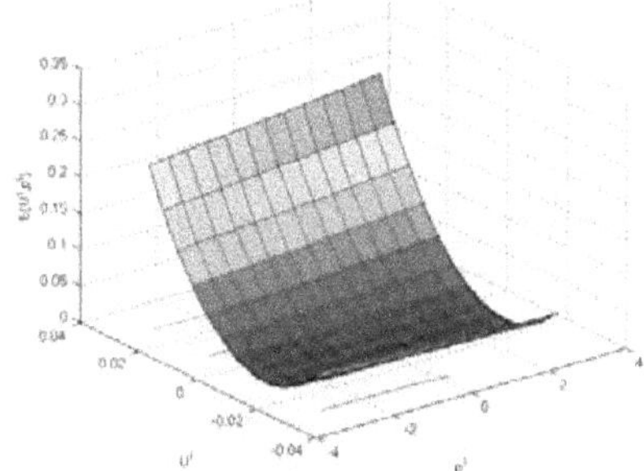
Figure 4: The energy function $E(U^i,p^i)$.

6. Conclusion

The key point here is to exploit the convex framework and to make an extensive use of the Fenchel-Legendre inequality in order to build the appropriate energy error. Two BC identification examples will illustrate the capabilities of the proposed approach. The first one involves a regular (C^2) potential modeling the behavior of an elastic material with nonlinear compressibility. The second one exhibits a non twice-differentiable potential modeling the asymmetric behavior of cracked geomaterials.

7. References

[1] Leitão A. An iterative method for solving elliptic Cauchy problems, Numer. Funct. Anal. Optim., 21, 715-742. 2000.

[2] Kügler P. and Leitão A. Mean value iterations for non linear elliptic Cauchy problems, Numerische Mathematik, 96, 269-293.2003.

[3] Andrieux S., Baranger T.N. and Ben Abda A. Solving Cauchy problem by minimizing an energy-like functional, Inverse Problems, 22, 115-133, 2006.

[4] Andrieux S., Baranger T.N. An energy error-based method for the resolution of the Cauchy problem in 3D linear elasticity, Computer Methods in Applied Mechanics and Engineering, 197, 9-12, 902-920, 2008.

[5] Andrieux S. and Baranger T. N. Energy methods for Cauchy problems for evolution equations, 6th International Conference on Inverse Problems in Engineering: Theory and Practice, Journal of Physics: Conference Series 135 (2008) 012007, doi:10.1088/1742-6596/135/1/012007, 2008.

[6] Baranger T.N., Andrieux S., An optimization approach for the Cauchy problem in linear elasticity, *J. Multidisciplinary optimization*, 35, 141-152, 2008.

[7] Escriva X., Baranger T.N., Hariga Tlatli N. Leak identification in porous media by solving the Cauchy problem, C.R. Mécanique, 335, 410-406, 2007.

[8] Ekeland I., Témam R. Convex analysis and variational problems, Siam, Philadephia, 1999.

[9] Ladevèze P. Nonlinear Structural Mechanics - New Approaches and Non-Incremental Methods of Calculation, Springer-Verlag, New York, 1998.

[10] Barbu V. and Kunisch K. Identification of non linear elliptic equations, Appl. Math. Optim., 33(2), 139-167, 1996.

[11] Optimization Toolbox, Matlab, http://www.mathworks.com.

USE OF ACTIVE CONSTRAINTS EQUATIONS IN INVERSE PROBLEMS ELASTOSTATIC IDENTIFICATION OF CRACKS

Marcus Alexandre Noronha de Brito[1,2], *Luciano Mendes Bezerra*[1], *William Taylor Matias Silva*[1]

[1]*Graduate Program and Structures and Construction*
University of Brasília- UnB,Brasília, Federal District, +55 61 3321-1000
Email: marcusanb@yahoo.com.br
lmbz@unb.br
taylor@unb.br

[2]*Federal Institute of Education, Science and Technology of Brasília – IFB*
Email: marcus.brito@ifb.edu.br

Abstract

The use of inverse problems in engineering has intensified in recent decades. In mechanics, there are basically two types of inverse problems: the inverse problem of identification and reconstruction. The latter, usually reconstructs boundary conditions while the first attempts to identify the problem domain. In problems of crack detection is the need to identify the domain of mathematical problem - which makes it an inverse problem. To identify a crack in a domain Ω of an object, some numerical methods as FEM and BEM associated with optimization methods can be used in some algorithms in solving an inverse problem of identification. To identify a crack in a mathematical domain, there is the need to establish the "equations of constraint" to indicate being a crack inside or outside the field of object. In the latter case, there is no physical sense to the inverse problem in question and therefore should be avoided. This article discusses a numerical algorithm, computational identification of cracks opened within objects using the ECM associated with optimization methods, but focuses on the restriction equations avoiding undesirable solutions of the inverse problem, outside the realm of the physical object. The constraint equations are used to penalize the object function during the optimization or minimization of a functional defined in this work. This article discusses the mathematical aspects of the theorem of Jordan, who built the solution algorithm of the inverse problem sought guarantees that the crack remains feasible within the domain of mathematical problem, within the domain of the object.

Keywords: Inverse problems, identification of cracks, constraint equations.

1. Introduction

The problems in engineering there are many, many years, however the use of inverse problems in solving these occurred from the last decades, in works such as: Analysis and solutions of inverse problems in heat conduction, Weber [1] and only one of many examples where de facto inaccessibility or difficulty in the trial of at least one of information such as: The domain Ω, the boundary conditions Γ of an object under study, the equations governing this problem, the properties involved in addressing them and their applications, makes an inverse problem.

In this study, the constraints, described by Eqn. (1) is the term z_i as components of the vector z.

$$C_j(z_i) \geq 0 \tag{1}$$

Equations are widely used in engineering problems so as to achieve a reduction extremize the purpose of a function that is being used, so as not to have their solutions in undesirable levels, ie without physical sense. Thus, in cases where the problems should have a significant solution within the body studied, such as fault detection and / or fractures within an object with your problem domain Ω, the restriction equations are then widely used for penalizing the objective function in a model where vector z can be maintained within the possible region in the domain boundary Γ and thereby avoiding meaningless solutions, like finding a failure or break out of the body being studied, for example.

2. Active constraint equations

The possible field is defined by the points inside $\overline{\Omega}$, studied the problem domain. For simple linear geometry of the body Ω in case of failure of discovery, is the restrictive function Cj(zi) that can be simply expressed by Eqn. (2) in the case of a rectangular domain.

$$C_j(z_i) \pm z_i \mp (\alpha \pm \varepsilon) \tag{2}$$

In this equation α is a constant that represents a coordinate of the corners of the rectangular field, and ε is a small number to ensure that failure is an ε-neighborhood within the limits of the provision of object. The provision of this ε-neighborhood is essential for the formulation of the BEM (Boundary Element Method) since the singularities arise as the point of placing on the approaches of the integration scheme limit. The restrictions given by Eqn. (2) are steps in the parameter values assumed by zi and when a constraint equation is violated, Eqn. (3) is applied to the field $\overline{\Omega}$ that can be done to determine the failure or restoration of the traction $\overline{\Gamma_1}$, where is the step length is taken when $\alpha^{(K)}$ is necessary, according to Eqn. (3).

$$\alpha^{(k)} = \begin{cases} 1.00x\alpha^{(k)}; z^{(k+1)} \in \overline{\Omega}; (z^{(k+1)} \in \overline{\Gamma}_1) \\ 0.90x\alpha^{(k)}; z^{(k+1)} \notin \overline{\Omega}; (z^{(k+1)} \notin \overline{\Gamma}_1) \end{cases} \tag{3}$$

The test of Eqn. (3) is repeatedly applied $\alpha^{(K)}$ to $z^{(K+1)} \in \overline{\Omega}$, for determining the failure or $z^{(K+1)} \in \overline{\Gamma_1}$, in the case of reconstruction of traction.

The steps in Eqn. (3) are only sufficient to maintain the failure and/or fracture within a body Γ, where the domain boundary is a convex polygon in a simple manner. This is not the case when the solid Ω is not convex and has a more complex shape with a contour Γ defined by non-linear. In this study, for such a general case, steps to limit the vector z are model indirectly imposed by Eqn. (4) an active constraint equation where ε was previously defined.

$$C_a(z_i) \leq \varepsilon \tag{4}$$

The active constraint is selected as the minimum distance between a point P^f, the contour Γ^f of the fault and another point P on the boundary Γ of domain Ω with the body. The active constraint represents the smallest computed distance between the edge of failure Γ^f and body contour Γ, where all points of failure are within Γ.

The task of verifying that a point P^f chosen at fault is inside or outside a convex polygon or non-convex, simply-connected or multiply-connected domain with boundary Γ, is based on the theorem of Jordan curve topology, Preparata et al.[2].

This theorem is illustrated in Figure 1 in which any given point of failure, called P^f, and a straight line that starts at any Q, starting in P^f, then the point where P^f is in Γ if and only if the line Q intersects Γ in a number of times.

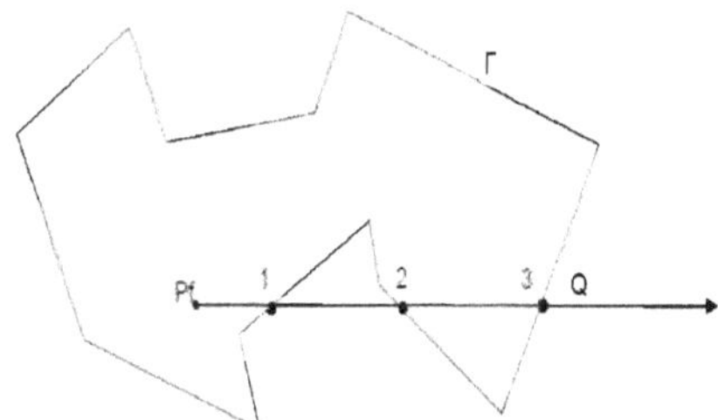

Figure 1: Determination of the point P^f located within a polygon boundary Γ by a ray passing

Based on this fact, the whole point P^f of failure Ω^f is first checked if it is inside or outside the boundary Γ of the body. P^f is defined by the vector model *z*. The active constraint equation of Eqn. (4) is computed and in the case of rape, then Eqn. (3) applies to bring *z* (and failure) into the contour Γ.

Given *z*, the algorithm checks whether the point of failure P^f is inside Γ is shown schematically in Figure 2. As Ω can have a non-linear boundary, each boundary is discretized into elements and where Γ is a extremize by passing a square box out by element nodes 1, 2 and 3. The points $(x_i^*; y_i^*)$ are defined by jumping each box as shown in Figure 2, and expressed by Eqn. (5) to Eqn. (13).

$$a_1 = \frac{(y_3 - y_1)}{(x_3 - x_1)} \tag{05}$$

$$b_1 = y_2 - a_1 x_1 \tag{06}$$

$$b_2 = y_2 - a_1 x_2 \tag{07}$$

$$c_1 = y_1 + \frac{x_1}{a_1} \tag{08}$$

$$c_2 = y_3 + \frac{x_3}{a_1} \tag{09}$$

$$(x_1^*; y_1^*) \equiv (x_1; y_1) \tag{10}$$

$$(x_a^*; y_a^*) \equiv (\frac{c_2 - b_2}{a_1 + 1/a_1}; a_1 \frac{c_2 - b_2}{a_1 + 1/a_1} + b_2) \tag{11}$$

$$(x_b^*; y_b^*) \equiv (\frac{c_1 - b_2}{a_1 + 1/a_1}; a_1 \frac{c_2 - b_2}{a_1 + 1/a_1} + b_2) \tag{12}$$

$$(x_3^*; y_3^*) \equiv (x_3; y_3) \tag{13}$$

A semi-infinite beam crossing, Q, started on node failure, P^f, finds the number of jumps out of the boxes with 3 points of the contour bouncing boxes intercepted, and properly solving the parabolic equation in natural coordinates, finding the intersection point. If the number of intersections is found, then point of failure that P^f is within the boundary Γ.

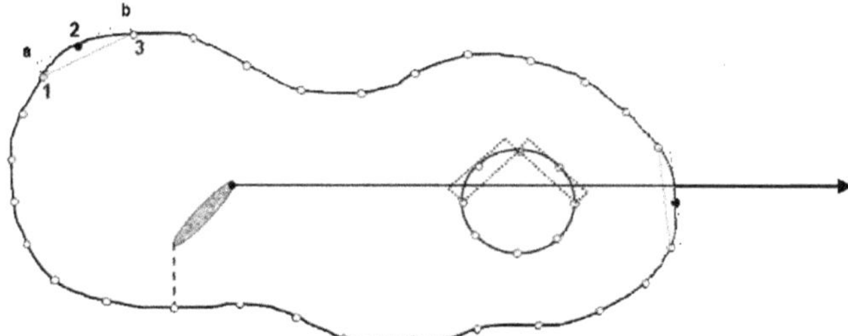

Figure 2: Determining the location of a point of imperfection or fracture inside the body.

Given a point, the definition of the failure is presented in accordance with z, $P_i^f \equiv (x_i^f; y_i^f)$ and the whole point is shown as $P_j \equiv (x_j; y_j)$ defining the boundary Γ, the constraint equation $C_a(z_i)$ can be expressed by Eqn. (14).

$$C_a(z) = Min\sqrt{[x_i^f(z) - x_j]^2 + [y_i^f(z) - y_j]^2} \tag{14}$$

Where $j = 1,2,...,\overline{N}$ and $i = \overline{N}+1, \overline{N}+2,..., \overline{N}_e$, being $\overline{N}$ the number of nodes defined by the outer boundary Γ, and $\overline{N}+1,...,N_e$ are the points that define the fracture. Its derivative with respect to the vector is represented by Eqn. (15).

$$\frac{\partial C_a(z)}{\partial z} = \frac{[x_i^f(z) - x_i]\dfrac{\partial x_i^f(z)}{\partial z} + [y_i^f(z) - y_i]\dfrac{\partial y_i^f(z)}{\partial z}}{\sqrt{[x_i^f(z) - x_j]^2 + [y_i^f(z) - y_i]^2}} \tag{15}$$

3. Objective Function Penalty

The Eqn. (14) and Eqn.(15) are used in evaluating the objective function F given by Eqn. (16) that allow the optimization of the constraint on the solution of the problem using methods unrestricted, which Θ is an internal function of penalty that is increased according with the objective function f given by Eqn. (17) and $\Re$ a penalty parameter used.

$$\mathcal{F}(z, \Re) = f(z) + \Theta(C_j(z_i), \Re) \tag{16}$$

$$f(z) = w\sum_{k=1}^{m}\sum_{i=1}^{2}(\psi_{ik} - \hat{\psi}_{ik})^2 \tag{17}$$

Eqn. (17) is an initial rating for solving problems in two-dimensional case, where w is a multiplicative parameter that significantly increases a numerical increase in the minimization process, ψ_{ik} is the vector which are leased to the amounts computed stresses and displacements along the direction corresponding i to the position k taken or failure to limit configuration traction set $i = 1,2$, with corresponding directions *x* and *y* respectively, *m* is the total number of positions, and $\hat{\psi}_{ik}$ is the vector of experimental measurements for inverse problems elastostáticos.

In the case of reconstruction of the limit of traction, the restriction equations are equations similar to Eqn. (2) and represent straightforward steps over *z*. In this case, the derivatives $C_j(z_i)$ *of also can be obtained.*

4. Mesh update

A boundary element mesh is defined by coordinates of nodes and connectivity of elements. In this article, for IESPs (Inverse Problems Elastostáticos) which makes the discovery of a failure and reconstruction of boundary traction, the connectivity boundary element is assumed unchanged. Only the coordinates of the nodes that define the fault position and extending the limit will change with lost traction vector, and modified in the minimization process.

On the discovery of the failure, it is free to move throughout the body domain $\overline{\Omega}$. As a possible *z*, the coordinates of the nodes of failure are updated according to Eqn. (18) and Eqn.(19), which are the functions of specifications to detect the fault.

Whereas, for example, a circular fault, we can define it in terms of parameters $z=\{x_0, y_0, r\}$, which (x_0, y_0) are the coordinates of the center point and radius of the circle. Parametric functions specifying the shape and position of limit points (x^f, y^f), of the circular fault.

$$x^f = x_0 + r\cos\varphi \tag{18}$$

$$y^f = y_0 + r\,\mathrm{sen}\,\varphi \tag{19}$$

Where $-\pi \leq \varphi \leq \pi$. In case of failure elliptical, $z=\{x_0, y_0, a, b, \theta\}$; and parametric functions specifying the shape and position of limit points, (x^f, y^f) are given by Eqn. (20) and Eqn.(21).

$$x^f = x_0 + a\cos\varphi\cos\theta - b\,\mathrm{sen}\,\varphi\,\mathrm{sen}\,\theta \tag{20}$$

$$y^f = y_0 + a\cos\varphi\,\mathrm{sen}\,\theta + b\,\mathrm{sen}\,\varphi\cos\theta \tag{21}$$

Where (x_0, y_0) are the coordinates of the center of the ellipse, *a* and *b* are the dimensions of the minor and θ the minor axis, respectively, and is the slope of the major axis with respect to horizontal x-axis, as show in Figure 3.

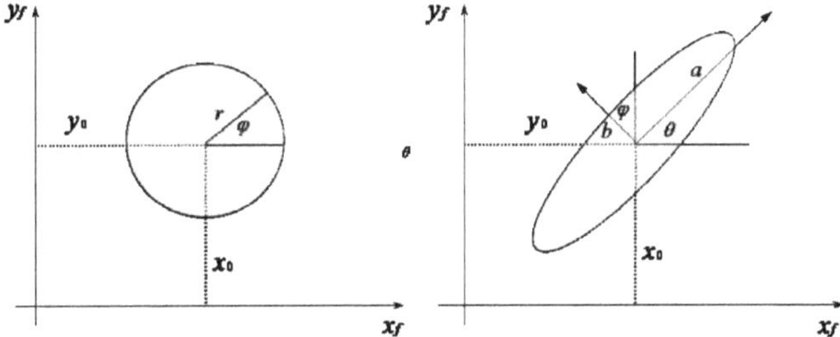

Figure 3: Parameters for the definition of circular and elliptical flaws.

For the reconstruction of boundary tractions, when the vector z changes, then the position should be updated and extent of the drifts. The alleged statements are only recently applied in the limit we can set out in the field $\overline{\Gamma_1}$.

5. Concluding Remarks

Optimization solutions for inverse problems elastostáticos by equations of constraints on active fault detection in structures using methods of evidence limits, has been widely used. The purpose of the functions in this procedure is to perform a comparative analysis between the displacements obtained computationally by assuming a fault configuration and displacements measured experimentally in the solid parsed purpose of the use of active **constraint equations is** augmented by the geometric constraints that reduce the field of object considered faulty inaccurate in its location through adjustment of the first order numerical solutions satisfactory, leading to a significant advancement in aiding the commission of an analysis estrutura.As objective functions are minimized by the algorithms used metric approaching the boundary elements. A variety of problems of numerical examples with circular and elliptical flaws is considered, the fault configuration is predicted in each case starting with different initial guesses for the initial failure, thus demonstrating the versatility of the formulations presented, stability and effectiveness of the method solving real problems with small errors in the measurement data was demonstrated by numerical data in realizadas.A analysis approach for detecting faults using boundary elements is more efficient compared to the nearest place with finite elements, for this it is difficult and somewhat cumbersome to modify the domain at each iteration, which would require that the failure to modify its settings so since the initial iteration to the final, which would have a high time for that operation and thus a high operational cost.

6. References

[1] Weber C. F. 1981. Analysis and solution of the ill-posed inverse heat conduction problem.Int. Heat Mass Transfer. 24: 1783-1792.

[2] Preparata, F. P., and M. I. Shamos. 1985.Computational Geometry An Introdution. New York: Spring-Verlag.

[3] Bezerra L. M. , and Saigal. 1991a. A boundary element formulation for the inverse elastostatics problem (IESP) of flaw detection. Int. J. Numer. Methods in Eng, in press.

[4] Bezerra L. M. Inverse Elastostatics Solutions with Boundary Elements, Ph.D. thesis, Carnegie Mellon University, Pittsburgh, PA, 1993.

NON-CONTACT TESTING OF ELASTIC MATERIALS BY INVERSE METHODS OF NONLINEAR ACOUSTICS

Oleg M. Alifanov[1] , Aleksey V. Nenarokomov[1], and Kirill A. Nenarokomov[1]

[1] Aerospace Department, Moscow Aviation Institute,
4 Volokolamskoe Hgw., Moscow, 125993, Russia,
Email: kir-nenar@yandex.ru

Abstract

Elastic properties of flaws differ from ones of base medium. Gradient of elastics properties on the border of flaws results in the appearance of non-classic nonlinearity in the region. Such nonlinearity exceeds significantly the physical nonlinearity of a medium under diagnostics. Determination of spatial distribution of structural acoustic nonlinearity in the sample under investigation allows spotting the defects. In the paper the experimental technique for remote diagnostics of subsurface flaws in elastic materials is considered.

Key Words: nonlinear acoustics, inverse problems, defects in elastic materials

1. Introduction

The purpose of our research was to develop an experimental technique and setup for remote (non-contact) diagnostics of structural defects in elastic materials. The performance of equipment is based on nonlinear interaction of two acoustic beams of finite amplitude in investigated material. The most promising direction in further development of non-destructive diagnostics (research methods) for the elastic composite materials is to use the procedure of inverse problems to estimate $\varepsilon(p)$. Such problems are of great practical importance in the study of properties of composite materials used as non-destructive elastic surface coating in objects of space technology, power engineering etc The experimental equipment and the method developed could be applied for determination of material's properties; the availability of corresponded experimental facilities allows us to provide uniqueness of the solution.

The mathematical model of wave propagation in specimen is

$$\frac{\partial^2 p}{\partial t^2} - c^2 \frac{\partial^2 p}{\partial x^2} = \frac{1}{c^2 \rho}\varepsilon(p)\frac{\partial^2 p}{\partial t^2} \quad (1)$$
$$x \in (0,d),\ t \in (0,t_{\max}]$$

$$p(x,0) = p_0,\ x \in [0,d] \quad (2)$$

$$\frac{\partial p}{\partial t}(x,0) = 0,\ x \in [0,d] \quad (3)$$

$$-\frac{\partial p}{\partial x}(0,t)=\rho_0\frac{\partial^2\xi_1}{\partial t^2},\ t\in(0,t_{max}) \tag{4}$$

$$p(d,t)=p_2(t),\ t\in(0,t_{max}) \tag{5}$$

In model (1)-(5) the coefficient $\varepsilon(p)$ is unknown.

The results of pressure measurements at the left side of the material's specimen are assigned as necessary additional information to solve an inverse problem

$$p^{exp}(0,t)=p_1(t),\ t\in(0,t_{max}) \tag{6}$$

The experimental equipment and the method presented below could be applied for estimating of material's property.

2. Inverse Problem Algorithm

In the inverse problem (1)-(6) it is necessary first of all to indicate as a temperature range $[p_{min}, p_{max}]$ of the unknown functions, which is general for all experiments, and for which the inverse problem analysis has a unique solution. For p_{min} the minimum value of initial pressure p_0 is used. Of much greater importance is a correct sampling of value p_{max}. Proceeding from the necessity to provide uniqueness of solution, it seems possible to sample for p_{max} a maximum pressure value gained on the measured tools positioned on the both surface at every testing specimen. Suppose then that the unknown function is searched in the parametric form. With this purpose introduce in the interval $[p_{min}, p_{max}]$ uniform difference grids with the number of nodes N_ε.

$$\omega_\varepsilon=\{p_k=p_{min}+(k-1)\Delta p,\ k=\overline{1,N_\varepsilon-1}\}$$

Then we approximate the unknown functions on grids using the cubic B-splines

$$\varepsilon(p)=\sum_{k=1}^{N_\varepsilon}\varepsilon_k\varphi_k(p) \tag{7}$$

where ε_k, k=1,..., N_ε are unknown. As a result of approximation, the inverse problem is reduced to the search of a vector of unknown parameters $\overline{\varepsilon}=\{\varepsilon_k\}$, k=1,..., N_ε. Writing down a least-square discrepancy of the calculated and experimental pressure values in points of measuring we receive the residual functional

$$J(\overline{\varepsilon})=\int_0^{t_{max}}(p(0,t)-p_1(t))^2\,dt \tag{8}$$

where $p(0,t)$ is determined from a solution of the boundary problem (1)-(6) using the approximation (7). So, proceeding from the principle of iterative regularization, the unknown vector can be determined through minimization of functional (8) by gradient methods of the first order prior to a fulfilment of the condition

$J(\bar{p}) \leq \delta_p$

where $\delta_p = \int_0^{t_{max}} \sigma_m(t)\,dt$ - integral error of temperature measurements $p_1(t)$, and σ_m - measurement variance. To construct an iterative algorithm of the inverse problem solving a conjugate gradient method was used. The successive approximation process is constructed as follows:

1) a-priori, an initial approximation of the unknown parameter vector $\overline{\varepsilon^0}$ is set
2) a value of the unknown vector at the next iteration are calculated

$$\bar{\varepsilon}^{s+1} = \bar{\varepsilon}^s + \gamma^s \bar{g}^s, \quad s = 1,\ldots,s^*,$$

(9)

$$\bar{g}^s = -J_\varepsilon' + \beta^s \bar{g}^{s-1}, \quad g^0 = 0,$$

$$\beta^0 = 0,$$

$$\beta^s = \left\langle \left(\bar{J}_\varepsilon'^{(s)} - \bar{J}_\varepsilon'^{(s-1)}\right), \bar{J}_\varepsilon'^{(s)} \right\rangle_{R^{N_\varepsilon}} \Big/ \left\| \bar{J}_\varepsilon'^{(s)} \right\|^2_{R^{N_\varepsilon}}$$

where $\bar{J}_p'^{(s)}$ - value of the functional gradient at current iteration.

The greatest difficulties in realizing the gradient methods are connected with calculation of the minimized functional gradient. In the approach being developed the methods of calculus of variations are used. Here an analytic notation for the minimized functional gradient can be obtained

$$\frac{\partial J}{\partial \varepsilon_k} = -\int_0^{t_{max}} \int_0^d \psi(x,t) \frac{1}{c^2 \rho} \frac{\partial^2 p}{\partial t^2}\,dx\,dt \qquad (9)$$

$$k = 1,\ldots N_\varepsilon$$

where $\psi(x,\tau)$ - solution of a boundary-value problem adjoint to a linearized form of the initial problem (1)-(5).

$$\frac{\partial^2 \varphi}{\partial t^2} - c^2 \frac{\partial^2 \psi}{\partial x^2} = \frac{1}{c^2 \rho} \varepsilon(p) \frac{\partial^2 \psi}{\partial t^2} + \frac{2}{c^2 \rho} \frac{d\varepsilon}{dp} \frac{\partial^2 p}{\partial t^2} \psi + 2 \frac{\partial \psi}{\partial t} \frac{1}{c^2 \rho} \frac{d\varepsilon}{dp} \frac{\partial p}{\partial t} + \psi \frac{1}{c^2 \rho} \frac{d^2 \varepsilon}{dp^2} \left(\frac{\partial p}{\partial t}\right)^2$$

$$x \in (0,d),\ t \in [0,t_{\max}) \qquad (10)$$

$$\psi(x,t_{\max}) = 0,\ x \in [0,d] \qquad (11)$$

$$\frac{\partial \psi}{\partial t}(x,t_{\max}) = 0,\ x \in [0,d] \qquad (12)$$

$$c^2 \frac{\partial \psi}{\partial x}(0,t) + 2(p(0,t) - p_1(t)) = 0,\ t \in [0,t_{\max}) \qquad (13)$$

$$\psi(d,t) = 0,\ t \in [0,t_{\max}) \qquad (14)$$

The descent parameter γ^s is determined from the condition

$$\gamma^s = arg \min_{\gamma^s \in R^+} \left(J\left(\bar{\varepsilon}^s + \gamma^s \bar{g}^s\right)\right) \tag{15}$$

A linear estimation is used for determination of the descent step. It can be calculated as

$$\gamma^s = \frac{\int_0^{t_{max}} \Delta p(0,t)\left(p(0,t) - p_1(t)\right)dt}{\int_0^{t_{max}} \Delta p^2(0,t)dt} \tag{16}$$

where $\Delta p(x,t)$ is the Frechet differential of $p(x,t)$ at point $\varepsilon(p)$, and is the solution of the boundary-value problem

$$\frac{\partial^2 \Delta p}{\partial t^2} - c^2 \frac{\partial^2 \Delta p}{\partial x^2} = \frac{1}{c\rho^2}\varepsilon(p)\frac{\partial^2 \Delta p}{\partial t^2} + \frac{1}{c\rho^2}\frac{d\varepsilon}{dp}\frac{\partial \Delta p}{\partial t}\frac{\partial^2 p}{\partial t^2} + \frac{1}{c\rho^2}\frac{\partial^2 p}{\partial t^2}\sum_{k=1}^{N_\varepsilon} g_k \varphi_k(p)$$

$$x \in (0,d),\ t \in (0,t_{max}] \tag{17}$$

$$\Delta p(x,0) = 0,\ x \in [0,d] \tag{18}$$

$$\frac{\partial \Delta p}{\partial t}(x,0) = 0,\ x \in [0,d] \tag{19}$$

$$-\frac{\partial \Delta p}{\partial t}(0,t) = 0,\ t \in (0,t_{max}] \tag{20}$$

$$\Delta p(d,t) = 0,\ t \in (0,t_{max}) \tag{21}$$

As an example, the results of data processing on experimental study of material specimens are used, which involves the thermo-vacuum facility developed in MAI at the Department of Space System Engineering. The test facility consists of a horizontally cylindrical set vacuum chamber with a water-cooling system.

3. Experimental Technique

The effectiveness of the process is defined with value of nonlinear parameter in an interaction area. If two plane waves of frequencies ω_1 and ω_2 (their pressure amplitudes are $P(\omega_1)$ and $P(\omega_2)$ correspondingly) fall on a defect of the material structure with typical size *d*, then on the border of defects the signal of heterodyne frequencies with pressure $P(\omega_1 \pm \omega_2)$ appears as a result of nonlinear interaction [3].

$$P(\omega_1 \pm \omega_2) = \frac{\varepsilon(p)d}{4c^3\rho}(\omega_1 \pm \omega_2)P(\omega_1)P(\omega_2), \tag{22}$$

here *c*, ρ are known sound velocity and material density correspondingly. As is clear from (22), under constancy of incident waves' amplitudes $P(\omega_1)$ and $P(\omega_2)$ the value of signal of heterodyne frequencies $P(\omega_1 \pm \omega_2)$ is defined with value of nonlinear parameter $\varepsilon(p)$.

Nonlinearity of a material with damaged structure is the result of elastic properties changes on the border of inhomogenuity and surrounding medium. Knowing the value of signal amplitude on heterodyne frequency during scanning makes it possible to restore the spatial distribution of structural damages in the investigated material [4].

The block scheme for the computer-controlled experimental setup is on Figure 1a. The ultrasound emitters are the two focusing ultrasound transmitters (Fig. 1b). Each transmitter consists of 60 discrete ultrasound Murata radiators with resonant frequency of ~ 40 kHz, placed in first four Fresnel zones. The transmitters have a focal distance of 0.25 m and peak sound pressure in focus 140 dB.

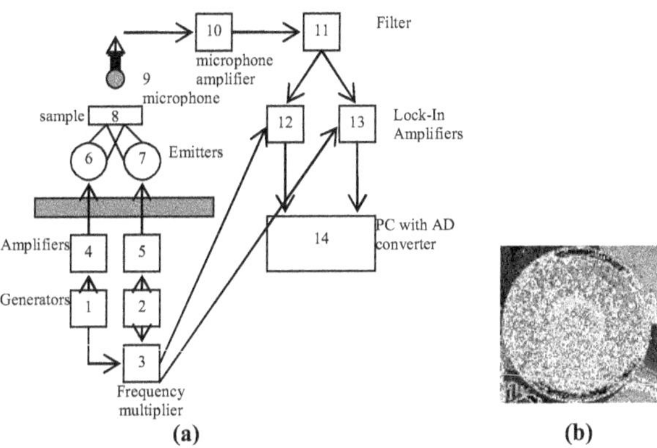

Figure 1: The block scheme of a setup (a) and used ultrasound transmitter (b)

Two signals with frequencies 38.5 kHz and 39.5 kHz respectively are taken from the generators (1), (2) and passed to the frequency multiplier (3) and to the power amplifiers (4), (5). From amplifiers (4) and (5) the signals go the on transmitters (6), (7). The signals with the sum and difference frequencies of $\omega_1 + \omega_2 = 80 kHz$ and $\omega_1 - \omega_2 = 1 kHz$ respectively, are produced by multiplier (3) and set as the reference signals on Lock-In Amplifiers (12) and (13).

The transmitters are fixed on the optical bench and are focused in a single area of investigated sample (8). As a result of nonlinear interaction of probe signals in the sample under study, besides signals of base frequencies emitted by antennas, the ones of sum $\omega_1 + \omega_2 = 80 kHz$ and difference $\omega_1 - \omega_2 = 1 kHz$ frequencies appear in the spectrum of reflected signals (Fig. 2). Nonlinear response is detected with capacitor microphone (9). Further, the signal is boing amplified (10) and passed through the HF and LF filters (11) to the inputs of two Lock-In Amplifiers (12) and (13). This allows us to measure both difference and sum frequencies. The signal with the sum frequency is 35 dB higher than the one with the difference frequency. Therefore all further results are presented for the measurements made with sum frequency.

The measurement of nonlinear response value dependence on coordinate allows to define spatial distribution of nonlinear parameter in the sample and to localize the damages of structure.

Background noise measured on an object without structural damages is used as a zero signal value in all of the experiments.

4. Experimental Results

For testing measurements the silicone sample was made (Fig. 2). Its density was 1200 kg/m^3. The sizes of the sample were 7x5x2.5 cm. The subsurface structural inclusions (a grain of millet, inner air shot of 0.3 cm in diametr, and a grain of buckweat) were placed 1.5 cm away from each other in the sample at one depth of 0.5 cm and imitated the structural damage between layers of a composite material.

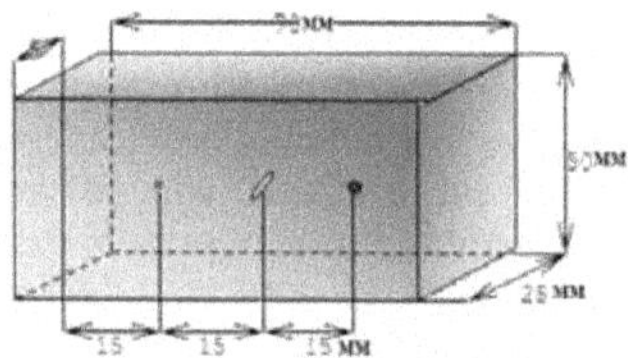

Figure 2: The sample with subsurface structural inclusions

Scanning of the sample with focused ultrasound beams along the horisontal axis were made with fixed-step-size, and nonlinear response dependence on coordinates was built (Fig. 3). Conducted experiments allowed defining its spatial distribution with an accuracy to 0.5 cm, set with focal spot size of the emitters. Nonlinear response exceeded background noise approximately by 12-18 dB.

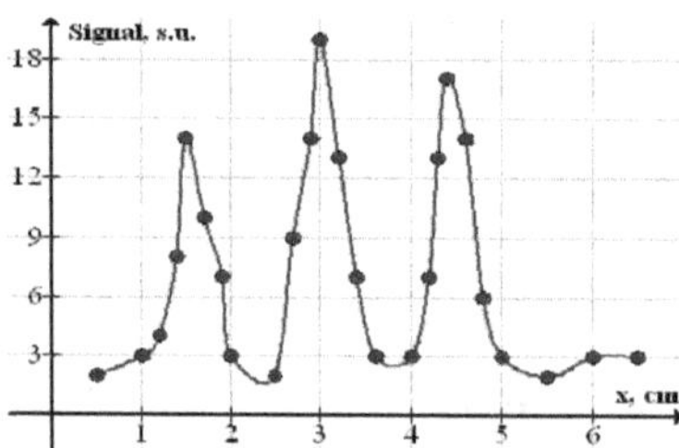

Figure 3: Sample scanning results

5. Conclusions

The sufficient complex of program and experimental facility, allowing with the high accuracy to define the defects in elastic structural and thermal engineering materials is developed. The deviations of the calculated pressure (using $\varepsilon(p)$ estimations) from the pressure measured in the experiments are insignificant showing sufficient accuracy in the estimations of $\varepsilon(p)$ of the analyzed materials.

6. References

[1] O.M. Alifanov, E.A. Artyukhin and S.V. Rumyantsev, Extreme Methods for Solving Ill-Posed Problems with Applications to Inverse Problems, Begell House, 1995.

[2] O.M. Alifanov, Inverse Heat Transfer Problems, Springer-Verlag, 1994.

[3] O.V. Rudenko, S.N. Soluyan, Theoretical basics of nonlinear acoustics. Nauka, 1975.

[4] Mostafa Fatemi, Lester E. Wold, Azra Alizad, James F. Greenleaf. *IEEE Trans Med. Imaging* 21(1), 1-8, 2002.

On Finite Element Analysis of An Inverse Problem in Elasticity

David W. Nicholson
Mechanical, Materials and Aerospace Engineering
University of Central Florida, Orlando, FL 32816
Email: nicholsn@mail.ucf.edu

Abstract

The present investigation concerns an inverse problem in elasticity, modeled by the finite element method. Even when the inverse problem possesses a unique solution when modeled "exactly", a given mesh may cause the finite element model to fail to do so. This investigation addresses a particular inverse problem and identifies a matrix *non-singularity criterion* assuring a unique solution for the FEA model. A numerical test is introduced to verify satisfaction of the criterion, exploiting Gram-Schmidt orthogonalization. The test also serves to identify how the mesh can be modified to obviate singularity. Examples are given demonstrating the effectiveness of the criterion and the test.

1. Introduction. Direct and Inverse Problems

In an elastic body under static loads, the finite element equation may be written as

$$\mathbf{K}\begin{Bmatrix}\mathbf{u}_1\\ \mathbf{u}_2\\ \mathbf{u}_i\end{Bmatrix}=\begin{Bmatrix}\mathbf{f}_1\\ \mathbf{f}_2\\ \mathbf{0}\end{Bmatrix} \qquad \mathbf{K}=\begin{bmatrix}\mathbf{K}_{11} & \mathbf{K}_{12} & \mathbf{K}_{13}\\ \mathbf{K}_{21} & \mathbf{K}_{22} & \mathbf{K}_{23}\\ \mathbf{K}_{31} & \mathbf{K}_{32} & \mathbf{K}_{33}\end{bmatrix} \tag{1}$$

Here the $n_i \times 1$ vector $\mathbf{u}_i$ denotes the displacement degrees of freedom at internal nodes, while the $n_1 \times 1$ vector $\mathbf{u}_1$ and the $n_2 \times 1$ vector $\mathbf{u}_1$ denote displacement degrees of freedom at two different sets of boundary nodes. It is assumed that external forces are applied only to boundary nodes, with the $n_1 \times 1$ force vector $\mathbf{f}_1$ and the $n_2 \times 1$ force vector $\mathbf{f}_2$ referred to the two sets. The purpose of using two sets will be evident in the subsequent discussion of an inverse problem. All matrices and vectors are real.

The system has n degrees of freedom in which $n = n_1 + n_2 + n_i$. It is assumed that Eq. (1) represents degrees of freedom remaining after any *simple constraints* on the body have been applied. Here, in a simple constraint a displacement degree of freedom is specified and the corresponding reaction force is an unknown. This contrasts with complex constraints appearing in inverse problems in which *both* the displacement *and* force are prescribed for a degree of freedom.

In finite element models of static problems in linear elasticity, the $n \times n$ stiffness matrix **K** is symmetric positive definite. In the direct problem, which is considered 'well posed', the force vectors $\mathbf{f}_1$ and $\mathbf{f}_2$ are prescribed, and the corresponding displacement vectors $\mathbf{u}_1$ and $\mathbf{u}_2$ are unknowns to be determined (after enforcing the simple constraints). Accordingly, at each boundary degree of freedom, only one quantity (the force) is specified. Positive definiteness of **K** implies that the solution of the direct problem exists and is unique.

In contrast, in the specific inverse problem under investigation, the first node set may be 'overspecified' in that displacement and force are specified at each nodal degree of freedom. Correspondingly, the second node set is 'underspecified' in that neither the displacement nor the force are specified at the degrees of freedom, and in fact are both unknowns to be determined by solving the inverse problem, *assuming the 'exact' solution exists and is unique.* Introducing conditions and a method for determining existence and uniqueness of the solution *in a finite element model* is the goal of the current investigation. Examples of investigations of inverse problems in mechanics are Dennis, Dulikravitch and Yoshimura [1] and in Gladwell [2].

2. Condition for the Existence of a Solution to the Inverse Problem

2.1 *On Nonexistence of a Unique Solution in the Finite Element Model*

We first illustrate that a unique solution to the finite element model may not exist even though the exact direct problem does possess a unique solution. Consider the following problem:

let $\mathbf{K}_{11}$ and $\mathbf{K}_{22}$ be two $n \times n$ symmetric positive definite matrices and let $\mathbf{K}_{12}$ denote a singular $n \times n$ matrix. Next introduce the matrix **H** given by

$$\mathbf{H} = \begin{bmatrix} \mathbf{K}_{11} & \sqrt{\mathbf{K}_{11}}\mathbf{K}_{12} \\ \mathbf{K}_{12}^T\sqrt{\mathbf{K}_{11}} & \mathbf{K}_{12}^T\mathbf{K}_{12}+\mathbf{K}_{22} \end{bmatrix} = \begin{bmatrix} \sqrt{\mathbf{K}_{11}} & \mathbf{0} \\ \mathbf{K}_{12}^T & \sqrt{\mathbf{K}_{22}} \end{bmatrix}\begin{bmatrix} \sqrt{\mathbf{K}_{11}} & \mathbf{K}_{12} \\ \mathbf{0}^T & \sqrt{\mathbf{K}_{22}} \end{bmatrix}.$$

Now suppose that the right hand side exhibits the vector $\begin{Bmatrix} \mathbf{f}_{k1} \\ \mathbf{f}_{u2} \end{Bmatrix}$ in which $\mathbf{f}_{k1}$ is known while $\mathbf{f}_{u2}$ is unknown. Also suppose that the left hand side now contains the vector $\begin{Bmatrix} \mathbf{u}_{k1} \\ \mathbf{u}_{u2} \end{Bmatrix}$ in which $\mathbf{u}_{k1}$ is known but $\mathbf{u}_{u2}$ is unknown. We are now confronted by the inverse problem

$$\begin{bmatrix} \mathbf{K}_{11} & \sqrt{\mathbf{K}_{11}}\mathbf{K}_{12} \\ \mathbf{K}_{12}^T\sqrt{\mathbf{K}_{11}} & \mathbf{K}_{12}^T\mathbf{K}_{12}+\mathbf{K}_{22} \end{bmatrix}\begin{Bmatrix} \mathbf{u}_{k1} \\ \mathbf{u}_{u2} \end{Bmatrix} = \begin{Bmatrix} \mathbf{f}_{k1} \\ \mathbf{f}_{u2} \end{Bmatrix}.$$

The upper block row serves to determine that

$$\sqrt{\mathbf{K}_{11}}\mathbf{K}_{12}\mathbf{u}_{u2} = \mathbf{f}_{k1} - \mathbf{K}_{11}\mathbf{u}_{k1}.$$

But $\mathbf{K}_{12}$ and hence $\sqrt{\mathbf{K}_{11}}\mathbf{K}_{12}$ are singular: either there is no solution for $\mathbf{u}_{u2}$ or there are many solutions.

2.2 *Sufficient Condition*

A sufficient condition is presented for the existence and uniqueness of the solution to the finite element model in the current inverse problem. In a given physical application, we will see that it is very possible that the stiffness matrix in one mesh will not satisfy the sufficient condition, while the stiffness matrix for another mesh for the same problem will satisfy the sufficient condition.

From the third row in Eq. (1), we have

$$\mathbf{K}_{31}\mathbf{u}_1 + \mathbf{K}_{32}\mathbf{u}_2 + \mathbf{K}_{33}\mathbf{u}_i = 0 \tag{2}$$

and now, recalling that $\mathbf{K}_{33} > 0$,

$$\mathbf{u}_3 = -\mathbf{K}_{33}^{-1}(\mathbf{K}_{31}\mathbf{u}_1 + \mathbf{K}_{32}\mathbf{u}_2) \tag{3}$$

Upon substitution, the two upper block rows provide the equations

$$\left[\mathbf{K}_{11} - \mathbf{K}_{13}\mathbf{K}_{33}^{-1}\mathbf{K}_{31}\right]\mathbf{u}_1 + \left[\mathbf{K}_{12} - \mathbf{K}_{13}\mathbf{K}_{33}^{-1}\mathbf{K}_{32}\right]\mathbf{u}_2 = \mathbf{f}_1 \tag{4}$$

$$\left[\mathbf{K}_{21} - \mathbf{K}_{23}\mathbf{K}_{33}^{-1}\mathbf{K}_{31}\right]\mathbf{u}_1 + \left[\mathbf{K}_{22} - \mathbf{K}_{23}\mathbf{K}_{33}^{-1}\mathbf{K}_{32}\right]\mathbf{u}_2 = \mathbf{f}_2 \tag{5}$$

Positive definiteness of the original matrix in the direct problem implies that the principal minors of **K** are all positive definite. It follows that

$$\begin{matrix} \mathbf{K}_{11} - \mathbf{K}_{13}\mathbf{K}_{33}^{-1}\mathbf{K}_{31} > 0 & \text{(a)} \\ \mathbf{K}_{22} - \mathbf{K}_{23}\mathbf{K}_{33}^{-1}\mathbf{K}_{12} > 0 & \text{(b)} \end{matrix} \quad \begin{bmatrix} \mathbf{K}_{11} - \mathbf{K}_{13}\mathbf{K}_{33}^{-1}\mathbf{K}_{31} & \mathbf{K}_{12} - \mathbf{K}_{13}\mathbf{K}_{33}^{-1}\mathbf{K}_{32} > 0 \\ \mathbf{K}_{21} - \mathbf{K}_{23}\mathbf{K}_{33}^{-1}\mathbf{K}_{31} & \mathbf{K}_{22} - \mathbf{K}_{23}\mathbf{K}_{33}^{-1}\mathbf{K}_{12} > 0 \end{bmatrix} > 0 \quad \text{(c)} \tag{6}$$

Now suppose that $\mathbf{u}_1 = \mathbf{u}_k$ and $\mathbf{f}_1 = \mathbf{f}_k$ are known, while $\mathbf{u}_2 = \mathbf{u}_u$ and $\mathbf{f}_2 = \mathbf{f}_u$ are unknown, thereby introducing the inverse problem of interest. Equations (4) and (5) are now rewritten as

$$[\mathbf{K}_{11} - \mathbf{K}_{13}\mathbf{K}_{33}^{-1}\mathbf{K}_{31}]\mathbf{u}_k + \left[\mathbf{K}_{12} - \mathbf{K}_{13}\mathbf{K}_{33}^{-1}\mathbf{K}_{32}\right]\mathbf{u}_u = \mathbf{f}_k \tag{7}$$

$$\left[\mathbf{K}_{21}-\mathbf{K}_{23}\mathbf{K}_{33}^{-1}\mathbf{K}_{31}\right]\mathbf{u}_k+\left[\mathbf{K}_{22}-\mathbf{K}_{23}\mathbf{K}_{33}^{-1}\mathbf{K}_{12}\right]\mathbf{u}_u=\mathbf{f}_u \tag{8}$$

Equation (7) now furnishes

$$\left[\mathbf{K}_{12}-\mathbf{K}_{13}\mathbf{K}_{33}^{-1}\mathbf{K}_{32}\right]\mathbf{u}_u=\mathbf{f}_k-\left[\mathbf{K}_{11}-\mathbf{K}_{13}\mathbf{K}_{33}^{-1}\mathbf{K}_{31}\right]\mathbf{u}_k. \tag{9}$$

For the solution of the finite element model expressed by Eqs (7,8) to exist and be unique it is necessary and sufficient that $\mathbf{K}_{12}-\mathbf{K}_{13}\mathbf{K}_{33}^{-1}\mathbf{K}_{32}$ be nonsingular. Note that the existence of the solution is unaffected by the actual values of $\mathbf{u}_k$ and $\mathbf{f}_k$. Also note that it is not necessary to attain a *simultaneous* solution for $\mathbf{u}_u$ and $\mathbf{f}_u$, as may be supposed. Instead, once $\mathbf{u}_u$ is obtained by solving Eq. (9), $\mathbf{f}_u$ is immediately found from Eq. (8).

One immediate *necessary* condition for $\mathbf{K}_{12}-\mathbf{K}_{13}\mathbf{K}_{33}^{-1}\mathbf{K}_{32}$ to be nonsingular is that the mesh must ensure that $n_1=n_2$, since otherwise it would be non-square and hence singular.

Of course the inverse $\mathbf{K}_{33}^{-1}$ appears in the foregoing relation. We will see in the next section that the evaluation of this matrix inverse can be avoided.

3. Test for Nonsingularity of $\mathbf{K}_{12}-\mathbf{K}_{13}\mathbf{K}_{33}^{-1}\mathbf{K}_{32}$

We now introduce a simple numerical test for determining whether $\mathbf{K}_{12}-\mathbf{K}_{13}\mathbf{K}_{33}^{-1}\mathbf{K}_{32}$ is nonsingular given that $n_1=n_2$. It is first demonstrated that nonsingularity of $\mathbf{K}_{12}-\mathbf{K}_{13}\mathbf{K}_{33}^{-1}\mathbf{K}_{32}$ is equivalent to requiring that the matrix $\begin{bmatrix}\mathbf{K}_{12} & \mathbf{K}_{23}\\ \mathbf{K}_{32} & \mathbf{K}_{33}\end{bmatrix}$ be nonsingular. In particular, $\begin{bmatrix}\mathbf{K}_{12} & \mathbf{K}_{23}\\ \mathbf{K}_{32} & \mathbf{K}_{33}\end{bmatrix}$ is *singular* if and only if the vector $\begin{Bmatrix}\mathbf{x}\\ \mathbf{y}\end{Bmatrix}$ exists which satisfies

$$\begin{bmatrix}\mathbf{K}_{12} & \mathbf{K}_{23}\\ \mathbf{K}_{32} & \mathbf{K}_{33}\end{bmatrix}\begin{Bmatrix}\mathbf{x}\\ \mathbf{y}\end{Bmatrix}=\begin{Bmatrix}\mathbf{0}\\ \mathbf{0}\end{Bmatrix} \tag{10}$$

Of course $\mathbf{x}$ is $n_1\times 1$ and $\mathbf{y}$ is $n_i\times 1$. But, from the bottom row, $\mathbf{K}_{32}\mathbf{x}+\mathbf{K}_{33}\mathbf{y}=\mathbf{0}$ and hence $\mathbf{y}=-\mathbf{K}_{33}^{-1}\mathbf{K}_{32}\mathbf{x}$ in the singular case. Substitution into the top row and some manipulations now furnish

$$(\mathbf{K}_{12}-\mathbf{K}_{23}\mathbf{K}_{33}^{-1}\mathbf{K}_{32})\mathbf{x}=\mathbf{0} \tag{11}$$

as anticipated. It follows that $\mathbf{K}_{12}-\mathbf{K}_{23}\mathbf{K}_{33}^{-1}\mathbf{K}_{32}$ is singular whenever $\begin{bmatrix}\mathbf{K}_{12} & \mathbf{K}_{23}\\ \mathbf{K}_{32} & \mathbf{K}_{33}\end{bmatrix}$ is singular , and is non-singular whenever $\begin{bmatrix}\mathbf{K}_{12} & \mathbf{K}_{23}\\ \mathbf{K}_{32} & \mathbf{K}_{33}\end{bmatrix}$ is nonsingular.

4. Method of Verifying Existence and Uniqueness of Inverse Solution

We now present a numerical test for the *nonsingularity* of an $n\times n$ matrix **A**, exploiting linear independence of its rows. The literature on tests for nonsingularity, and more generally eigenvalues, identifies several major methods contrasting with the current method. One is determinant search, while a second is based on the Singular Value Decomposition (cf Dahlqvist and Bjork [3]). Of course the latter is very effective and otherwise plays a major role in inverse problems owing to its relative insensitivity to errors (eg, Chan [4]). However, its numerical implementation typically depends on using transformations to convert the matrix to triangular or bidiagonal form, and is not selective in that it produces all singular values and not just the minimum value. In contrast the test introduced below, based on Gram-Schmidt orthogonalization (cf Dahlqvist and Bjork [3]) shows how entries of the untransformed matrix can be modified to avoid singularity in the finite element model of interest and more generally to avoid eigenvalue determination.

First, the matrix is written in the form

$$\mathbf{A} = \begin{bmatrix} \mathbf{a}_1^T \\ \mathbf{a}_2^T \\ . \\ \mathbf{a}_{n-1}^T \\ \mathbf{a}_n^T \end{bmatrix} \tag{12}$$

in which the ith row of the matrix **A** is written as the row vector $\mathbf{a}_j^T$. We now construct a set of orthonormal base vectors $\mathbf{e}_i$ as follows (via Gram-Schmidt orthogonalization). The n base vectors $\mathbf{e}_j$ are given by

$$\begin{aligned} & \mathbf{e}_1 = \mathbf{a}_1 / |\mathbf{a}_1| \\ \hat{\mathbf{e}}_2 = \mathbf{a}_2 - (\mathbf{a}_2^T \mathbf{e}_1)\mathbf{e}_1 \qquad & \mathbf{e}_2 = \hat{\mathbf{e}}_2 / |\hat{\mathbf{e}}_2| \\ . \qquad & . \\ \hat{\mathbf{e}}_{n-1} = \mathbf{a}_{n-1} - \sum_{j=1}^{n-1} (\mathbf{a}_{n-1}^T \mathbf{e}_j)\mathbf{e}_j \qquad & \mathbf{e}_{\text{n-1}} = \hat{\mathbf{e}}_{n-1} / |\hat{\mathbf{e}}_{n-1}| \\ \hat{\mathbf{e}}_n = \mathbf{a}_n - \sum_{j=1}^{n-1} (\mathbf{a}_n^T \mathbf{e}_j)\mathbf{e}_j \qquad & \mathbf{e}_{\text{n}} = \hat{\mathbf{e}}_n / |\hat{\mathbf{e}}_n| \end{aligned} \tag{13}$$

If the matrix **A** is nonsingular, the jth row vector cannot be a linear combination of the foregoing *j-1* vectors. Accordingly, the jth row vector $\mathbf{a}_j$ exists in the *j*-dimensional subspace spanned by the base vectors $\mathbf{e}_1, \mathbf{e}_2, ...\mathbf{e}_j$. For the moment suppose that the matrix **A** is singular with rank deficiency unity, and that $\mathbf{a}_n^T$ is a linear combination of the foregoing row vectors and *hence exists in an n-1 dimensional subspace spanned by the base vectors* $\mathbf{e}_1, \mathbf{e}_2, ..\mathbf{e}_j..\mathbf{e}_{n-1}$. However, if $\mathbf{e}_n$ lies in the n-1 dimensional subspace and is also orthogonal to the base vectors of the subspace, it follows that $\mathbf{e}_n = \mathbf{0}$. More generally, if the matrix is singular, then at least one row vector $\mathbf{a}_k^T$ is a linear combination of the foregoing row vectors $\mathbf{a}_1^T, \mathbf{a}_2^T ...., \mathbf{a}_{k-1}^T$, and 'spans' the corresponding $(k-1)$ dimensional subspace. The vector $\mathbf{e}_k$ likewise is orthogonal to the base vectors of the space it lies in, and hence $\hat{\mathbf{e}}_k = \mathbf{0}$.

Accordingly, the condition for the matrix **A** to be nonsingular is

$$\hat{\mathbf{e}}_k \neq 0, \quad k = 1, 2, 3, ..., n \tag{14}$$

If any of the vectors $\hat{\mathbf{e}}_k$ vanish, **A** is singular.

We now illustrate the application and performance of the nonsingularity test expressed in Eq. (14). Consider the matrix

$$\mathbf{A} = \begin{bmatrix} 1 & 2 & 3 \\ 2 & 4 & 6+\varepsilon_1 \\ 3 & 6 & 9+\varepsilon_2 \end{bmatrix} \tag{15}$$

We seek to determine whether the current test correctly indentifies values of ε_1 and ε_2 for which A is singular.

$$\mathbf{e}_1 = \frac{1}{\sqrt{14}} \begin{Bmatrix} 1 \\ 2 \\ 3 \end{Bmatrix} \qquad \hat{\mathbf{e}}_2 = \begin{Bmatrix} 2 \\ 4 \\ 6+\varepsilon_1 \end{Bmatrix} - \{2 \;\; 4 \;\; 6+\varepsilon_1\} \frac{1}{\sqrt{14}} \begin{Bmatrix} 1 \\ 2 \\ 3 \end{Bmatrix} \frac{1}{\sqrt{14}} \begin{Bmatrix} 1 \\ 2 \\ 3 \end{Bmatrix} = \frac{1}{14} \begin{Bmatrix} -3 \\ -6 \\ 5 \end{Bmatrix} \varepsilon_1 \tag{16}$$

It is evident that $\mathbf{A}$ is singular if $\varepsilon_1 = 0$. The current procedure furnishes that $\hat{e}_2 = 0$ if $\varepsilon_1 > 0$, but $\hat{e}_2 \to 0$ as $\varepsilon_1 \downarrow 0$. Assume for the moment that $\varepsilon_1 > 0$, such that

$$\mathbf{e}_2 = \frac{1}{\sqrt{70}}\begin{Bmatrix} -3 \\ -6 \\ 5 \end{Bmatrix}. \tag{17}$$

The third base vector is now found as

$$\hat{e}_3 = \begin{Bmatrix} 3 \\ 6 \\ 9+\varepsilon_2 \end{Bmatrix} - \left[\{3 \;\; 6 \;\; 9+\varepsilon_2\}\frac{1}{\sqrt{14}}\begin{Bmatrix} 1 \\ 2 \\ 3 \end{Bmatrix}\right]\frac{1}{\sqrt{14}}\begin{Bmatrix} 1 \\ 2 \\ 3 \end{Bmatrix} - \left[\{3 \;\; 6 \;\; 9+\varepsilon_2\}\frac{1}{\sqrt{70}}\begin{Bmatrix} -3 \\ -6 \\ 5 \end{Bmatrix}\right]\frac{1}{\sqrt{70}}\begin{Bmatrix} -3 \\ -6 \\ 5 \end{Bmatrix} = \begin{Bmatrix} 0 \\ 0 \\ 0 \end{Bmatrix}. \tag{18}$$

The test in Eq. (14) indicates that the matrix $\mathbf{A}$ is singular regardless of ε_2. That this result is correct is easily seen by recognizing that the second column in $\mathbf{A}$ is proportional to the first column, regardless of ε_2 (or ε_1).

As a second example, let $\mathbf{A}=\begin{bmatrix} 1 & 2 \\ 2 & 4+\varepsilon \end{bmatrix}$. Application of the foregoing procedures reveals that $\hat{e}_2 \neq \mathbf{0}$ if $\varepsilon > 0$, and that $\hat{e}_2 \to \mathbf{0}$ as $\varepsilon \downarrow 0$, which of course is correct.

5. Mesh Dependence of Solution: Mesh Modification

As previously stated, it is quite possible that a unique solution exists in an inverse problem when treated 'exactly' in the theory of elasticity, but that a finite element model fails Eq. (14). Such a situation may reflect mesh dependence, and if so may be remedied by altering the mesh details. An obvious example of mesh dependence arises when $n_2 \neq n_1$. An example in which $n_1 = n_2$ is now presented involving mesh dependence. Consider an actively controlled beam, modeled using *two beam elements* in which the displacement of the mid-node, w_1, is sensed and an additional shear force on the end node arises according to $-V_2 = -\kappa w_1$. Of course κ is a 'gain'. The lengths of the elements are L_1 and L_2, and the bending rigidity in both elements is denoted by *EI*.

We now suppose that the slope and the moment are both known at the first node, while the displacement and the shear force (without the control system contribution) are known at the second node. Of course these conditions give rise to an inverse problem of the type of interest here. Ignoring inertial forces, the modified beam equation is now

$$EI\begin{bmatrix} 12/L_1^3+12/L_2^3 & 6/L_1^2-6/L_2^2 & -12/L_2^3 & -6/L_2^2 \\ 6/L_1^2-6/L_2^2 & 4/L_1+4/L_2 & 6/L_2^2 & 4/L_2 \\ -12/L_2^3 & 6/L_2^2 & 12/L_2^3 & 6/L_2^2 \\ -6/L_2^2 & 4/L_2 & 6/L_2^2 & 4/L_2 \end{bmatrix}\begin{Bmatrix} w(L_1) \\ -w_k'(L_1) \\ w_k(L_1+L_2) \\ -w'L_1+L_2) \end{Bmatrix} = \begin{Bmatrix} V_k(L_1) \\ M_k(L_1) \\ V_k(L_1+L_2) \\ M(L_1+L_2) \end{Bmatrix} - \begin{bmatrix} 0 & 0 & 0 & 0 \\ 0 & 0 & 0 & 0 \\ \kappa & 0 & 0 & 0 \\ 0 & 0 & 0 & 0 \end{bmatrix}\begin{Bmatrix} 0 \\ 0 \\ \kappa w(L_1) \\ 0 \end{Bmatrix} \tag{19}$$

in which the subscript k refers to known (prescribed) quantities. Equation (1.19) is now rewritten to incorporate the gain in the stiffness matrix, and the subscript k implies that the quantity is prescribed.

$$EI\begin{bmatrix} 6/L_1^2-6/L_2^2 & 4/L_2 \\ -12/L_2^3+\kappa/EI & 6/L_2^2 \end{bmatrix}\begin{Bmatrix} w(L_1) \\ -w'(L_1+L_2) \end{Bmatrix} = \begin{Bmatrix} M_k(L_1) \\ V_k(L_1+L_2) \end{Bmatrix} - EI\begin{bmatrix} 4/L_1^1+4/L_2^2 & 6/L_2^2 \\ 6/L_2^2 & 12/L_2^3 \end{bmatrix}\begin{Bmatrix} -w_k'(L_1) \\ w_k(L_1+L_3) \end{Bmatrix} \tag{20}$$

Equation (20) has no solution in the specific case in which $\frac{\kappa}{EI} = \frac{9}{L_1^2 L_2} + \frac{3}{L_2^3}$, shown equally well by taking the determinant in this example or by applying Eq. (14). However, with a mesh such that $\frac{\kappa}{EI} \neq \frac{9}{L_1^2 L_2} + \frac{3}{L_2^3}$, the existence of a solution is assured.

6. Conclusion

A mathematical relation is derived which serves to determine the existence of the solution to an inverse problem modeled using the finite element method. The inverse of a matrix is present in the relation. However, the need to evaluate the inverse is avoided. To enable application of the relation, a simple test based on Gram-Schmidt orthogonalization is introduced and illustrated with an example. Using a simple inverse problem consisting of an actively controlled two element beam, it is demonstrated that it is possible for an unfortunate FEA mesh design to render the finite element matrices singular even though the underlying inverse problem possesses a solution when solved using exact mathematics. The relation and the test will enable identifying a suitable FEA mesh for an inverse problem for which a solution exists.

7. References

[1] Dennis, B.H., Dulikravitch, G.S., and Yoshimura, S., "A Finite Element Formulation for the Determination of Unknown Boundary Conditions for Three Dimensional Steady Thermoelastic Problems", J. Heat Transfer-Trans ASME, Vol 126, pp. 110-118, 2004.

[2] Gladwell, G.M.L., "Inverse Vibrations Problems for Finite Element Models", Inverse Problems, Vol 13, pp., 311-322, 1997.

[3] Dahlqvist, G., and A. Bjork, A. Numerical Methods, Prentice-Hall, Inc., Englewood Cliffs, N.J., 1974.

[4] Chan, T.F., "An Improved Algorithm for Computing the Singular Value Decomposition", ACM Transactions of Mathematical Software, Vol 8, No. 1, pp 73-83, 1982.

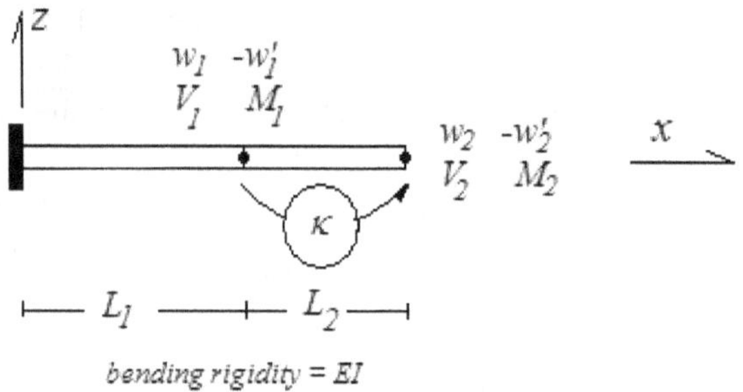

Figure 1: Two Element Model of an Actively Controlled Beam

An Inverse POD-RBF Network Approach to Parameter Estimation in Mechanics

Craig A. Rogers[1], Alain Kassab[1], Eduardo Divo[2], Ziemowit Ostrowski[3], and Ryszard Bialecki[3]

[1]Department of Mechanical, Materials, and Aerospace Engineering
University of Central Florida, Orlando, FL
Email: Alain.Kassab@ucf.edu
[2]School of Engineering Technology
Daytona State College, Daytona Beach, FL
Email: divoe@daytonastate.edu
[3]Institute of Thermal Technology, Faculty of Power and Environmental Protection
Silesian University of Technology, Gliwice, Poland
Email: Ryszard.Bialecki@polsl.pl

Abstract

In applied mechanics it is always necessary to know the fundamental parameters of a system in order to generate an accurate numerical model or to predict future operating conditions. These fundamental parameters include but are not limited to, material properties of a specimen, boundary conditions governing a system or essential dimensional characteristics that define the system or body. In certain instances, there may be little knowledge about the systems conditions or properties; as a result the problem cannot be accurately modeled using standard numerical methods. Thus, it is critical to define an approach that is capable of identifying such characteristics of the problem at hand. In this paper, an inverse approach is formulated using proper orthogonal decomposition (POD) with an accompanying radial basis function (RBF) network to estimate various physical parameters of a specimen with little prior knowledge of the system. Specifically an elasticity problem is considered and modeled using the finite elements method (FEM). In order to create the truncated POD-RBF network to be utilized in the inverse approach, a series of direct FEM solutions is used to generate a data set of deformations within the system or body, each produced for a unique set of physical parameters. The data set is then transformed via POD to generate an orthonormal basis to accurately solve for the desired material characteristics using the Levenberg-Marquardt (LM) algorithm. For now, the LM algorithm can be simply defined as a direct relation to the minimization of the Euclidean norm of the objective Least Squares function(s). The goal of this paper is to provide a flexible inverse technique which can be applied into various fields of engineering and/or mechanics. This approach is designed to offer an inexpensive, yet accurate, way to estimate material characteristics or fundamental properties using nondestructive evaluations. While the POD-RBF inverse approach outlined in this paper focuses primarily in applications to elasticity, this technique is designed to be directly applicable to other realistic conditions and/or industries.

Key Words: Proper orthogonal decomposition, inverse problem, parameter estimation, heat conduction, elasticity, fracture mechanics.

1. Introduction

The concept of proper orthogonal decomposition (POD) began over a century ago as a statistical tool developed by Pearson. Since that time, this method has been redeveloped under various names and in vastly different applications. Depending on how the input data is utilized POD is also similarly known as Karhunen-Loéve decomposition (KLD), principal component analysis (PCA) or singular value decomposition (SVD) [6]. Furthermore, this technique has been implemented in various applications from signal processing and control theory, human face recognition, data compression, parameter estimation and many others. However, to the knowledge of the authors POD has never been extended into the field of fracture mechanics and additionally has little evidence of its application in elasticity.

The application of POD as an inverse method arose due to the demanding task of solving an ill-posed inverse problem. Inverse methods have been studied for decades, ever since the mainstream use of FEM began to emerge. Different concepts have been outlined by many authors with techniques to retrieve the best solution [4,7,8]. The main techniques developed for solving inverse problems is primarily regularization; however, other methods like model reduction by decreasing the degrees of freedom and/or filtering excess error also benefit the solution. The method of POD is utilized by finding the correlation between the known direct problem and the solution to be desired [6]. The application of POD can be used to produce a low-order, but high quality, approximation of the solution field. More specifically POD is capable of capturing dominant components (called principal components) of the data with typically only a few modes. This is due to the ability of POD to approximate a set of vectors using a rotated coordinate frame, where the angles of rotation are denoted as the POD basis [5,9]. Of course, the primary reason POD is a favorable in solving inverse problems, is that it provides many features of the desired methods for solving inverse problems, such as model reduction, error filtration and regularization.

Specifically in this paper, the method of POD is applied to three different engineering problems. First, a basic two dimensional heat conduction problem is modeled to estimate a spatially dependent thermal conductivity. As will be demonstrated later in this paper, this approach has a direct correlation to the analytical eigenfunctions governing the system. The next example deals with a linear elastic bar in tension in order to estimate the isotropic material coefficients within a steel sample. Finally, a compact tension specimen, relevant in fracture mechanics, is used to estimate the crack length of a sample under a constant Mode 1 loading [2,3].

2. Method

The first step in the implementation of POD is the creation of the *snapshot* which is the collection of *N* sampled values of **u** - the field under consideration. In heat conduction problems the vector ***u*** stores the discrete temperature field while in elasticity the snapshot is a sampled deformation field. Next, a collection of *M* snapshots denoted as $\mathbf{u}^j$ (for $j = 1, 2 \ldots M$) are generated by altering the parameter(s) upon which the field depends on. In the current scope, the altered parameters refer to the parameters describing the spatial distribution of the conductivity, Young modulus and Poisson ratio as well as the crack length. Generally the altered parameters can be any selection of material properties and/or boundary conditions. Each $\mathbf{u}^j$ is then stored inside rectangular *N x M* matrix **U** denoted as the snapshot matrix. The snapshot field may be created by numerical modeling of the system, say FEM or BEM, sampling an analytical solution or from actual empirical data. The goal of POD is to establish a set of orthonormal vectors $\mathbf{\Phi}^j$ (for $j = 1, 2 \ldots M$) resembling the snapshot matrix **U** in an optimal way. The matrix $\mathbf{\Phi}$ is commonly referred to as the POD *basis* and follows:

$$\mathbf{\Phi} = \mathbf{U} \cdot \mathbf{V} \tag{1}$$

V represents the eigenvectors of the covariance matrix **C** and can easily be derived using the nontrivial solution of the general eigenvalue problem denoted as:

$$\mathbf{C} \cdot \mathbf{V} = \mathbf{\Lambda} \cdot \mathbf{V} \tag{2}$$

$\mathbf{\Lambda}$ represents a diagonal matrix that stores the eigenvalues λ of the covariance matrix **C**, which is defined as:

$$\mathbf{C} = \mathbf{U}^{\mathrm{T}} \cdot \mathbf{U} \tag{3}$$

It may also serve to note that **C** is symmetric and positive definite and λ is always real and positive. Typically λ is sorted in a descending order and can often be attributed to the energy of the POD mode (base vector). This energy decreases rapidly with the increasing mode number. Since higher modes hold little energy (or data) of the system they can be discarded without influencing the accuracy of representation (1). This is known as the truncation of the POD basis and is accomplished by deciding which fraction of the energy of the system can be neglected in later calculations. The resulting POD basis $\hat{\mathbf{\Phi}}$, referred to as the *truncated* POD basis consists of $K < M$ vectors and is:

$$\hat{\mathbf{\Phi}} = \mathbf{U} \cdot \hat{\mathbf{V}} \tag{4}$$

This also corresponds to the truncation of the eigenvector matrix, denoted as $\hat{\mathbf{V}}$, which stores the first K^{th} eigenvectors of **C**. The truncated POD basis (4) is also known to be orthogonal $\hat{\mathbf{\Phi}}^{\mathrm{T}} \cdot \hat{\mathbf{\Phi}} = \mathbf{I}$ and presents optimal approximation properties. Once $\hat{\mathbf{\Phi}}$ is known, the snapshot matrix **U** can be regenerated and approximated as:

$$\mathbf{U} = \hat{\mathbf{\Phi}} \cdot \mathbf{A} \tag{5}$$

Where **A** stands for the amplitudes associated with $\mathbf{u}^j$. Now referring to the orthogonality of $\hat{\mathbf{\Phi}}$, the amplitudes can be determined from:

$$\mathbf{A} = \hat{\mathbf{\Phi}}^{\mathrm{T}} \cdot \mathbf{U} \tag{6}$$

At this time, data may begin to be extrapolated for information on the current problem. To do this, consider a vector **p** which stores the parameters on which the solution depends. The transient derivation is not further described in this paper, for more information refer to [4,5].

Next, the amplitudes **A** are defined as a nonlinear function of the parameter vector **p**. The unknown constant coefficients of the current combination are gathered in a matrix **B**, shown as:

$$\mathbf{A} = \mathbf{B} \cdot \mathbf{F} \tag{7}$$

F is defined as the matrix of interpolation functions, where the set of interpolation functions $f_i(\mathbf{p})$ can be chosen arbitrarily. However, some choices of interpolation functions may lead to an ill-conditioned system of equations for the coefficient matrix **B**. In this paper, radial basis functions (RBF's) have been used as the interpolating function of choice due to their nice approximation and smoothing properties. Here the Hardy inverse multi-quadric radial basis function has been employed as:

$$f_i(\mathbf{p}) = f_i(|\mathbf{p} - \mathbf{p}^i|) = \frac{1}{\sqrt{|\mathbf{p} - \mathbf{p}^i|^2 + r^2}} \tag{8}$$

Where r is defined as the RBF smoothing factor and $\mathbf{p}^i$ corresponds to the same parameter **p** used to generate $\mathbf{u}^i$ (for $i = 1, 2 \ldots M$). It should be seen that the argument of the ith RBF is the distance $|\mathbf{p} - \mathbf{p}^i|$ between its current parameter **p** and the reference parameter $\mathbf{p}^i$.

To use (7), the matrix of coefficients **B** needs to be evaluated. This can be done by simple inversion:

$$\mathbf{B} = \mathbf{A} \cdot \mathbf{F}^{-1} \tag{9}$$

As stated previously, **F** is the matrix of interpolation functions defined as set of M identical vectors $\mathbf{f}(\mathbf{p})$ defined as $\{\mathbf{f}\}_j = f_j\left(|\mathbf{p} - \mathbf{p}^j|\right)$. Requiring that (9) is exact for all vectors used to generate the snapshots, leads to a definition of the **F** matrix as:

$$\mathbf{F} = \begin{bmatrix} f_1(|\mathbf{p}^1 - \mathbf{p}^1|) & \cdots & f_1(|\mathbf{p}^j - \mathbf{p}^1|) & \cdots & f_1(|\mathbf{p}^M - \mathbf{p}^1|) \\ \vdots & & \vdots & & \vdots \\ f_i(|\mathbf{p}^1 - \mathbf{p}^i|) & \cdots & f_i(|\mathbf{p}^j - \mathbf{p}^i|) & \cdots & f_i(|\mathbf{p}^M - \mathbf{p}^i|) \\ \vdots & & \vdots & & \vdots \\ f_M(|\mathbf{p}^1 - \mathbf{p}^M|) & \cdots & f_M(|\mathbf{p}^j - \mathbf{p}^M|) & \cdots & f_M(|\mathbf{p}^M - \mathbf{p}^M|) \end{bmatrix} \tag{10}$$

With $\mathbf{p}^i$ and $\mathbf{p}^j$ vectors of parameters used to generate i^{th} or j^{th} snapshot respectively.

At this point it should be stressed that the matrix of amplitudes **A** and the matrix of coefficients **B** are known using the above relations. Now equating (6) and (7) yields the following.

$$\hat{\mathbf{\Phi}}^{\mathrm{T}} \cdot \mathbf{U} = \mathbf{B} \cdot \mathbf{F} \tag{11}$$

Using the orthogonality of $\hat{\mathbf{\Phi}}$, it can easily be seen that the snapshot matrix **U** can be approximated as:

$$\mathbf{U}(\mathbf{p}) \approx \hat{\mathbf{\Phi}} \cdot \mathbf{B} \cdot \mathbf{F}(\mathbf{p}) \tag{12}$$

Such that after the coefficient matrix **B** is evaluated, a low dimensional model of (5), now defined as (12), can be set in vector form as:

$$\mathbf{u}(\mathbf{p}) \approx \hat{\mathbf{\Phi}} \cdot \mathbf{B} \cdot \mathbf{f}(\mathbf{p}) \tag{13}$$

This model will now be referred to as the *trained* POD-RBF network and is completely capable of reproducing the unknown field that corresponds to any arbitrary set of parameters **p**. It must be noted, that extrapolation outside the range of **p** used to generate the initial snapshots $\mathbf{u}^i$ can lead to poor accuracy of the model.

Finally, the trained POD-RBF network in (13) is used to retrieve the values of the unknown parameter vector **p**. This is done in a least squares sense by taking the sum of the squares of the data obtained from (13) and subtracting it from the actual experimental data ***y***. To avoid additional interpolation, the sampling points of the field should coincide with the sensor locations. Finally the nonlinear least squares equation reads:

$$\Psi(\mathbf{p}) = \sum_{i=1}^{N} \left(u(\mathbf{p}^i) - y_i \right)^2 \tag{14}$$

The least-squares functional is augmented with the aid of a regularization term and minimized with respect to the variable **p**, leading to:

$$J(\mathbf{p}) = \left[\sum_{i=1}^{N} \left(u_i(\mathbf{p}) - y_i \right) \frac{\partial u_i(\mathbf{p})}{\partial \mathbf{p}} \right] + \alpha \left[\sum_{i=1}^{N} \left(u_i(\mathbf{p}) - \bar{y} \right) \frac{\partial u_i(\mathbf{p})}{\partial \mathbf{p}} \right] \tag{15}$$

Which is minimized using the LM algorithm. First order regularization is performed with respect to $\bar{y}$ denoting the mean of y.

3. Results

A compact tension C(T) specimen is modeled using FEM software and the above POD-RBF inverse approach is applied to determine the unknown crack length. The C(T) specimen was modeled following ASTM E399 standards for plain strain fracture toughness and can be seen in Fig. 1.

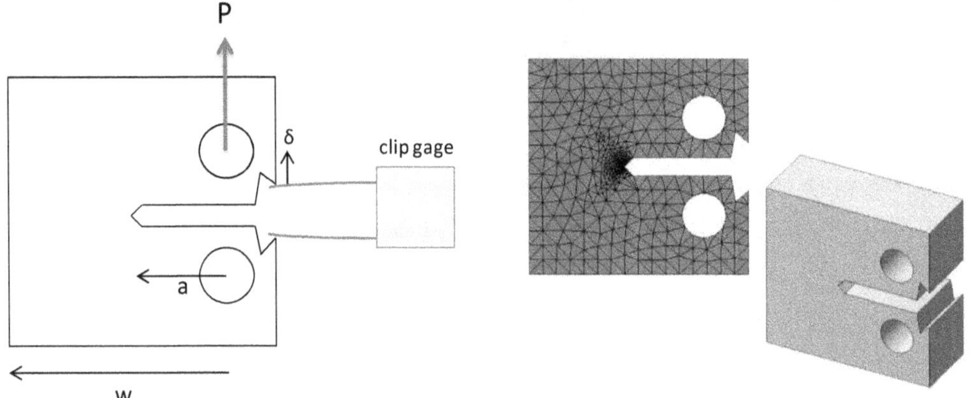

Figure 1: Model of compact tension specimen

The snapshots were generated by measuring the deformations at the notch opening of the C(T) specimen as to replicate a standard fracture experiment with a clip gage. Various crack length sizes were then implemented via FEM that ranged from 0.35 - 0.55 in to create the snapshot matrix **U**, with a total of 21 snapshots *M*. Next, the eigenvalues of the covariance matrix **C** were calculated and truncated after the 5th eigenvalue of a possible 21, as shown in Table 1. In this example, the inverse crime was avoided by calculating the initial snapshot deformations using linear elements within the FEM software. Likewise, the experimental measurements were estimated using parabolic tetrahedral elements for higher experimental accuracy; this will also allow for a more conservative estimate of the crack length.

Table 1: Truncated eigenvalues of fracture mechanics application

	λ1	λ2	λ3	λ4	λ5
λ	2.17E+08	1.81E+01	3.23E+00	3.21E+00	1.25E+00

The simulated experimental data was then obtained by adding a noise of ± 10 % of the mean value to the FEM solution. A plot of the deformation and error can be seen in Fig. 2 and Table 2.

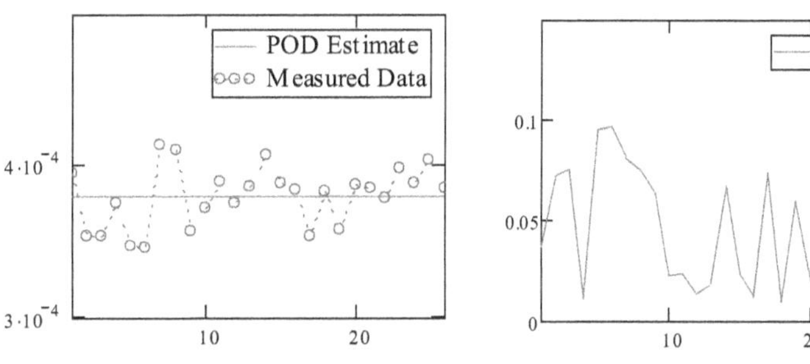

Figure 2: Deformation (left) and error (right) for ± 10% noise solution under Mode 1 loading

Table 2: POD-RBF estimated crack lengths at various amounts of added noise

	Actual (in.)	No Noise	± 5 % Noise	± 10 % Noise
Crack Length	0.416	0.43749	0.43760	0.43766

By observing the deformations, it is easy to see the least squares fit goes through the mean of data, despite the noisy solutions. This allows the POD-RBF inverse routine to optimally pick the crack length with minimal error in regard to the initial snapshot matrix developed

4. Conclusions

The POD-RBF inverse technique applied in this paper provided an excellent solution with respect to the unknown parameters to be desired. The numerical example provided help to show that the POD inverse technique outlined is quite insensitive to measurement errors, even in the presence of somewhat unlikely pessimistic conditions. Moreover, the POD-RBF inverse approach provides an efficient means of reducing the size and degrees of freedom of the problem while also optimizing accuracy of the solution to be determined. With the addition of a regularization parameter presented inside the least squares objective function(s), the solution converges effortlessly and quickly.

The overall purpose of this paper is to show that the POD-RBF technique can easily be applied to much more complicated systems and domains, as well as in other fields of research [4-6,9,10]. Ideally, the method can be further proven with applications with real experimental data. Nevertheless, this technique provides an inexpensive and nondestructive way to efficiently estimate a set of unknown parameters that have a critical role in the development of a system.

5. References

[1] Press, W.H, Teukolsky, S.A., Vetterling, W.T., & Flannery, B.P. (1992). Numerical recipes in FORTRAN 77. New York, NY: Cambridge University Press.

[2] Sanford, R.J. (2003). Principles of fracture mechanics. Upper Saddle River, NJ: Pearson Education, Inc.

[3] Anderson, T.L. (2005). Fracture mechanics: fundamentals and applications. Boca Raton, FL: CRC Press.

[4] Bialecki, R.A., Kassab, A.J., & Fic, A. (2004). Proper orthogonal decomposition and modal analysis for acceleration of transient fem thermal analysis.International Journal for Numerical Methods in Engineering, (62), 774-797.

[5] Fic, A, Bialecki, R.A., & Kassab, A.J. (2005). Solving transient nonlinear heat conduction problems by proper orthogonal decomposition and the finite-element method. Numerical Heat Transfer, (48), 103-124.

[6] Ostrowski, Z, Bialecki, R.A., & Kassab, A.J. (2005). Estimation of constant thermal conductivity by use of proper orthogonal decomposition. Computational Mechanics, (37), 52-59.

[7] Comino, L, & Gallego, R. (2005). Material constants identification in anisotropic materials using boundary element techniques. Inverse Problems in Science & Engineering, 13(6), 635-654.

[8] Znaidia, S, Mzali, F, Sassi, I, Mhimid, A, & Jemni, A. (2005). Inverse problem in a porous medium: estimation of effective thermal properties. Inverse Problems in Science & Engineering, 13(6), 581-594.

[9] Klimanek, A. (2010). Numerical modeling of heat, mass and momentum transfer in natural draft wet cooling tower. PhD thesis, Gliwice: Silesian University of Technology, Poland.

[10] Ostrowski, Z, Klimanek, A, & Bialecki, R.A. (2010). CFD two-scale model of a wet natural draft cooling tower. Numerical Heat Transfer, (57), 119-137.

[11] Feeney, B. F., & Kappangantu R. J., (1998) On the Physical Interpretation of Proper Orthogonal Modes in Vibrations. Sound Vibrations, 211, 607-616.

[12] Sirovich, L. (1995). Empirical Eigenfunctions and Low Dimensional Systems. In L. Sirovich, New Perspective in Turbulence. Springer-Yerlag.

[13] Rogers, C. (2010). Parameter Estimation in Heat Transfer and Elasticity using trained POD-RBF network inverse methods. MSc thesis,: University of Central Florida, Orlando, USA.

Numerical investigation of ductile damage parameters identification: benefit of local measurements

Emile Roux and Pierre-Olivier Bouchard

MINES ParisTech, CEMEF – Centre for material forming,
CNRS UMR 7635, BP 207, 1 rue Claude Daunesse, 06904 Sophia Antipolis cedex, France
Email: emile.roux@mines-paristech.fr

Abstract

Identification of material parameters is an important issue to improve the accuracy of finite element computations. Identification of these parameters by inverse analysis is based on experimental observables coming from mechanical experiments. In this paper, a simple tensile test is used. Two types of observables are investigated to identify ductile damage law parameters. The first one is a global measurement, such as the load-displacement curve. The second is a local observable based on full field measurements. Our approach, based on response surfaces, allows an efficient analysis of identification issues. Ill conditioned problems and multi-minima can be obtained using only a global observable. Full field measurements are a good way to improve the identification of plastic hardening and damage law parameters. In fact, local measurements combined with global ones, lead to a better formulation of the inverse problem.

Key Words: Parameters identification, inverse analysis, global optimization, sensitivity analysis, ductile damage, full field measurements

1. Introduction

Identification of material parameters is an important issue to improve the accuracy of finite element computations. Some mechanical joining processes, such as riveting or clinching are based on materials plastic deformation ability. The mechanical strength of these joining points is linked to the plastic hardening and damage history of materials during the joining process [1]. To predict the assembly final mechanical strength, reliable damage law parameters have to be idnetified.

Identification by inverse analysis of damage law parameters can be achieved using different kind of mechanical tests. Local or global measurements can be done on these tests for,the identification procedure. These different choices - local or global measurements - can be illustrated by two papers. Abendroth et al. [2] make the identification of damage law parameters using global measurement on small punch tests, and Springmann et al. [3] make the identification of damage law parameters using local field measurement on tensile tests. Therefore, in this paper, we present an approach based on response surfaces to detail the benefit of local measurements.

First, we describe the inverse analysis issue. This description details the direct model, the mechanical test, the objective function definition and briefly the minimization algorithm. The minimization algorithm used here is able to construct a response surface. This response surface gives the evolution of the objective function all over the design space. In the second part of this paper, we present response surfaces constructed with different objective functions, computed using different observables. The final goal is to evaluate the benefit of each kind of observable to solve the identification issue for ductile damage parameters.

2. Inverse analysis issues

Experimental observable - Virtual measurement

Identification by inverse analysis of material parameters is done using a tensile test. In our study, this tensile test is a virtual tensile test. In other words, experimental measurements come from a tensile test finite element modeling. This methodology allows controlling all parameters and avoids experimental uncertainties. Material parameters values which are used to generate the virtual measurement are given in Table 1.

Two different measurements are done on this tensile test: the tensile load-displacement curve, and the displacement field on one side of the sample. Our goal is to show the ability of each observable to make the identification procedure of ductile damage parameters as accurate as possible.

Direct Model

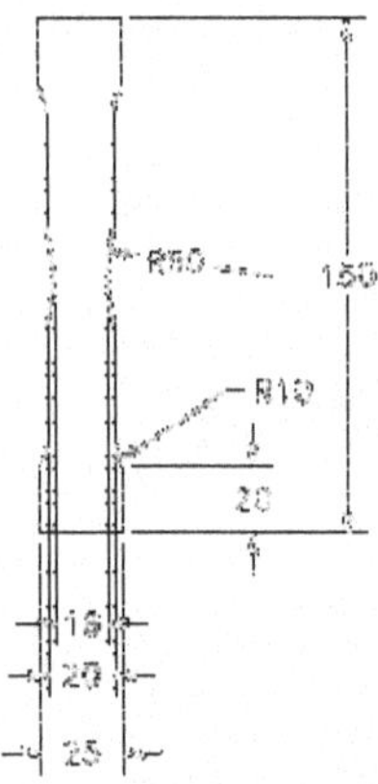

Figure 1 : Sample dimensions (mm), thickness is 2 mm - The sample is notched in order to localize strain in the centre

Parameters	Value
σ_y	46 MPa
K	430MPa
n	0.34
b	1
S_0	0.7
$\bar{\varepsilon}_d$	0.16

Table 1 : value of material parameters used to generate the virtual tensile test observable

The direct model is obtained using the finite element library CIMLib® developed at CEMEF.. Dimensions of the sample are given in Figure 1. To describe the material behavior used for cold material forming applications, an elastic-plastic law coupled with a ductile damage model has been chosen. The main equations describing the elastic-plastic behavior and ductile Lemaitre damage model are given here:

- Hardening law:

$$\sigma_0 = \sigma_y + K\,\bar{\varepsilon}^n \tag{1}$$

- Damage evolution law:

$$\dot{w} = \begin{cases} 0 \quad if \quad \bar{\varepsilon} < \bar{\varepsilon}_d \\ \dfrac{\lambda^{pl}}{1-w}\left(\dfrac{-Y}{S_0}\right)^b \quad if \quad \bar{\varepsilon} \geq \bar{\varepsilon}_d \end{cases} \tag{2}$$

The flow stress σ_0 is described by a hardening law equation(1), based on the yield stress σ_y, the consistency *K*, the equivalent plastic strain $\bar{\varepsilon}$, and the hardening exponent *n*. The evolution law for the damage parameter w is given by equation (2), where $\bar{\varepsilon}_d$ is the plastic strain threshold for damage growth, λ^{pl} is the plastic multiplier, *Y* the strain energy release rate, S_0 and *b* are material damage parameters. More details about this model are given in [4]. The mechanical behavior is coupled with the damage parameter by computing an effective stress $\tilde{\sigma}$, as shown in equation (3).

$$\tilde{\sigma}=\frac{\sigma}{1-w} \tag{3}$$

w equals zero for an undamaged material and grows toward 1 which is reached for complete fracture.

The objective of the identification procedure is to determine the parameters values for the plastic hardening and damage laws.

Objective functions

To solve the identification issue, an objective function has to be minimized. The formulation of this objective function is specific to each observable. Equation (4) defines how the objective function is defined using the load-displacement observable. *P* is the set of parameters to identify, F^{num} and $F^{\exp}$ are respectively the numerical and experimental forces, and d is the tool displacement. More details about this formulation and about how to deal with the softening part of the load-displacement curve can be found in [5]. Accounting accurately for this softening part and for the breaking instant (Load and displacement at fracture) is a key point to obtain a smooth and valuable cost function to identify damage parameters.

Equation (5) evaluates the objective function when the observable is the displacement field in the tensile direction. Ux^{num} and $Ux^{\exp}$ are respectively the numerical and experimental displacement fields in the tensile direction *x*. The same formulation is used to compute the objective function $fc_y(P)$ for transverse displacement fields.

$$\begin{cases} fc_F(P)=\sqrt{\dfrac{\sum_{i=1}^{n}\left[\left(F_i^{num}(P)-F_i^{\exp}\right)\right]^2\Delta d_i}{\min\left(\sum_{i=1}^{n}\left(F_i^{\exp}\right)^2\Delta d_i,\sum_{i=1}^{n}\left(F_i^{num}(P)\right)^2\Delta d_i\right)}} \\ \Delta d_i=d_i-d_{i-1} \end{cases} \tag{4}$$

$$fc_x(P)=\sum_{t=1}^{T}\left(\sqrt{\frac{\sum_{i=1}^{n}\left[\delta_i.\left(Ux_i^{num}(P)-Ux_i^{\exp}\right)\right]^2}{\sum_{i=1}^{n}\delta_i}}\right) \tag{5}$$

At the end, we obtain 3 objective functions. This 3 objective functions can be added up in order to obtain one single global objective function (equation (6)). Each objective function is weighted by a weight coefficient ω.

$$fc(P)=\omega_x\, fc_x(P)+\omega_y\, fc_y(P)+\omega_F\, fc_F(P) \tag{6}$$

Resolution strategy / Sensitivity analysis

To minimize the objective function, a minimization algorithm is used. A flowchart of the parallel Efficient Global Optimization [6] algorithm is presented in Figure 2. This algorithm is implemented in the MOOPI (MOdular Optimization software for Parameters Identification) software developed in the CIMLIB library. This algorithm deals with an iterative building of a kriging response surface thanks to the maximization of the expected improvement technique. In this paper, we analyze these response surfaces to evaluate the efficiency of each kind of observable.

The Efficient Global Optimization (EGO) algorithm is a global algorithm. The EGO algorithm is dedicated to time consuming objective functions. In fact, each evaluation of the objective function usually needs a finite element computation. This is extremely time consuming, in particular for highly non linear problems such as elastic-plastic and ductile damage mechanical behavior. So an adapted algorithm must be used. In order to increase the efficiency of this algorithm, we have developped a parallel version of the EGO algorithm (Figure 2). Our developments are based on a virtual enrichment method. This virtual enrichment technique sets, temporarily, the objective function value to a virtual value. This virtual value is set by the kriging response surface.

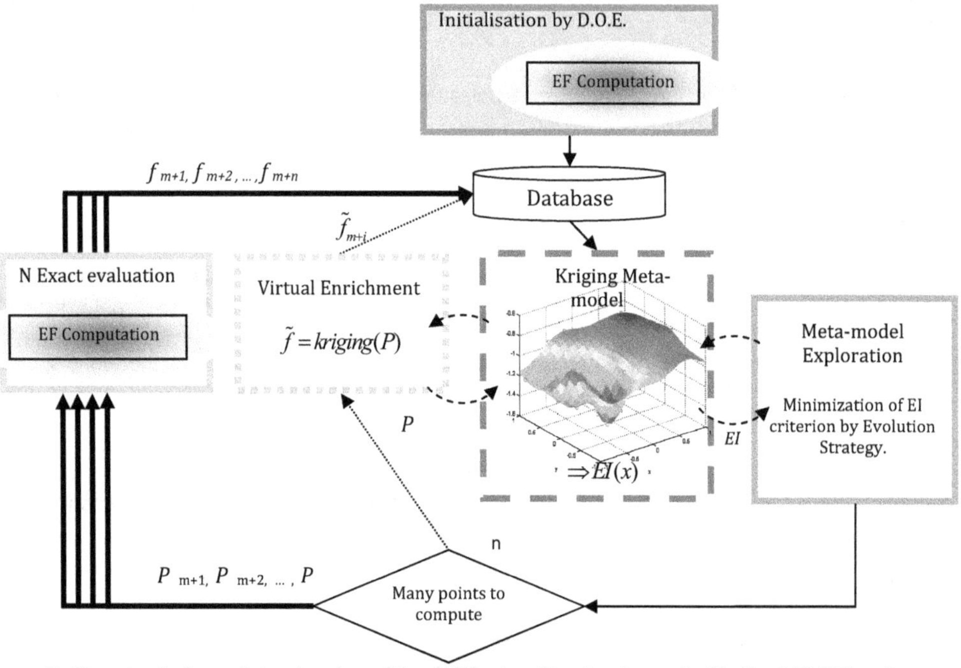

Figure 2: Flowchart of parallel extension of the EGO algorithm implemented in the MOOPI software - This algorithm is well suited to time consuming identification issues, and is able to take advantages of parallel computation capabilities

3. Results and discussions

The whole identification of materials parameters require the identification of six parameters (equation (1) and (2)).However, to illustrate the methodology and the influence of the observables choice, only two couples of parameters are selected. For these two couples, response surfaces are plotted in Figure 3 and Figure 4.

Hardening parameters identification

Figure 3 shows the evolution of objective functions evaluated using different observables. To plot the surfaces, parameter *K* ranges from 200 to 1000 MPa, and parameter *n* ranges from 0.1 to 0.6. Figure 3a exhibits an ill conditioned problem. According to the load-displacement curves, a set of *K-n* couples represent a good solution of the identification issue; i.e. objective function value is low. This set of couples values is illustrated by the red-dash line on Figure 3a. Figure 3c exhibits the same specificity; a set of K-n couples gives good results according to the objective function computed using transverse displacement fields (red-dash line on Figure 3c). Therefore, this two figures show an ill conditioned problem. However, the most interesting remark is that the two valleys (red-dash lines) have not the same orientation. So, adding these 2 objective functions enables to obtain a well conditioned identification problem. An illustration of the added up objective function is shown in Figure 3d. On this surface the minimum is well defined. Note that the use of the longitudinal displacement field (Figure 3b) also enables to converge towards a single minimum.

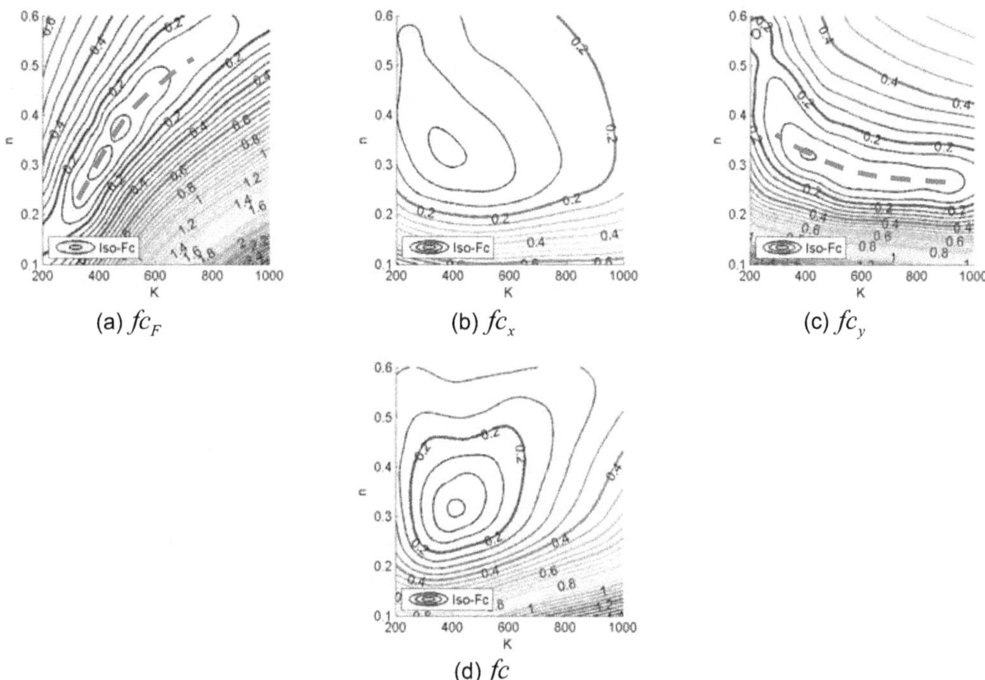

(a) fc_F (b) fc_x (c) fc_y

(d) fc

Figure 3 : sensitivity analysis of hardening parameters by kriging response surfaces, evolution of the four objective functions based on: (a) load-displacement, (b) displacement field Ux in the tensile direction, (c) displacement field Uy in the transverse direction, (d) total objective function

Damage law parameters identification

Figure 4 shows the evolution of the objective functions evaluated with different observables. To plot the surfaces, parameters *b* of the Lemaitre damage model ranges from 0.5 to 3, and parameter S_0 ranges from 0.1 to 1.5. The surface presented in Figure 4a is computed using the load-displacement curve objective function. This surface exhibits two minima. The global one (b=1, S_0=0.7) and a local one (red cross on Figure 4a). Same minima appear on the surface computed with the transverse displacement field (Figure 4c). However, the local wrong minimum is not observed on the surface computed with the tension direction displacement field (Figure 4b). of the use of local measurements, such as full field measurements, allow the suppression of local minima for ductile damage parameters identification.

Figure 4d shows the surface obtains by adding the three objective functions (equation(6)). This surface still exhibit two minima. This highlights one remaining difficulty with this methodology: the choice of the weight coefficients ω in equation(6) is important to obtain a single minimum.

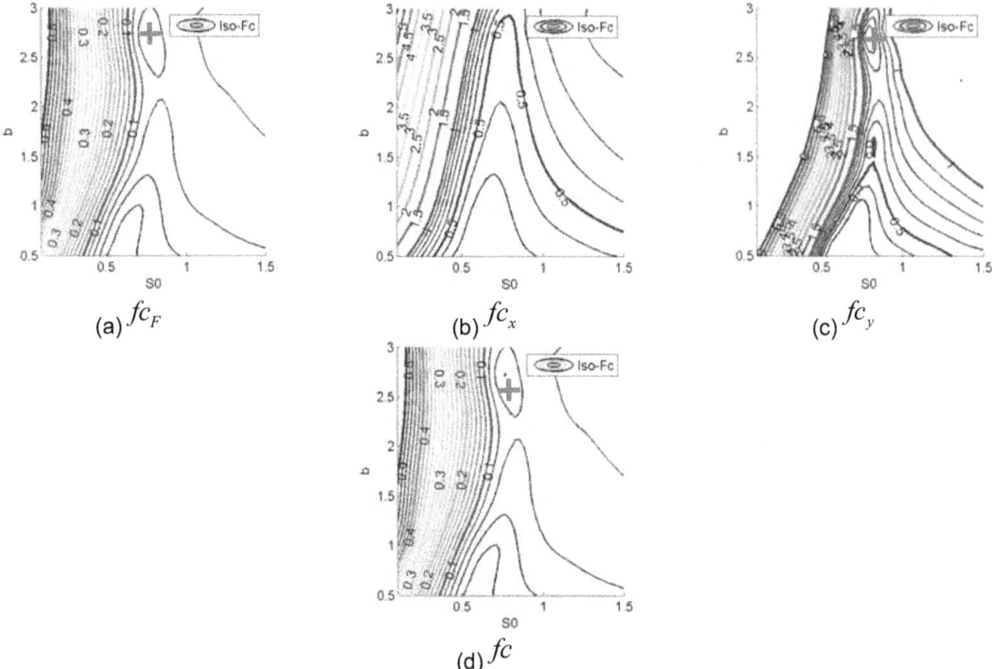

Figure 4 : sensitivity analysis of damage law parameters by kriging response surfaces, evolution of the three objective functions based on: (a) load-displacement, (b) displacement field Ux in the tensile direction, (c) displacement field Uy in the transverse direction, (d) total objective function

4. Conclusion

Response surfaces obtained with a parallel version of the EGO algorithm show that the use of local measurements is very efficient to improvet the identification of material parameters. Two couples of parameters have been studied in order to show this improvement. On the one hand, full field measurements suppress ill conditioned problems that can be observed for the hardening law identification. On the other hand, full field measurements limit the numbers of spurious local minima.

One difficulty of this methodology is to set the weight parameters of equation (6) in order to construct a mono-objective minimization problem. However, plotting the different response surfaces is a good way to evaluate and set these weight parameters.

Acknowledgement : This work has been done within the Mona Lisa project funded by CETIM.

5. References

[1] P.O. Bouchard, T. Laurent, L. Tollier, " Numerical modeling of self-pierce riveting - From riveting process modeling down to structural analysis", *J. Mater. Process. Technol.,* Vol. 202(1-3), pp. 290-300, 2008.

[2] M. Abendroth, M. Kuna, "Identification of ductile damage and fracture parameters from the small punch test using neural networks", Engineering fracture mechanics, Vol.73, pp.710-725, 2006.

[3] M. Springmann, M. Kuna, "Determination of ductile damage parameters by local deformation fields: Measurement and simulation", Archive of Applied Mechanics, Vol.75, pp.775-797, 2006.

[4] P.O. Bouchard, L. Bourgeon, S. Fayolle and K. Mocellin, "An enhanced Lemaitre model formulation for materials processing damage computation", *Int. J. Mater. Form.,* accepted, DOI 10.1007/s12289-010-0996-5.

[5] E. Roux, M. Thonnerieux and P.O. Bouchard, "Ductile damage material parameters identification: Numerical investigation", Proceedings of the Tenth International Conference on Engineering Computational Technology, paper 135, 2010, Valencia, Spain.

[6] D.R. Jones, "A Taxonomy of Global Optimization Methods Based on Response Surfaces", Journal of global optimization, Vol.21, pp.345-383, 2001.

A Stochastic-based Meshless Method for Modeling Heat Transfer

D W Pepper* and A F Emery+

* Nevada Center for Advanced Computational Methods, University of Nevada Las Vegas, Las Vegas, NV, 89154
+Department of Mechanical Engineering, University of Washington, Seattle, WA 98195-2600
email:emery@u.washington.edu

Abstract

A meshless method is applied to evaluate the effect of a stochastic boundary condition for a 'nominally' one dimensional conduction problem. The isothermal boundary temperature is perturbed in a spatially random, but correlated, fashion. The standard deviation of both the heat flux normal to the surface, q_x, and parallel to the surface, q_y, are shown to decay monotonically with depth. Both are strong function of the correlation length but do not show the amplification found for the case of a rough surface in which there was a maximum effect at an intermediate correlation length.

Keywords: Meshfree conduction, stochastic boundary conditions

1. Introduction

Although many numerical and analytical schemes exist for solving engineering problems, the meshless method is a particularly attractive method that is receiving attention in the engineering and scientific modeling communities. Finite difference (FDM), finite volume (FVM), and finite element (FEM) methods have been historically used to model a wide variety of engineering problems in complex geometries that may require extensive meshing. The meshless method is simple, accurate, and requires no meshing.

The need to accurately simulate various physical processes in complex geometries is important, and has perplexed modelers utilizing conventional numerical schemes for many years. Today, advances in numerical schemes and enhanced hardware have led to many commercial codes that can employ Herculean efforts to solve complex stress-strain, heat transfer, fluid flow, and other nearly intractable problems. Advances in the development and application of meshless techniques show they can be strong competitors to the more classical finite difference/volume and finite element approaches. Textbooks by Liu [1] and Fasshauer [2] discuss meshfree methods, implementation, algorithms, and coding issues for stress-strain problems; Liu includes Mfree2D, an adaptive stress analysis software package available for free from the web, and Fasshauer include MATLAB modules. Atluri and Shen [3] also produced a textbook that describes the meshless method in more detail, including much in-depth mathematical basis.

2. Meshless Methods

Meshless methods are uniquely simple, yet provide solution accuracies for certain classes of equations that rival those of finite elements and boundary elements without requiring the need for mesh connectivity. Ease in programming, no domain or surface discretization, no numerical integration, and similar formulations for 2-D and 3-D make these methods very attractive. There exist several types of meshless methods. The more common techniques include kernel methods, moving least square method, meshless Petrov-Galerkin, partition of unity methods, smooth-particle hydrodynamics, and radial basis functions. Each technique has particular traits and advantages for specific classes of problems. Generally the simplest and easiest to implement is the radial basis function approach. The examples illustrated here are based on radial basis functions (RBFs) and Kansas approach [4].

Over the last 10 years, development in using RBFs as a meshless method approach for approximating partial differential equations has accelerated. Kansas method, a domain-type meshless method, directly collocates RBFs using multiquadric approximations (MQ). Franke [5] published a review paper on 2-D interpolation methods and ranked MQ as the best based on its accuracy, speed, storage requirements, and ease of implementation. Meshless methods utilizing RBFs create mesh-free algorithms that are significantly simpler to employ than FDM/FVM, FEM, and BEM approaches, and truly eliminate the need for meshes requiring connectivity and optimization.

More recently, the use of the Method of Fundamental Solutions (MFS) with the dual reciprocity method (DRM) and RBFs has appeared [6]. The MFS method is also meshless, fast, and can be extended to multiple dimensions and is especially attractive for problems involving complex geometry. The MFS is equivalent to the more widely known Boundary Element Method (BEM). Previous use of the MFS limited its application to homogeneous equations; however, by coupling the MFS with the DRM and RBFs has led to the creation of an effective boundary-type meshless method for solving a wide range of partial differential equations. The MFS-DRM is slightly more computationally expensive than Kansas method, more difficult to implement, and requires the existence of fundamental solutions. However, it is more mathematically robust.

3. Radial Basis Functions

RBFs are the natural generalization of univariate polynomial splines to a multivariate setting. The main advantage of this type of approximation is that it works for arbitrary geometry with high dimensions and it does not require a mesh at all. An RBF is a function whose value depends only on the distance from some center point. Using distance functions, RBFs can be easily implemented to reconstruct a plane or surface using scattered data in 2-D, 3-D or higher dimensional spaces. Due to the use of the distance functions, the RBFs can be easily implemented to reconstruct the surface using scattered data in 2D, 3D or higher dimensional spaces.

In Kansas method, it is assumed that a variable (assume temperature) can be expressed as an approximation in the form

$$T(x) = \sum_{j=1}^{N} \phi_j(x) T_j \tag{1}$$

where T_j are the unknown temperature values to be determined, and $\phi_j(x) = \phi||x - x_j||$ is some form of RBF where $x_i, i = 1, \ldots, NI$, are interior points and $x_i, i = NI + 1, \ldots, N$, are boundary points. Some popular choices of RBFs include linear (r), cubic (r^3), multiquadrics (MQ) ($(r^2 + c^2)1/2$), polyharmonic splines ($r^{2n+1} log\ r$ in 2-D, r^{2n+1} in 3-D), Gaussian ($exp(-cr^2)$). The theory of RBFs interpolation is discussed in Powell [7]; Fasshauer [4] lists many RBFs and their derivatives. To illustrate the application of the meshless method with RBF, consider the 2-D Poisson's Equation

$$\nabla^2 T = f(x), \quad x \in \Omega \qquad\qquad T = g(x), \quad x \in \Gamma \tag{2}$$

where x=(x,y). Now approximate T(x) assuming

$$T(x) = \sum_{j=1}^{N} \phi(r_j) T_j \tag{3}$$

where $\phi(r_j$ is defined as a MQ basis function

$$\phi(r_j) = \sqrt{r_j^2 + c^2} = \sqrt{(x - x_j)^2 + (y - y_j)^2 + c^2} \tag{4}$$

where c is a predetermined shape parameter. Likewise, the derivatives are expressed as

$$\frac{\partial \phi}{\partial x} = \frac{x - x_j}{\sqrt{r_j^2 + c^2}}, \qquad \frac{\partial \phi}{\partial y} = \frac{y - y_j}{\sqrt{r_j^2 + c^2}}, \tag{5}$$

Substituting into the original equation set, one obtains

$$\sum_{j=1}^{N} \nabla^2 \phi(r_j) T_j = f(x_i) \qquad i = 1, 2, \ldots, NI$$
$$\sum_{j=1}^{N} \phi(r_j) T_j = g(x_i) \qquad i = NI + 1, NI + 2, \ldots, N \tag{6}$$

For a parabolic equation, an implicit scheme can be used, i.e.,

$$\frac{T^{n+1} - T^n}{\delta t} - \alpha(\frac{\partial^2 T}{\partial x^2} + \frac{\partial^2 T}{\partial y^2}) = f(x, y, t, T^n, \frac{\partial T^n}{\partial x}, \frac{\partial T^n}{\partial y}) \tag{7}$$

where δt denotes the time step and superscript $n+1$ is the unknown (or new) value to be solved. Substituting into Eq. (1), one obtains

$$\sum_{j=1}^{N} I(\frac{\phi_j}{\Delta t} - \alpha(\frac{\partial^2 \phi}{\partial x^2} + \frac{\partial^2 \phi}{\partial y^2}))T_j^{n+1} = \frac{\phi_j}{\Delta t} T^n(x_i, y_i) + f(x_i, y_i, t^n, T^n(x_i, y_i)), \qquad i = 1, \ldots, NI$$
$$\sum_{j=1}^{N} \phi(x_i, y_i) T_j^{n+1} = g(x_i, y_i, t^{n+1}), \qquad i = NI, \ldots, N \tag{8}$$

which produces an NxN linear system of equations for the unknown T^{n+1}.

4. Examples of Meshless Calculations

To illustrate the use of meshless methods, let us begin with a simple heat transfer problem, a two-dimensional plate subjected to prescribed temperatures applied along each boundary. The temperature at the mid-point (1,0.5) is used to compare the numerical solutions with the analytical solution. The analytical solution is given as

$$\Theta(x, y) = \frac{T - T_1}{T_2 - T_1} = \frac{2}{\pi} \sum_{n=1}^{\infty} \frac{(-1)^n + 1}{n} sin(\frac{n\pi x}{L}) \frac{sinh(n\pi y/L)}{sinh(n\pi W/L)} \tag{9}$$

which yields $\Theta(1, 0.5) = 0.445$, or $T(1, 0.5) = 94.5oC$. Table 1 lists the final temperatures at the mid-point using a finite element method, a boundary element method, and a meshless method.

Table 1

Method	mid point (° C)	Elements	Nodes
Exact	94.512	0	0
FEM	94.605	256	289
BEM	94.471	64	65
Meshless	94.514	0	325

As a second example problem, a two-dimensional domain is prescribed with Dirichlet and Neumann boundary conditions applied along the boundaries, as shown in Fig. 1. This problem, described in Huang and Usmani [8], was used to assess an h-adaptive FEM technique for accuracy. A fixed temperature of 100C is set along side AB; a surface convection of 0C acts along edge BC and DC with h = 750 W/mC and k = 52 W/mC. The temperature at point E is used for comparative purposes. The severe discontinuity in boundary conditions at point B creates a steep temperature gradient between points B and E. Figures 3(b,c) show the initial and final FEM meshes after two adaptations using bilinear triangles. The analytical solution for the temperature at point B is T = 18.2535C. Table 2 lists the results for the three methods compared with the exact solution. The initial 3-noded triangular mesh began with 25 elements and 19 nodes.

Table 2

Method	mid point (° C)	Elements	Nodes
Exact	18.2535	0	0
FEM	18.1141	256	155
BEM	18.2335	32	32
Meshless	18.2531	0	83

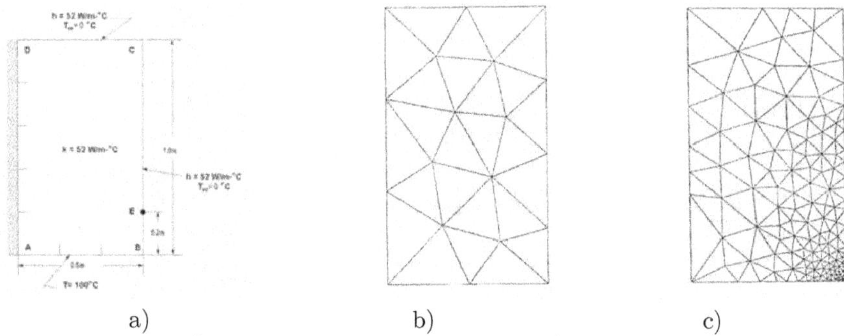

a) b) c)

Figure 1 a) geometry-boundary conditions, b) initial FEM mesh, c) final FEM adapted mesh [8]

Results obtained using a meshless method based on the Kansa's approach compare very well to solutions obtained using an h-adapting finite element and boundary elements for heat transfer in two-dimensions. All three techniques provide accurate results for the two example cases. The meshless method was clearly the fastest, simplest, and least storage demanding method to employ. Advances now being made in meshless methods will eventually enable the scheme to compete directly with the FEM and BEM on a much broader range of problems. Recent work using a meshless method for incompressible fluid flow with turbulence and heat transfer, including compressible flows with shock capture as well as bioengineering applications, are discussed in Pepper et al. [9].

5. *Stochastic Techniques*

Consider the two dimensional conduction of heat through a square slab of height and width W for which the surface at $x = 0$ is presumed to be randomly rough or whose temperature is random. In both cases the randomness has a correlation length $\mathcal{L}$. Because of this random boundary condition both x and y components of flux will exist, particularly close to the surface. For a value of $\mathcal{L}$ that is small in relation to the slab height W, i.e., approaching white noise, the effects are expected to attenuate quickly with increasing distance into the slab. For larger values of $\mathcal{L}$, implying smoother variations with respect to y, the effects on q_x are expected to propagate with little change through the thickness.

Let the variation in the spatial definition or in the temperature be defined as

$$x(y) = \Sigma\xi(y) \qquad \text{or } T(0,y) = 100 + \Sigma\xi(y) \tag{10}$$

where $\xi(y)$ is a random field whose mean at any specific value of y is zero with a standard deviation of unity and Σ is the standard deviation of the field. $\xi(y)$ has a spatial correlation length of $\mathcal{L}$. If $\mathcal{L} = 0$, i.e., $\xi(y)$ is characterized by white noise, we expect that the region in which the conductive flux is two dimensional will be very thin and approach zero in extent. If $\mathcal{L}$ is large in comparison to W, $\xi(y)$ will be nearly constant with respect to y, i.e., be a random variable with $\xi(y) = \overline{\xi}$ and the heat transfer will approach that of one dimensional conduction with values per unit depth of

$$\overline{q}_y = 0, \qquad \overline{q}_x = \frac{k\,\overline{\Delta T}}{\overline{W}} \qquad \text{with} \qquad \sigma(q_x) \approx \overline{q}_x \frac{\sigma(\xi)}{\overline{W}} \tag{11}$$

where the overbars represent averages.

Because of the random nature of the surface, this system is stochastic in nature and the evaluation of the net heat transfer will involve the solution of a partial differential equation over a random region or if the boundary condition is stochastic, solving the pde with a stochastic bc. In most of the reported literature, the stochasticity is associated with a property such as the thermal conductivity or a boundary condition. The two problems, that of a random roughness and that of a random boundary condition, are fundamentally different and require different approaches.

6. Meshing and Representation of a Random Field

To solve the problem using finite elements/volumes requires resolving the questions of the mesh size and how to represent the random field within the element. Unfortunately there are no definitive answers to either question. Several common approaches are: a) integrating $\xi(x,y)$ over the element, b) using a centroidal value, c) evaluating $\xi(x,y)$ at the nodal points and interpolating. In general, centroidal values tend to over predict and spatial averaging under predict the variability of $\xi(x,y)$ over the element. The approach depends upon the scale of fluctuations [10] and Li and Der Kiureghian [11] summarize several approaches.

The question of the appropriate mesh size is more difficult to resolve. One suggestion is that the correlation length $\mathcal{L}$ should span 2-3 elements [11]. If $\xi(y)$ is a zero mean, wide sense stationary, random field, some guidance can be found by considering how $\xi(y)$ is represented through the Karhunen-Loeve expansion in eigenfunctions given by [12]

$$\xi(y) = \sum_{i=1}^{\infty} f_i(y) u_i \tag{12a}$$

where u_i are uncorrelated random variables and $f_i(y)$ are orthogonal eigenfunctions that satisfy

$$f_i(y) = \int C(y,y') f_i(y') dy' \tag{12b}$$

$$\int f_i(y) f_j(y) dy = \delta_{ij}\sqrt{\lambda_i \lambda_j} \tag{12c}$$

where $C(y,y')$ is the covariance function and δ_{ij} is the Kronecker delta. Eq. (12a) defines a realization of $\xi(y)$ as the sum of the deterministic functions $f_i(y)$, that are ordered in terms of their eigenvalues, multiplied by multiple random variables, u_i that satisfy

$$E[u_i] = 0, \qquad E[u_i u_j] = \delta_{ij} \tag{12d}$$

While u_i are uncorrelated, they are not independent, and the typical procedure involves

1 representing a random field in terms of uncorrelated random variables, u_i, Eq. 3a
2 converting all random variables into Gaussian variables [13].

Inasmuch as a lack of correlation does not imply independence, the second step is necessary for both practical and theoretical reasons since uncorrelated Gaussian random variables are independent. Li [11] and Phoon [14] describe techniques for converting the random variables to Gaussian variables, usually in the form of multidimensional Hermite polynomials involving several Gaussian random variables, $\eta_i, i = 1, \ldots, G$. If G is large, the dimensionality of the problem can grow to an unacceptable size.

The series, Eq. 12a, is usually terminated at N terms and the combined variance of the first N vectors is given by

$$\sigma^2(\xi) = \sum_{i=1}^{N} \lambda_i^2 \tag{13}$$

The number of such random variables, u_i, needed will depend upon the desired accuracy. For a mesh with Δy=1/40 over $0 \leq y \leq 1$, the number of vectors needed to achieve 99.9% of the total variance is shown in Figure 2. Our computed results did not show adequate convergence unless more than 99% of the variance of the rough edge was accounted for.

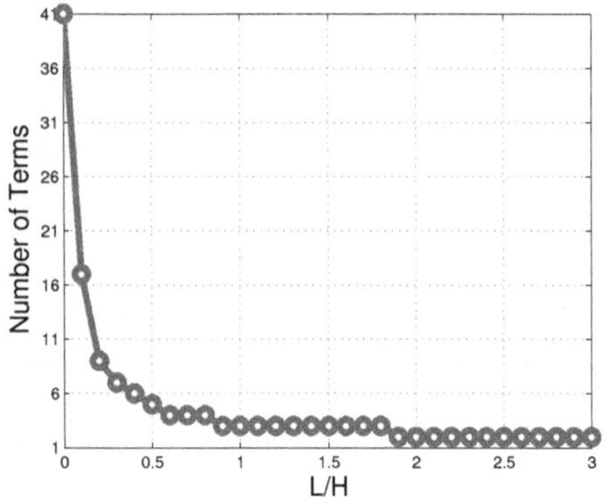

Figure 2 Number of Vectors, N, using Eq. 12a needed to achieve 99.9% of the true variance as a function of the correlation length $\mathcal{L}/H$

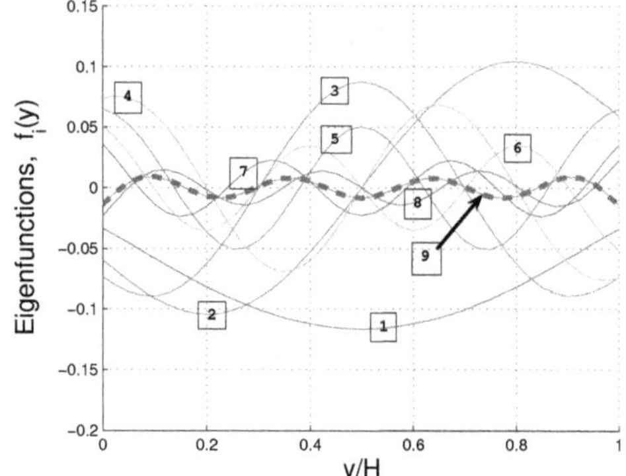

Figure 3 Principal Component Analysis for $\mathcal{L}/H = 0.2$

About the only analytical solution is for the covariance $C(y, y') = exp(|y - y'|/\mathcal{L})$. Unfortunately this possesses some undesirable characteristics, e.g., a discontinuous slope at $y = y'$ and is not a convenient covariance. For other covariances Eq. 12a must be solved numerically and the eigenfunctions are replaced by eigenvectors. In this case, the K-L expansion is usually referred to as Principal Components [13]. Figure 3 depicts the eigenvectors for the covariance $C(y, y') = exp((y - y')^2/\mathcal{L}^2)$. Each succeeding vector contributes less to the total variance. To achieve 99.9% of the total variance requires 9 terms in the expansion, Eq. 12a. The 9th vector, indicated by the heavy dashed vector, has a wave length of approximately $1.5\mathcal{L}$. This proved to be the case for all values of $\mathcal{L}$ examined from 0.01W to 3W. Since boundary conditions usually require of the order of 3 elements per wave length for satisfactory results, this suggests that a adequate mesh would be one in which$\Delta y \approx 0.5\mathcal{L}$.

7. Results for a Rough Edge

The problem was solved for a unit thermal conductivity and a roughness with a standard deviation, Σ, of 2% of the slab width, $\overline{W}$. The value of 2% was chosen as representing a reasonable degree of roughness since when using a Gaussian distribution truncated at $\pm 4\sigma$, it produces a maximum deviation of 0.08% of W, giving a range of thermal resistance from 0.92 to 1.08. A full description of the analysis is given in Reference [17].

Figure 4 shows the contours for q_x and q_y for an edge roughness of Σ =2%. Near the right edge, the conduction is essentially one dimensional since the flux in the y direction at all values of y has approached zero and as a consequence the standard deviation of the x flux approaches from below the value obtained from Eq. 11. Near the rough edge, there is a substantial disruption of the heat flow with large values of q_y that are of the order of q_x. Consequently, the standard deviations of both fluxes are significantly larger than that of the edge roughness. In the central portion, $0.2H \leq y \leq 0.8H$ the y flux is unconstrained, but near the insulated edges flow lines of the flux are forced to become aligned with the edge and the result is that q_x must vary more than it does in the central region. Consequently, the standard deviation near the insulated boundaries is of the order of 1.5 times that in the central core.

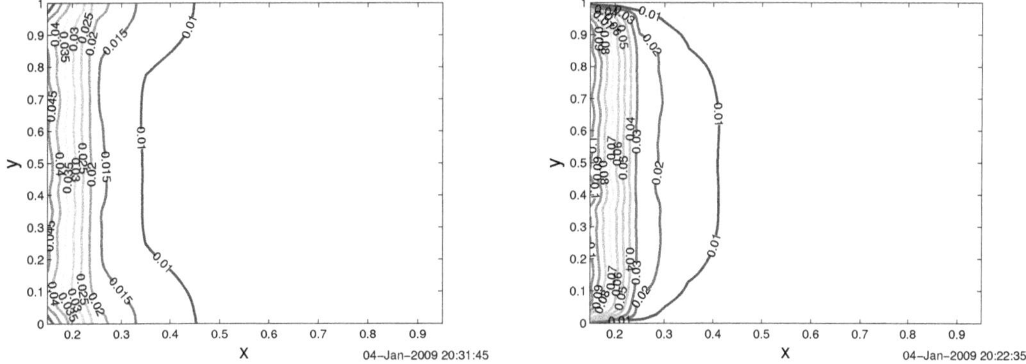

Figure 4 Contours for an Edge Roughness with Σ = 2% and $\mathcal{L}$=0.05W

Figure 5 shows the variation of $\sigma(q_x)$ at $x = 0.1W, y = H/2$ and $x = 0.15W, y = H/2$ as functions of $\mathcal{L}$. As expected, when $\mathcal{L} \rightarrow 0$ the variation goes to zero and as $\mathcal{L} \rightarrow \infty$ it approaches 2%, the value expected from Eq. 11. What is surprising is how the interaction between q_x and q_y results in a standard deviation at modest values of $\mathcal{L}$ of the order of 7%, or about 3 times the value expected for large values of $\mathcal{L}$.

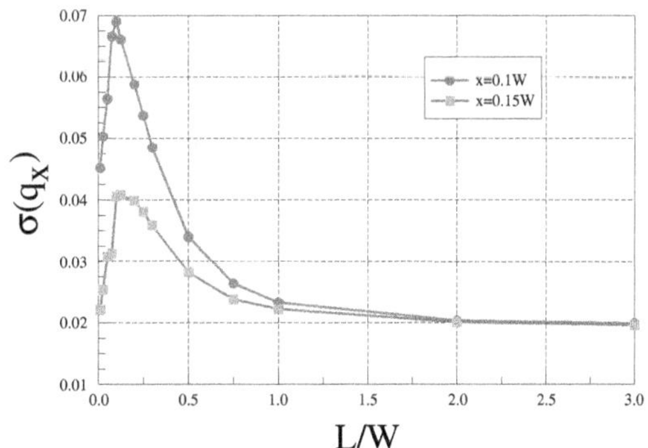

Figure 5 Effect of $\mathcal{L}$ upon the behavior of q_x for an Edge Roughness of 2%

7. Results for a Random Boundary Temperature

The case of a random boundary temperature was solved using Kansa's meshless approach with the grid shown in Figure 6 giving the variation in temperature shown in Figure 7.

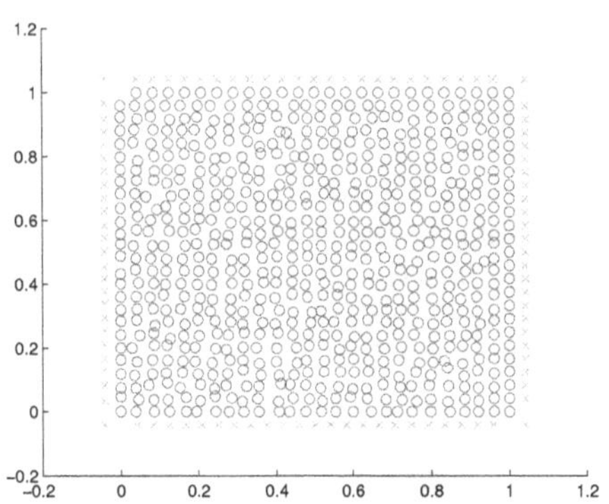

Figure 6 Points used in the meshfree analysis

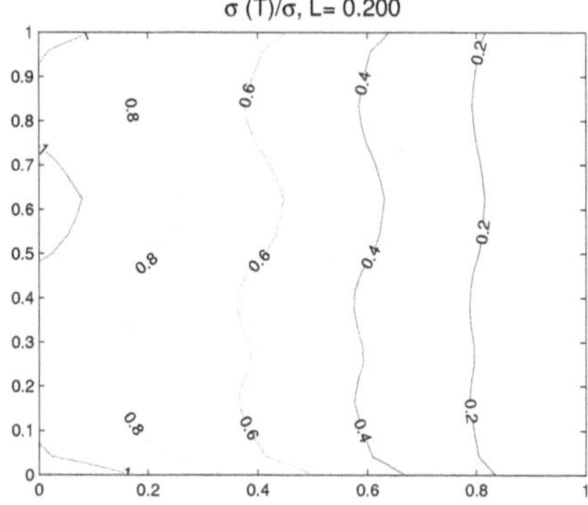

Figure 7 Standard Deviation of T(x,y) normalized by Σ for $\mathcal{L}$/W=0.2

The flux normal to the boundary, q_x, and that parallel to the boundary, q_y, were approximately equally affected for small values of $\mathcal{L}$. For larger values, the boundary temperature is nearly isothermal and $q_y \to 0$. q_x on the other hand approaches the values given by Eq. 11, ie., the random field reduces to a random variable. Figure 8 shows the effect on q_y for $\mathcal{L}/W = 0.2$.

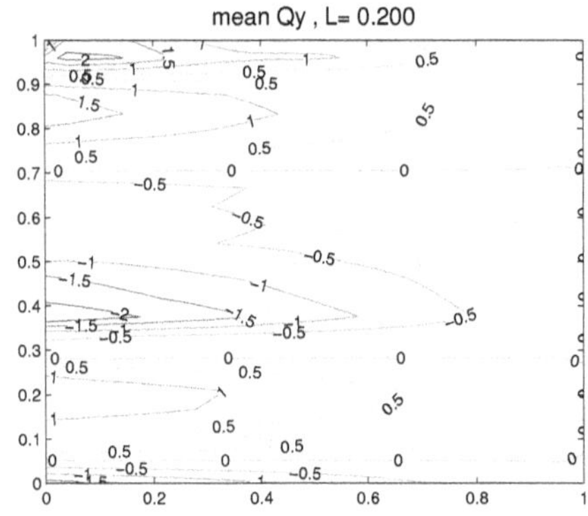

Figure 8a mean values of q_y for L/W=0.2%

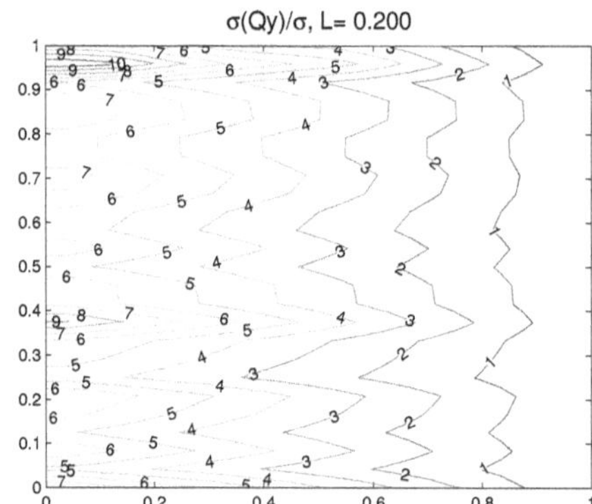

Figure 8b $\sigma(q_y)/\Sigma$ for $\mathcal{L}$/W=.2

Figure 9a shows the behavior of the standard deviations as a function of depth. As expected, $\sigma(q_y) = 0$ at x=W since there the temperature is isothermal. $\sigma(q_x)$ is given approximately by the Eq. 11. In contrast to the behavior of the rough edge we see: a) the depth of penetration is monotonic with $\mathcal{L}$ and does not result in a thin zone; b) there is no amplification of $\sigma(q)$ at a shallow depth, Figure 5. It smoothly reduces as $\mathcal{L}$ diminishes. Figure 9b compares the effects on $\sigma(q_x)$ and $\sigma(q_y)$ at the boundary, the point of maximum effect.

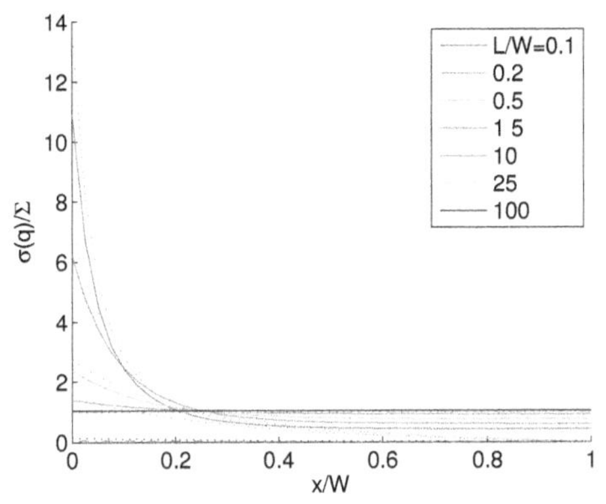

Figure9a $\sigma(Q)$ versus x/W (solid lines, q_x, dashed, q_y)

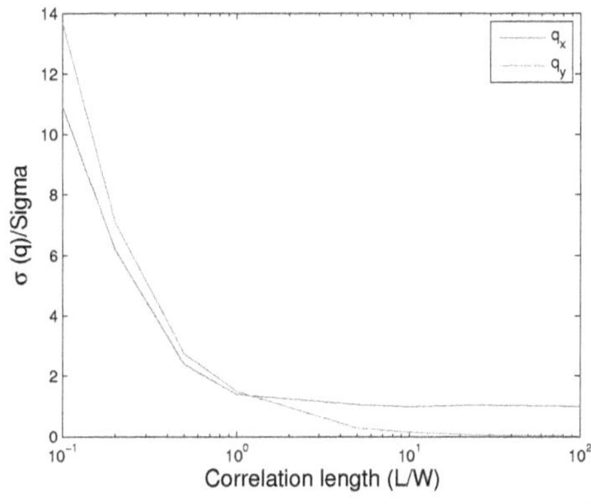

Figure 9b σ at $x/W = 0$ as functions of L/W

8. Conclusions

The requirement that the mesh size be related to the correlation length of the random noise, either for a boundary condition or for a domain property such as the conductivity, means that as the field approaches white noise the mesh density must approach unacceptably high values. As a consequence both FEM and BEM methods are computationally very expensive when treating realistic cases of stochasticity. Meshfree methods are ideally suited for these problem. For the case treated here, one simply adds a sufficient number of boundary nodes to capture the variation while retaining a reasonable number of internal nodes. If a field property is stochastic, one can use Atluri's locally meshless Petrov-Galerkin approach and assigns one set of point to represent the stochastic property and another set of nodes to represent the temperatures similar to the approach described by Rahman and Rao who used a perturbation approach, i.e., a one term expansion of the KL series, Eq. 12a. While the idea of adding sufficient control points appears simple, we observed that adding too many points led to ill-conditioned matrices and it was necessary to move both the boundary and internal points in a random fashion, but only with a slight displacement, to stabilize the solution. Since low values of $\mathcal{L}$ will require many more points, further work remains in studying the stability of the method.

9. Acknowledgments

These results were obtained as part of the research supported by the National Science Foundation through Grant 0626533 and the United States Federal Aviation Administration Cooperative Agreement 08-C-AM-UW.

10. References

1. Liu, G. R. (2002): Mesh Free Methods: Moving Beyond the Finite Element Method, CRC Press, Boca Raton, FL.
2. Fasshauer, G. E. (2007), Meshfree Approximation Methods with MATLAB, World Scientific Pub. Co., Hackensack, NJ.
3. Atluri, S. N. and Shen, S. (2002): The Meshless Local Petrov Galerkin (MLPG) Method, Tech Science Press, Encino, CA.
4. Kansa, E. J. (1990): Multiquatric A scattered data approximation scheme with applications to computational fluid dynamics II, Computers Math. Appl., 19, 8/9, pp. 147-161.
5. Franke, R. (1982), Scattered Data Interpolation: Tests of Some Methods, Math. Comput., 48, pp. 181-200.
6. Alves, C. J. S. and Chen, C. S. (2005), A new method of fundamental solutions applied to nonhomogeneous elliptic problems, Adv. Comput. Math., 23, 1-2, pp. 125-142.

7. Powell, M. J. D. (1992): The theory of radial basis function approximation in 1990, in Advances in Numerical Analysis, Vol. II, W. Light (Ed.), Oxford Sci. Pub., Oxford, UK, pp. 105-210.
8. Huang, H-C and Usmani, A.S. (1994), Finite Element Analysis for Heat Transfer, *Springer-Verlag,* London, UK
9. Pepper, D. W., Carrington, D. B., and Gewali, L. (2000): A Web-based, Adaptive Finite Element Scheme for Heat Transfer and Fluid Flow. ISHMT/ASME 4th Conf. on Heat and Mass Transfer, Jan. 12-14, Pune, India.
10. Vanmarcke, E. and Grigoriu, M. "Stochastic Finite Element Analysis of Simple Beams," *J. Engr. Mech.,* Vol 109, No 5, pp. 1203-1214, 1983
11. Li, C-C and Der Kiureghian, A. ,"Optimal Discretization of Random Fields," *J. Engr. Mech.,* Vol 119, No 6, pp. 1136-1154, 1993
12. Ghanem, R. G. and Spanos, P. D., *Stochastic Finite Elements,* Dover, Mineola, NY, 2003
13. Polynomial Chaos Decomposition for the Simulation of Non-Gaussian Nonstationary Stochastic Processes, Sakamoto, S. and Ghanem, R., *J. Engineering Mechanics,* Vol. 128, No 2, pp. 190-201, 2002
14. Phoon, K.. K., Huang, H. W., Quek, S. T., "Simulation of Strongly Non-Gaussian Processes using Karhunen-Loeve Expansion," *Probabilistic Engineering Mechanics,* Vol 20, pp. 18-198, 2005
15. Lin, G., Su, C-H, and Karniadakis, G. E, "Random Roughness Enhances Lift in Supersonic Flow," *Physical Review Letters,* Vol. 99, pg. 104501-1, 2007
16. Jackson, J. E., *A User's Guide to Principal Components,* J. Wiley and Sons, N.Y., NY, 1991
17. Emery, A. F., Dillon, H. and Mescher, A., 2010, "The effect of spatially correlated roughness and boundary conditions on the conduction of heat through a slab," *J. Heat Transfer,* v 132, n 5, p 1-11
18. Rahman, S. and Rao, B.N., 2001, "A Perturbation method for Stochastic Meshless Analysis in Elastostatics," *Intl. J. Numerical Methods in Engineering,* v. 50, pp 1969-1991

Practical Considerations when Using Sparse Grids with Bayesian Inference for Parameter Estimation

A F Emery and K. C. Johnson

Department of Mechanical Engineering, University of Washington, Seattle, WA 98195-2600, USA
email:emery@u.washington.edu

Abstract

Estimating parameters via Least Squares often involves a high computational cost in evaluating the model of the system response. One approach is to sample the response surface and interpolate. When done with a tensor grid the cost of developing an accurate representation of the surface often exceeds that associated with the least squares solution. Sparse grids offer an alternative that under some conditions may be of less cost. However, because sparse grids are inherently approximations and at any level are not complete polynomials their errors may lead to less efficient calculations in estimating parameters and may distort the marginal probability distributions of the parameters.

Keywords: Sparse Grids, Hierarchical Bayesian Inference, Parameter Estimation

1. *Introduction*

The standard approach to estimating parameters is the Least Squares analysis in which the estimated parameters are those that minimize the norm of the residuals based upon the fundamental assumptions that the model is correct and that any deviation of the model from the data is due to normally distributed zero mean errors. An important aspect of the inverse problem is the estimation of the standard deviation of the estimated parameters. In the least squares approach we define the model as $M(\Theta)$, where Θ represent the set of parameters. Let the true value of the parameters be denoted by Θ and the estimated values by $\hat{\Theta}$. The measurements, D, are presumed to be corrupted by the noise ϵ where $E[\epsilon] = 0$ and $cov[\epsilon] = \Sigma$. For non-linear problems, the parameters are found using a solution that is effected by an iterative procedure based upon linearization. The estimated property, $\hat{\Theta}$, is that which minimizes the functional $L(\hat{\Theta})$

$$L(\hat{\Theta}) = r^T(\hat{\Theta})\Sigma^{-1}r(\hat{\Theta}) \quad (1a) \qquad \text{where } r(\hat{\Theta}) \equiv D - M(\hat{\Theta}) \quad (1b)$$

Linearizing Eq. 1b about an estimate Θ_i gives

$$r(\hat{\Theta}_i) = D - M(\Theta_i) - \frac{dM}{d\Theta}\Big|_{\Theta_i}(\hat{\Theta}_i - \Theta_i) \tag{2a}$$

and $\hat{\Theta}_i$ is given by the iterative procedure

$$\hat{\Theta}_i - \Theta_i = (A_j^T\Sigma^{-1}A + \beta I)^{-1}A^T\Sigma^{-1}[D - M(\hat{\Theta}_i)] \tag{2b}$$

where $A = \partial M/\partial\Theta$. For N measurements and d parameters, D and M are $[Nx1]$ vectors, Θ is a $[1xd]$ row vector and A is a $[Nxd]$ matrix and β, the Levenberg-Marquardt parameter [1], is used for ill conditioned problems. Upon convergence of the iterations, the estimate $\hat{\Theta}$ satisfies

$$E[\hat{\Theta}] = \Theta \quad (3a) \qquad cov[\hat{\Theta}] = (A^T\Sigma^{-1}A)^{-1} \quad (3b)$$

A better approach is based upon hierarchical Bayesian inference [2] that allows us to incorporate knowledge about the reasonable values of the parameters sought and information about any other parameters involved in the model. Figure 1 compares the probability distribution (pdf) of the specific heat, c, obtained from Least Squares based on Laplace's assumption (namely that the distribution is assumed to be normally distributed about $\hat{c}$ with the standard deviation given by Eq. 3b) with the marginal distribution obtained from Bayesian inference when estimating the specific heat from the transient conduction problem described later in the paper. Note that the breadth of the distribution based upon the Bayesian approach is substantially greater and that an estimate of the uncertainty of $\hat{c}$ based upon least squares is a significant under estimate.

Unfortunately for many real problems, the computational expense for Bayesian inference is prohibitively high. One solution is to use an interpolation of the response function, but this is usually only realistically possible if the interpolation is cheap. Sparse grids offer a possible approach, but they have problems of their own. This paper describes the use of sparse grids and offers some practical advice for their effective use in Bayesian inference.

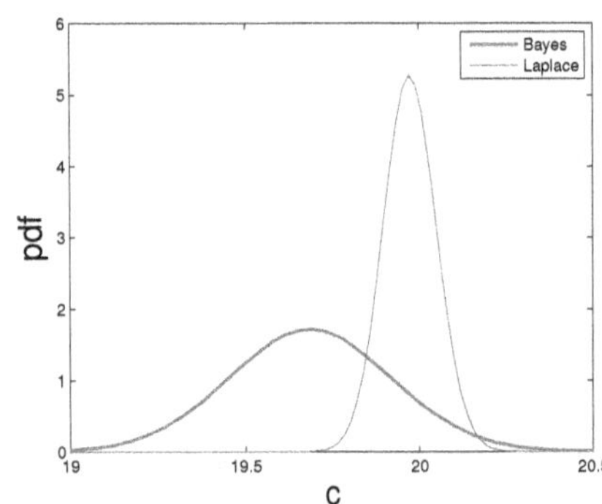

Figure 1. Comparison of the pdf (c) for Example Problem 1

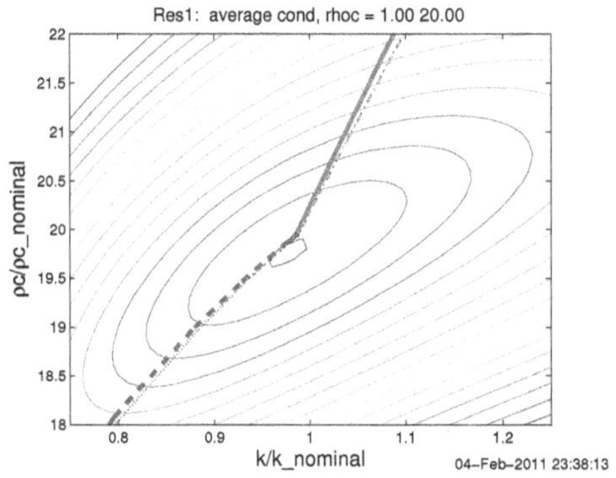

Figure 2 Contours of $L(\Theta)$ and Solution trajectories for Problem 1 using exact solutions and a sparse grid of level 3 for interpolation

2. Bayesian Inference

The Bayesian approach consists of determining the marginal distribution of a single parameter under assumptions for any prior knowledge of the parameters and of any other unknown elements [2]. Consider the case of two parameters to be estimated, Θ_1, Θ_2, from a set of data D that is presumed to be contaminated with a normally distributed error of standard deviation σ. From Bayes' relationship we have

$$p(\Theta_1, \Theta_2, \sigma|D) \propto p(D|\Theta_1, \Theta_2, \sigma)\ \pi(\Theta_1)\pi(\Theta_2)\pi(\sigma) \tag{4a}$$

where D represents the data, in our examples the measured temperatures. The probabilities, $\pi(\Theta_1), \pi(\Theta_2), \pi(\sigma)$ represent our prior knowledge about the parameters sought and the noise in the measurements. In general, these prior probabilities are often functions of other statistical quantities. For example, Θ may be presumed to follow a log normal distribution with a standard deviation of Σ_1 and Σ_1 may itself be a function of other parameters. When this occurs, the representation on the right side of Eq. 4a is said to be a hierarchical distribution. The most common priors for the parameters Θ are normal, lognormal, beta, and uniform distributions. Restricting the prior distributions to only one hierarchical parameter we have

$$p(\Theta_1, \Theta_2, \sigma|D) \propto p(D|\Theta_1, \Theta_2, \sigma)\ \pi(\Theta_1, \Sigma_1)\pi(\Theta_2, \Sigma_2)\pi(\sigma) \tag{4b}$$

The marginal posterior distributions of Θ_1 are obtained by integrating out $\Theta_2, \Sigma_1, \Sigma_2$ and σ from Eq. 4b to give Eq. 5.

$$p(\Theta_1) = \frac{\int_\sigma \int_{\Theta_2} \int_{\Sigma_2} \int_{\Sigma_1} p(D|\Theta_1, \Theta_2, \sigma)\ \pi(\Theta_1|\Sigma_1)\pi(\Sigma_1)\pi(\Theta_2|\Sigma_2)\pi(\Sigma_2)\pi(\sigma)d\sigma d\Theta_2 d\Sigma_1 d\Sigma_2}{\int_{\Theta_1} p(\Theta_1)d\Theta_1} \tag{5}$$

The integration can be done either by using Gaussian quadrature or by using a Markov Chain Monte Carlo approach with the Gibbs sampler [3].

For simple models, the evaluation of the right hand side of Eq. 5 for the integration usually does not entail substantial computational costs. However, for complex models, e.g., involving CFD or large scale FEA analysis, the computational cost is often too high to be acceptable. It has been our experience that for nearly normal distributions that 7-9 Gaussian quadrature points are needed for each parameter being integrated. Thus the evaluation of $p(\Theta_1)$involves integrating over $\Theta_2, \Sigma_1, \Sigma_2, \sigma$ in the numerator and over Θ_1 in the denominator and will require of the order of 7^5= 16807 evaluations.

During the last several years the use of sparse grids has proven to be a reasonable alternative to the usual tensor grid. For relatively smooth variations of the model behavior with respect to the parameters being sampled, a sparse grid for 5 parameters with a level of 4 requires 801 sample points contrasted to the 16807 points needed for the tensor grid.

While conceptually simple, the sparse grid approach must be applied carefully if good results are to be obtained. This paper describes the fundamentals of the sparse grid approach and demonstrates its value when applied to the estimation of thermal properties from temperatures measured in a simple 1D experiment. The method is then applied to a complex real problem with measured temperatures that is computationally demanding.

3. Problems Considered

Two problems were analyzed to assess the usefulness of sparse grids. The first problem was transient conduction in a bar of length L initially at a temperature of 1, insulated at x=0 and convecting to an ambient fluid of temperature 0 with a convection coefficient h at x=L. The aim was to estimate the values of h, the conductivity k, and the specific heat c. Temperatures were computed at x/L=[0, .5, 1.0] at increments of the Fourier number of 1 for $h = k = 1$ and ρc=20. While this sounds like a reasonable inverse problem, the temperature is a function only of $Bi = hL/k = 1$ and $\alpha = k/\rho = 0.05$, consequently only two properties can be determined. If all three are to be estimated, the problem is ill conditioned and may not be recognized as such. In this case solutions are often attempted using some form of regularization or the Levenberg-Marquardt (LM) approach. Using the LM approach, convergence, if achieved at all, is reached very slowly. We bring this point up here because many heat transfer problems often behave as though they are functions of only a few groupings of variables, fewer than the apparent number of parameters in the model and are not well suited for inverse estimation because if convergence is obtained, it is only through a very large numbers of iterations and large values of β. Such conditions are computationally expensive and appear to be ideal candidates for using sparse grids because of their low computational costs, misleading the analyst as to their usefulness.

Figure 2 shows the search trajectory when estimating only k and c from simulated data with 1% zero mean, random, normally distributed noise. Two different starting points were used and the temperatures and sensitivities needed in Eq. 2b were computed from the closed form product solution [4] and by using a sparse grid. The paths for the sparse grid process are only slightly different than those from the exact solution and reach the same estimated values.

The second problem was a "real" experiment in which a composite panel was heated with an electric blanket placed on its upper surface and cooled by free convection from the top and bottom surface and by forced convection from the bottom of the stringer. Figure 3 is a cross section of the panel. The goal was to estimate the heat losses from the upper and lower surfaces using temperatures measured with thermocouples embedded in the composite, at the location shown, and from infrared thermograms of the upper surface.

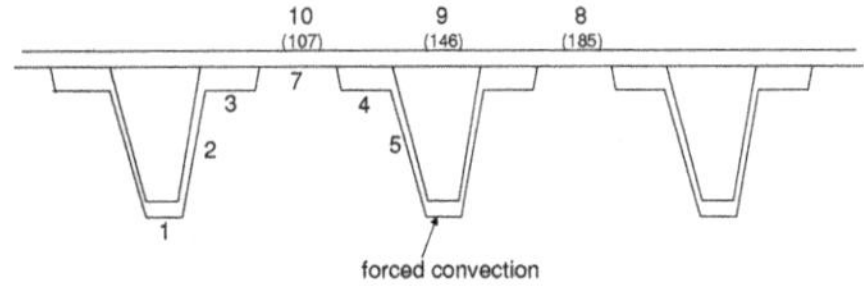

Figure 3: Schematic of the panel thermocouples are at the numbered points

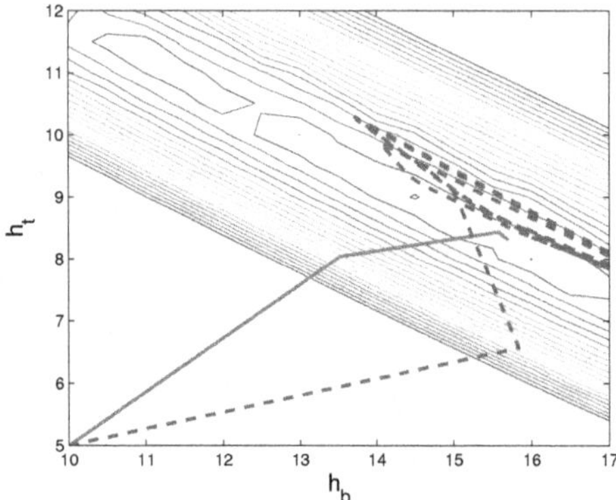

Figure 4 Solution trajectories for Problem 2 using exact solutions and a sparse grid of level 3

The panel temperatures needed in Eq. 2 for the model $M(h_t, h_b, h_s, t)$ were computed using the finite element multiphysics program Comsol [5]. Figure 4 shows the trajectories of the least squares solutions obtained using temperatures and sensitivities computed directly from Comsol and those from a sparse grid representation of the response surface.

From the contours of $L(\Theta)$, Eq. 1a, we see that the problem is ill conditioned – the surface has many local minima and is too flat to unequivocally identify a global minimum. The ill conditioning arose because the high in-plane conductivity led to sensitivities that resembled each other too much and because in the center of the panel the heat loss was related to the sum $h_t + h_b$, not to the individual values. In this case the sparse grid interpolated values are so inexact that the sparse grid estimation never converges and simply wanders around the surface in an unpredictable way.

4. *Sparse Grid Fundamentals*

The idea behind sparse grids is very simple. Consider a function of two variables, $f(x, y)$. The usual representation is through the Taylor series

$$\begin{aligned} f(x,y) &= f(0,0) + (\frac{\partial f}{\partial x}x + \frac{\partial f}{\partial y}y) + \frac{1}{2}(\frac{\partial^2 f}{\partial x^2}x^2 + 2\frac{\partial f}{\partial x}\frac{\partial f}{\partial y}xy + \frac{\partial^2 f}{\partial y^2}y^2) + \; ... \\ &= a_0 + (a_{11}x + a_{22}y) + (b_{11}x^2 + b_{12}xy + b_{22}y^2) + ... \end{aligned} \quad (6)$$

Each term in this expansion represents a "complete polynomial", i.e., the powers of x and y sum to the single value, e.g., the first term is a complete polynomial of power 0, the second of power 1, the third of power 2 et seq. Consider a tensor grid of sample points to represent the 2nd order expansion. In each direction, this grid has 3 points, giving a total of 9 points. However, there are only 6 constants that need to be evaluated. Thus one needs only 6 sample points. Of the four points on the diagonals, only one is needed, the remainder are superfluous. The question is which diagonal point to choose and how to define an automatic procedure for selecting it. In other words we can use a sparser grid than the tensor grid. Smolyak [6] developed a method that led to high quality when interpolating and integrating functions. The method consists of an interpolating function that is a set of linear combinations of products of the univariate polynomials. Consider three sample points in each of the x and y directions, i.e., 5 points. This will allow us to represent the univariate functions, $f(x) = [1, x, x^2], f(y) = [1, y, y^2]$. Smolyak's method represents $f(x, y)$ by products of these univariate representations, i.e., $f(x, y) = f(x)f(y)$. In this case we obtain $f(x, y) = [1, x, y, x^2, xy, y^2, x^2y, xy^2, x^2y^2]$. Obviously with the 5 sample points the product cannot represent all 9 of these terms and some of the terms are eliminated to give $f(x, y) = [1, x, y, x^2, y^2]$, that is we filter the product solution.

Smolyak's method did not attract much attention until Barthelmann et al. [7] demonstrated that the method was optimal. If we denote Smolyak's formula as $\mathcal{A}(l, d)$where d is the number of dimensions and l denotes the level of interpolation, they proved that $\mathcal{A}(k + d, d)$ is exact for all polynomials of complete degree k, that it is exact for some polynomials of degree up to $k + d$, that the process is efficient in use of the number of sample points if nested grids (i.e., grids where points of one level are intermediate to points of the preceding level) are used and that the approximation is 'best' in that the error bounds are least.

A sparse grid can be characterized by its 'level', essentially the degree of the complete polynomial that can be interpolated. When using nested grids, each level contains the contributions of the lower levels. A variety of sample point spacings may be used in the grids, but it is well known that the Chebyshev points lead to the min-max fitting which is almost as good as the best approximation possible.

Since Barthelmann's paper sparse grids have enjoyed increasing usage. Klimke [8] developed the program "Spinterp" which is available as a Matlab toolbox. A number of papers exist in which sparse grids are used for Bayesian inference and other problems of high dimensions [9-14].

5. Appropriate Level of the Sparse Grid

Because the number of sample points grows rapidly as the level is increased (for 3 parameters, level 2= 25 pts, 3= 64 pts, 5= 441 pts) it is important to determine the appropriate level. Unfortunately this cannot be done in advance. One approach is based upon Klimke's sparse grid method that uses 'surpluses' [8]. A surplus is the amount each sample point contributes to the basis function, i.e., equal to the difference between the value of the function at the sample point, $f(x_j)$, and the interpolated value at that point based upon the preceding level of the sparse grid, $A(l - 1, d)$. However the surplus is not a measure of a single sample point's contribution to the interpolation over the entire response surface. Furthermore, the surplus can be determined only for scalar quantities, thus eliminating its use for the vectors and matrices used in the Bayesian inference. It is possible to determine the surpluses for a specific scalar, in our case the temperature at a given point in space and time. For the transient bar problem the temperature and sensitivities at $x/L = 0.5$ and $\alpha t/L^2 = 21$ were calculated on the tensor grid defined by each parameter, h, k, c ranging from one half to twice its nominal value, i.e., $0.5 \leq h, k \leq 2$ and $10 \leq c \leq 40$. using 11 sample points in each dimension. The maximum and rms errors in the sparse grid interpolation at each of these tensor points were evaluated. The reduction in error as the sparse grid level was increased are shown in Figure 5a along with the estimated error from the program "spinterp". Similar comparisons made for other values of x and t gave comparable results. In almost all cases the estimated reduction in the maximum error based on surpluses was a reasonable estimate of the increased accuracy as the sparse grid level was increased.

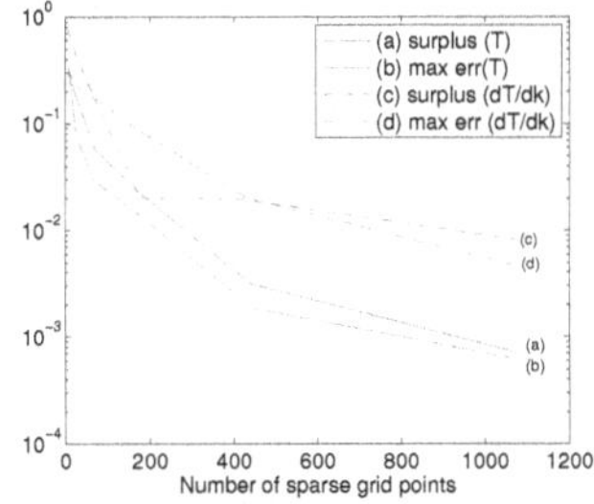

5a. For T and $\partial T/\partial k$, Problem 1

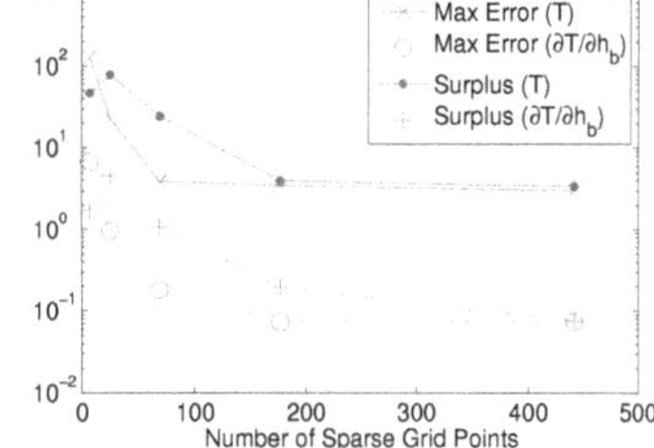

5b. For T and $\partial T/\partial h$, Problem 2

Figure 5. Reduction in Maximum Errors as a function of Sparse Grid Level

Although Figure 5 suggests that in general there was a monotonic behavior as the sparse grid level increased, this was not always the case for every point studied. Figure 6 shows how the higher level sparse grid interpolations for values of T and $\partial T/\partial k$ at x=0.5L converge as a function of level. While the temperature is well represented for a grid of level 3, $\partial T/\partial k$ requires a minimum of level 5 for good representation over entire range of times.

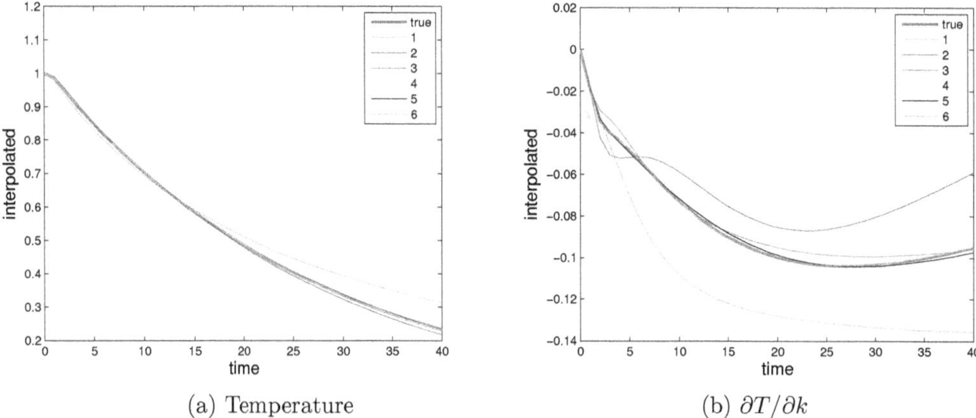

(a) Temperature (b) $\partial T/\partial k$

Figure 6. Interpolated Values for Sparse Grids of Increasing Levels for Problem 1 Using the Chebyshev Grid

6. Conclusions

Calculating the marginal probability distribution, Eq. 5, requires a fairly dense sampling of the integrand near the point of maximum probability and, if using a sparse grid a reasonable level of accuracy. For the problems considered here, and true in general, the response surface shows some strong variations near the vertices of the parameter space. Consequently it is important to choose the smallest parameter ranges possible. Although the maximum a-posteriori probability may not coincide with the estimated value of from the least squares analysis, $\hat{\theta}$, we have found that defining the range of each parameter to be $\hat{\theta} \pm n\ \sigma(\hat{\theta})$ where $\sigma(\hat{\theta})$ is given by Eq. 3b and n is a small number, usually 4 or 5, suffices. The advantage of the sparse grid (or of any other interpolation) is that the effect of assuming different priors in Eq. 5 can be examined with minimal computational expense. For the panel, computing the time history for a given set of parameters required in the order of 65 seconds using Comsol as compared to 0.8 seconds when using a sparse grid of level 5.

Our recommended procedure is:

1) determine $\hat{\Theta}$ with the least squares approach using the model $M(\Theta, x, t)$
2) define the parameter space based upon $\sigma(\hat{\theta})$
3) develop the sparse grid interpolation for levels l_1 to l_m
    a) at each level choose some representative responses and use "spinterp" to evaluate the maximum surplus as a measure of the accuracy of the sparse grid
    b) compare the behavior of the sparse grid representation of these points with their actual performance, as in Figure 6
    c) choose the appropriate maximum level, l_M based upon the surpluses and 3b)
4) use the sparse grid to calculate the marginal pdf

The choice of the representative points in 3a) must be made with some care to ensure that a good agreement is not fortuitous or unrepresentative of the interpolation accuracy over the entire parameter space. If the comparison is of the transient behavior, then although N responses at T times will involve NT calls to "spinterp", these computations are cheap.

Besides reporting the surpluses "spinterp" can provide information about adaptive grids, i.e., grids with different numbers of sample points in each dimension. While this may reduce the number of sample points needed, especial care must be taken to ensure that the interpolation is accurate over the entire parameter space. Although Eq. 5 involves only interpolation of the response, we have found that good interpolation of the sensitivities rather than just of the response is a better measure of the sparse grid acceptability.

These conclusions are valid only for well conditioned problems. For the panel test, the values of $\sigma(\hat{\theta})$ were of the order of 15% and an adequate coverage of the parameter space needed for an accurate evaluation of Eq. 5 required interpolation over such a wide range of parameters and responses that reasonable levels of sparse grid were not a viable option. In fact, for such ill conditioned problems it is questionable whether evaluating the marginal pdf is of any value unless one is specifically interested in the tails of the distribution, e.g., reliability studies.

7. Acknowledgements

A portion of this research was conducted under the sponsorship of FAA and NSF. We wish to thank Professor N. R. Aluru and Doctoral candidate, N. Agarwal for the initial Matlab version of the sparse grid basis program.

8. References

[1] Press, W. H., Flannery, B. P., Teukoslky, S. A. and Vetterling, W. T., 1986, *Numerical Recipes,*, Cambridge Univ. Press, Cambridge, UK

[2] O'Hagan, A and Forster, J, 2004, *Bayesian Inference, Kendall's Advanced Theory of Statistics,* Vol. **2b**, Oxford Univ. Press, Oxford, UK

[3] Gelman, A, Carlin, J B, Stern, H S and Rubin, D B, 2004, *Bayesian Data Analysis,* Chapman & Hall/CRC

[4] Carslaw, H. S. and Jaeger, C. 1959, *Conduction Of Heat In Solids,* Oxford, Clarendon Press

[5] COMSOL Version 3.5a, *Heat Transfer Module User's Guide,*COMSOL Multiphysics, 2008

[6] Smolyak, S., 1963, "Quadrature and Interpolation Formulas for Tenser Product of Certain Classes of Functions," *Soviet Math. Dokl.,* 4, pp 240-243

[7] Barthelmann, V., Novak, E. and Ritter, K., 2000, "High Dimensional Polynomial Interpolation on Sparse Grids," *Advances in Computational Mathematics,*, Vol. 12, pp 273-287

[8] Klimke, A., 2008, "Sparse Grid Interpolation Toolbox User's Guide,"
www.ians.uni-stuttgart.de/spinterp/download.html

[9] Achatz, S., 2003, "Higher Order Sparse Grid Methods for Elliptic Partial Differential Equations with Variable Coefficients," *Computing,* 71, 1-15

[10] Garcke, J. and Hegland, M., 2009, "Fitting multidimensional data using gradient penalties and the sparse grid combination technique," *Computing,* 84, pp 1-25

[11] Gerstner, T. and Griebel, M., 2008, "Sparse Grids," *Encyclopedia of Quantitative Finance,*, J. Wiley and Sons, N.Y,., NY

[12] Ma, X. and Zabaras, N., 2009, "An Adaptive Hierarchical Sparse Grid Collocation Algorithm for the Solution of Stochastic Differential Equations," *Journal of Computational Physics* 228, pp 3084-3112

[13] Ganapathysubramanian, B. and Zabars, N., 2007, "Sparse Grid Collocation Schemes for Stochastic Natural Convection Problems," *J. Computational Physics*, 225, pp 652-685

[14] Lin, B. and Tartakovsky, A. M., 2009, "An Efficient, High-Order Probabilistic Collocation Method on Sparse Grids for Three-Dimensional Flow and Solute Transport in Randomly Heterogeneous Porous Media," *Advances in Water Resourses,* 32, pp 712-722

[15] Agarwal, N. and Aluru, N. R. , N. R., 2009, "Stochastic Analysis of Electrostatic MEMS Subjected to Parameter Variations," *Journal of Microelectromechanical Systems,* VOL. 18, pp. 1454-1468

Identification of Thermophysical Properties of Nanocomposites via Integral Transforms, Bayesian Inference and Infrared Thermography

Diego C. Knupp[1], Carolina P. Naveira Cotta[2], João Vitor C. Ayres, Renato M. Cotta[3], Helcio R. B. Orlande[4]

Mechanical Engineering Department, Laboratory of Transmission and Technology of Heat - LTTC
Universidade Federal do Rio de Janeiro, POLI & COPPE/UFRJ, Rio de Janeiro, RJ, Brazil
Email: [1]diegoknupp@gmail.com, [2]cpncotta@hotmail.com, [3]cotta@mecanica.ufrj.br, [4]helcio@mecanica.ufrj.br

Abstract

Simultaneous estimation of space variable thermal conductivity and heat capacity in heterogeneous samples of nanocomposites is dealt with by employing a combination of the Generalized Integral Transform Technique (GITT), for the direct problem solution, Bayesian inference as implemented with the Markov Chain Monte Carlo (MCMC) method, for the inverse analysis, and infrared thermography, for the temperature measurements. Another aspect of the proposed approach is the integral transformation of the thermographic experimental data along the space variable, which allows for a significant data compression since the inverse analysis is undertaken within the transformed field. Results are presented for the validation of the experiment with a homogeneous polyester plate, as well as for a polyester - alumina nanoparticles composite with abrupt variation of the filler concentration.

Keywords: Nanocomposites, Variable Thermophysical Properties, Infrared Thermography, Bayesian Inference, Integral Transforms, Hybrid Methods

1. Introduction

Composite materials have been providing engineers with increased opportunities for tailoring structures to meet a variety of property and performance requirements. For the accurate determination of local variations in physical properties within heterogeneous media, one must seek a technique that provides the required information of spatially distributed measurements, for a successful solution of the appropriate inverse problem [1]. In addition, as the morphology of the heterogeneous medium directly influences the spatial behavior of the physical properties, the use of the non-intrusive technique of infrared thermography becomes of major interest, for being capable of providing measurements with high spatial resolution and high frequency [1].

In general, inverse methods require accurate and computationally fast direct problem solution methodologies, in order to handle the large amount of data that might be available and to allow for the computationally intensive iterative analysis often required by the inverse problem solution [2]. Solution techniques for partial differential equations that exploit the available analytical knowledge and rely on modern symbolic computation platforms offer advantages over classical numerical approaches in a number of applications, such as the Generalized Integral Transform Technique (GITT) for the hybrid numerical-analytical solution of convection-diffusion problems [3,4]. This approach is based on the extension of the classical integral transform method, making it sufficiently flexible to handle problems that are not a priori transformable, such as with arbitrary space-dependent and nonlinear coefficients in either the governing equation or the boundary conditions. This class of hybrid numerical-analytical methods was quite successfully employed in the direct and inverse analysis of heat conduction in heterogeneous media [5-7], including a novel proposition of performing the inverse analysis with the integral transformed experimental data [7]. Also, the effort to integrate the knowledge on the GITT application into a general purpose computational code resulted in an open source mixed symbolic-numerical algorithm called UNIT (**UN**ified **I**ntegral **T**ransforms) [8], for hybrid integral transform solutions of convection-diffusion problems.

In this paper, we present results obtained with an experiment involving a partially heated thin plate made of a nanocomposite composed of polyester resin and alumina nanoparticles, with an abrupt variation in the thermophysical properties. The inverse analysis procedure employs the infrared thermography as the measurement technique and the integral transform approach for the direct problem solution. Based on Bayesian inference, the inverse analysis methodology advanced in [7] introduces the use of integral transformed temperature measurements and the estimation within the transformed domain, providing a significant reduction on experimental data handling and computational effort.

2. Direct Problem Formulation and Solution

We consider the one-dimensional version of the general formulation on transient diffusion solved in [5,9], for the potential $T(x,t)$ dependent on position $x \in [0,L]$ and time $t > 0$. The diffusion equation and respective initial and boundary conditions are given by:

$$w(x)\frac{\partial T(x,t)}{\partial t} = \frac{\partial}{\partial x}\left[k(x)\frac{\partial T(x,t)}{\partial x}\right] - d(x)T(x,t) + P(x,t,T), \quad 0 < x < L,\ t > 0 \tag{1a}$$

$$T(x,0) = f(x), \quad 0 \le x \le L \tag{1b}$$

$$\alpha_0 T(x,t) - \beta_0 k(x)\frac{\partial T(x,t)}{\partial x} = \phi_0(x,t,T), \quad x = 0,\ t > 0 \tag{1c}$$

$$\alpha_1 T(x,t) + \beta_1 k(x)\frac{\partial T(x,t)}{\partial x} = \phi_1(x,t,T), \quad x = L,\ t > 0 \tag{1d}$$

A formal exact solution of problem (1) is achievable accounting for the exact integral transformation when the specific eigenvalue problem with all the space variable coefficients is chosen, such as in [5-7]. This solution is based on the direct application of the Classical Integral Transform Method to the linear version of problem (1) and the Generalized Integral Transform Technique [3,4] to the eigenvalue problem with space variable coefficients. An alternative approach is also available [9], by applying the GITT directly to the nonlinear problem (1), such as in the UNIT algorithm [8] implemented in the symbolic computational platform *Mathematica* [10]. In either one of the employed direct problem solutions, we start with the inverse formula below:

$$T(x,t) = T_f(x;t) + \sum_{i=1}^{\infty}\tilde{\psi}_i(x)\bar{T}_i(t), \quad \text{where} \quad \bar{T}_i(t) = \int_0^L w^*(x)\tilde{\psi}_i(x)T^*(x,t)dx \tag{2a,b}$$

where $T_f(x;t)$ is a proposed analytical filtering solution, and $\bar{T}_i(t)$ the transformed potentials in the above eigenfunction expansion, defined with the integral transformation operation above.

The eigenvalues μ_i and normalized eigenfunctions $\tilde{\psi}_i(x)$, are obtained from the chosen eigenvalue problem in each case, which specifies the weighting function to be adopted, $w^*(x)$, as described in further details in [5,9].

3. Inverse Problem Solution

This work provides an experimental verification of the inverse analysis methodology advanced in [7], in which Bayesian inference was used for the estimation of spatially variable equation and boundary condition coefficients in heat diffusion problems, by employing the Markov chain Monte Carlo (MCMC) method, implemented with the Metropolis-Hastings algorithm [11]. With this method, the sampling procedure used to recover the posterior distribution is in general the most expensive computational task, since the direct problem is calculated for each state of the Markov chain. Therefore, the use of a fast, accurate and robust computational implementation of the direct solution is extremely important. Thus, the integral transformation approach becomes very attractive for the combined use with the MCMC method. The coefficients $w(x)$, $k(x)$ and $d(x)$ can be themselves expanded in terms of eigenfunctions, together with a filtering solution to enhance their convergence, in the following form, exemplified for $w(x)$:

$$w(x) = w_f(x) + \sum_{k=1}^{\infty}\tilde{\Gamma}_k(x)\bar{w}_k, \qquad \text{inverse} \tag{3a}$$

$$\bar{w}_k = \int_0^{L_x}\hat{w}(x)[w(x) - w_f(x)]\tilde{\Gamma}_k(x)dx, \ \text{transform} \tag{3b}$$

where $\hat{w}(x)$ is the weighting function for the chosen normalized eigenfunction $\tilde{\Gamma}_k(x)$ [6,7].

In the Bayesian approach, inference is drawn by constructing the joint probability distribution of all unobserved quantities based on what is known about them. This knowledge incorporates previous information about the phenomena under study and is also based on values of observed quantities when they are available [11]. This approach is based on Bayes' theorem, which can be written as

$$p(\mathbf{P}|\mathbf{Y}) = \frac{p(\mathbf{Y}|\mathbf{P})p(\mathbf{P})}{p(\mathbf{Y})} \tag{4}$$

Solving an inverse problem within the Bayesian framework basically consists of the following three subtasks: (i) Based on all information available for the unknown $\mathbf{P}$ prior to the measurements $\mathbf{Y}$, model a prior probability density $p(\mathbf{P})$; (ii) Model the likelihood function $p(\mathbf{Y}\mid\mathbf{P})$ that describes the interrelation between the observations and the unknowns; (iii) Develop methods to explore the posterior probability density $p(\mathbf{P}\mid\mathbf{Y})$. When it is not possible to analytically obtain the corresponding posterior distributions, one needs to use a method based on simulation. The inference based on simulation techniques uses samples from the posteriori $p(\mathbf{P}\mid\mathbf{Y})$ to extract information about this distribution. Several sampling strategies are proposed in the literature, including the Monte Carlo method via Markov Chain (MCMC) implemented with the Metropolis-Hastings algorithm [11], which is adopted in this work.

The unknown quantities in this work are the variable thermal properties and the effective heat transfer coefficient, which are expressed as eigenfunction expansions, as well as the applied the heat flux, which is parameterized as explained below. Thus, the total number of parameters N_P to be estimated is given by the sum of parameters in each expansion, including the number of parameters in each filter function, and the number of parameters in the heat flux expression.

By following the methodology advanced in [7], the solution of the inverse problem was obtained within the transformed temperature field. The experimental temperature data is integral transformed by using equation (5a), thus compressing the experimental measurements in the spatial domain into a few transformed temperature modes.

Transform: $$\overline{T_{\exp,i}}(t) = \int_0^{Lx} w(x)\tilde{\psi}_i(x)[T_{\exp}(x,t) - T_f(x;t)]dx \tag{5a}$$

Inverse: $$T_{\exp}(x,t) = T_f(x;t) + \sum_{i=0}^{Ni} \tilde{\psi}_i(x)\overline{T_{\exp,i}}(t) \tag{5b}$$

4. Experimental Setup and Procedure

The experimental setup presented in Fig. 1a employs temperature measurements obtained from the infrared camera FLIR SC660, with 640x480 image resolution, and -40°C to 1500°C temperature range. The main components of the setup are: (a) IR camera (FLIR SC660); (b) camera stand for vertical experiment configuration; (c) frame with the sample and the heater; (d) sample support; (e) data acquisition system (Agilent 34970-A); (f) microcomputer for data acquisition. Fig. 1b illustrates an image produced by the FLIR SC660 camera after some elapsed heating time. As the test sample we have manufactured a nanocomposite plate, which is composed of polyester resin as matrix and alumina nanoparticles as filler, in such a way that ¾ of the plate's length has 28.5% of alumina nanoparticles in mass and the other ¼ of the plate's length is composed only of polyester resin, with no addition of filler, promoting an abrupt change of thermophysical properties. The plate's thickness is 1.51 mm and its lateral and vertical dimensions are 40 mm and 80 mm, respectively. For verification purposes, another homogeneous sample was used in this study, which is composed of polyester resin only, with no addition of filler. The thickness of this plate is 1.7 mm and the lateral and vertical dimensions are also 40mm and 80 mm, respectively.

An electrical resistance (38.2 Ω) was employed for the heating of the plate. This resistance has the same lateral dimensions as of the plate but half of its length (40x40 mm), and is located at the upper half of the plate. The resistance is attached to the plate with the aid of a thermal compound paste. Two experimental arrangements were investigated by varying the position of the plate with respect to the electrical resistance: first the portion with no addition of filler has been placed in contact with the electrical resistance and for the second case the plate has been turned upside down. These two cases are schematically represented in Figs. 2. For the homogeneous polyester resin sample, as two identical plates were available, we have set up a plate-heater-plate sandwich, which was attached to the sample support with insulated corners. This is thus the third case investigated in this work. These experimental setups are hereafter called Cases 1, 2 and 3, respectively. In order to reduce uncertainty in the IR camera readings, the plate surfaces were painted with a graphite ink, which brought its emissivity to around $\varepsilon = 0.97$, as stated by the ink manufacturer.

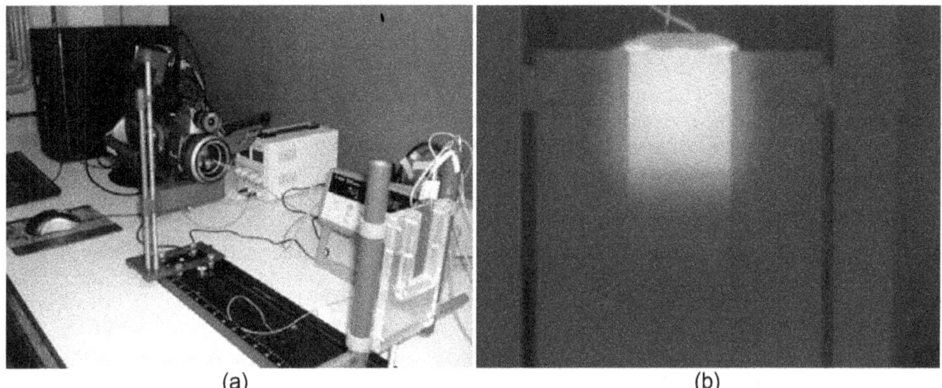

(a) (b)

Figure 1: (a) General view of the experimental setup for the infrared thermography analysis; (b) Image produced by the FLIR SC660 camera after some elapsed heating time.

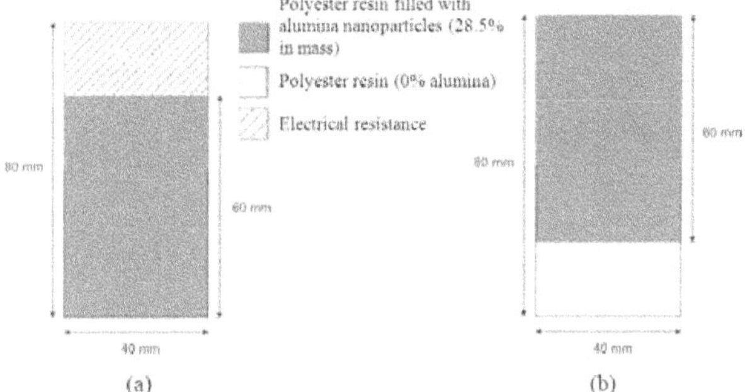

Figure 2: (a) Schematic representation of the experimental setup for (a) Case 1 and (b) Case 2.

5. Results and Discussion

Considering a lumped formulation across the sample thickness, the direct problem is represented by the heat conduction equation with initial and boundary conditions given by eqs.(1), and with the following parameters:

$$d(x) = \frac{h_{eff}(x)}{L_z}; \quad P(x,t,T) = -\frac{h_{eff}(x)T_\infty}{L_z} + \frac{q_w(x,t)}{L_z}; \quad f(x) = T_\infty; \quad \alpha_0 = \alpha_1 = 0; \quad \beta_0 = \beta_1 = 0; \quad \phi_0 = \phi_1 = 0; \qquad \text{(7a-f)}$$

where $h_{eff}(x)$ is the effective heat transfer coefficient, $q_w(x,t)$ is the applied heat flux and L_x and L_z are the plate's length and thickness, respectively. In [9] it has been verified that the appropriate identification of thermophysical properties through the proposed experimental setup requires the simultaneous estimation of the heat transfer coefficient and the time variation of the applied heat flux in light of the apparent border effects and thermal capacitance of the heater. For the time variation of the applied heat flux the following parameterization is employed:

$$q_w(x,t) = q(x)f(t) \qquad f(t) = c - ae^{-bt} \qquad \text{(8a,b)}$$

For the thermal conductivity and heat capacity, as the filter in the inverse analysis procedure for Cases 1 and 2 we have considered a step function, with a guessed position along the nanocomposite plate's length where the alumina nanoparticles concentration transition occurs, while for Case 3, we have considered a linear filter. For these parameters we have adopted normal priors, with 15% standard deviation, centered in literature values for the polyester resin [12] and in Lewis-Nielsen's formula prediction [13] for the region filled with alumina nanoparticles. Also, normal priors were employed for the two values of the parameters in the filter of the effective heat transfer coefficients, h_{x0} and h_{xL}, provided by available correlations of natural convection and linearized radiation. For the remaining parameters, non-informative priors have been adopted.

Table 1 presents the estimated thermophysical properties in the three cases examined, where 120,000 states in the MCMC method have been generated, being the first 40,000 states neglected in order to achieve the equilibrium of the chain. One may observe that the estimates obtained for the polyester resin properties of the nanocomposite plate in Case 1 are very close to those obtained in Case 3, where the experimental setup involved homogeneous polyester resin plates. On the other hand, the estimated parameters for the polyester in Case 2 are not in good agreement with those estimated for case 1. That's probably because for case 2 the portion of the nanocomposite plate composed with polyester resin without addition of filler has been placed away from the applied heat flux. This result was expected, since this homogeneous portion of the plate suffers a smaller variation of temperature during the experiment, yielding locally low sensitivity coefficients for the parameters. For the portion of the nanocomposite plate corresponding to the polyester resin filled with alumina nanoparticles, it may be observed that Cases 1 and 2 yield estimates very close to each other. The values estimated for the thermal conductivity are slightly higher than those provided by the Lewis-Nielsen formula [13], and similar results have been observed in [14].

Table 1: Estimated thermophysical properties.

Property	Material	Prior	Estimates [99% confidence intervals]
Case 1			
k [W/m°C]	polyester	N(0.16, 15%)	0.1617 [0.1605, 0.1629]
	polyester + alumina	N(0.193, 15%)	0.2042 [0.2022, 0.2062]
w [J/m^3°C]	polyester	N(1.595x10^6, 15%)	1.59x10^6 [1.58x10^6, 1.6x10^6]
	polyester + alumina	N(1.736x10^6, 15%)	1.760x10^6 [1.758x10^6, 1.763x10^6]
Case 2			
k [W/m°C]	polyester	N(0.16, 15%)	0.1486 [0.1466, 0.1506]
	polyester + alumina	N(0.193, 15%)	0.2025 [0.1995, 0.2055]
w [J/m^3°C]	polyester	N(1.595x10^6, 15%)	1.529x10^6 [1.526x10^6, 1.532x10^6]
	polyester + alumina	N(1.736x10^6, 15%)	1.743x10^6 [1.738x10^6, 1.748x10^6]
Case3			
k [W/m°C]	polyester	N(0.16, 15%)	0.159 [0.157, 0.161]
w [J/m^3°C]	polyester	N(1.595x10^6, 15%)	1.566x10^6 [1.558x10^6, 1.574x10^6]

Figures 3a,b depict the estimated thermal conductivity of the nanocomposite plate and the residuals between the measurements and the calculated temperatures, respectively, for Case 1. Figures 3 and Table 1 show that the present approach was capable of accurately recovering the two levels of the thermal conductivity step variation, thus resulting in small residuals. On the other hand, the residuals presented in figure 3b are correlated. Anyhow, we note that the inverse problem solution was obtained with the transformed temperature modes, while the residuals presented in Fig. 3b correspond to the actual temperature measurements.

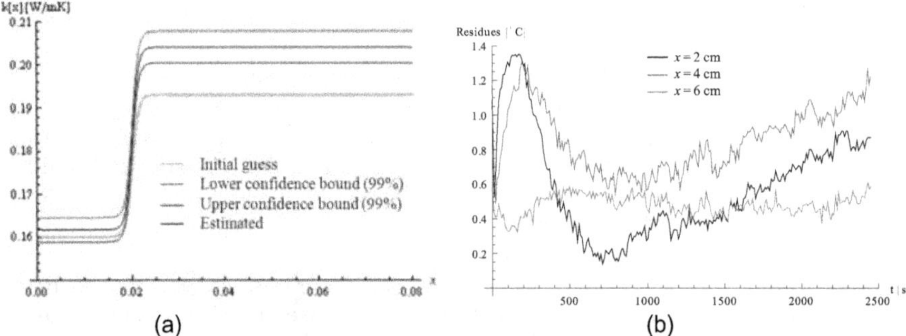

(a) (b)

Figure 3: (a) Estimated thermal conductivity of the nanocomposite plate and (b) Residuals between the experimental data and the calculated temperatures for Case 1.

Finally, we present in Figs.4 a comparison of the experimental data and the simulated results obtained by means of the GITT via the UNIT code, with the estimated parameters. Figs. 4 present such a comparison for the spatial distribution of temperature at t=400 sec (a), and for the time evolution of the temperatures at x=2 cm (b), for Case 1. Both figures show an excellent agreement between the simulated results and the experimental data.

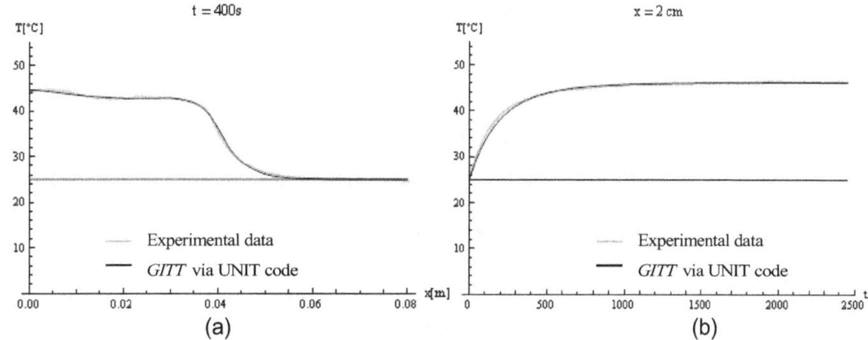

(a) (b)

Figure 4: Case 1 - (a) spatial distribution of temperature at $t = 400$ s and (b) time evolution of the temperature at (a) $x = 2$ cm using the estimated parameters in the UNIT code.

6. References

[1] Fudym, O., Orlande, H. R. B., Bamford, M., Batsale, J. C., "Bayesian Approach for Thermal Diffusivity Mapping from Infrared Images Processing with Spatially Random Heat Pulse Heating", J. Physics. Conference Series (Online), Vol. 135, pp 12-42, 2008.

[2] Ozisik, M. N. and Orlande, H. R. B., Inverse Heat Transfer: Fundamentals and Application, Taylor and Francis, New York, 2000.

[3] Cotta, R. M., Integral Transforms in Computational Heat and Fluid Flow, CRC Press, Florida, 1993.

[4] Cotta, R. M., Mikhailov, M. D., Heat Conduction: Lumped Analysis, Integral Transforms, Symbolic Computation, Wiley-Interscience, 1997.

[5] Naveira Cotta, C. P., Cotta, R. M., Orlande, H. R. B., Fudym, O., "Eigenfunction expansions for transient diffusion in heterogeneous media", Int. J. Heat & Mass Transfer, Vol. 52, pp 5029-5039, 2009.

[6] Naveira Cotta, C. P., Orlande, H. R. B., Cotta, R. M., "Integral Transforms and Bayesian Inference in the Identification of Variable Thermal Conductivity in Two-Phase Dispersed Systems", Num. Heat Transfer – part B Fundamentals, Vol. 57, No. 3, pp 173-203, 2010.

[7] Naveira-Cotta, C.P., R.M. Cotta, and H.R.B. Orlande, "Inverse Analysis with Integral Transformed Temperature Fields for Identification of Thermophysical Properties Functions in Heterogeneous Media", *Int. J. Heat & Mass Transfer*, V.54, no.7-8, pp.1506-1519, 2011.

[8] Sphaier, L.A., R.M. Cotta, C.P. Naveira-Cotta, and J.N.N. Quaresma, "The UNIT Algorithm for Solving One-Dimensional Convection-Diffusion Problems via Integral Transforms", *Int. Comm. Heat & Mass Transfer,* (in press).

[9] Knupp, D. C., Naveira Cotta, C. P., Ayres, J. V. C., Orlande, H. R. B., Cotta, R. M., "Experimental-theoretical analysis of a transient heat conduction setup via infrared thermography and unified integral transforms", International Review of Chemical Engineering, Vol. 2, pp 736-747, 2010.

[10] Wolfram, S., The Mathematica Book*, version 5.2*, Cambridge-Wolfram Media, 2005.

[11] Kaipio, J. and Somersalo, E., Statistical and Computational Inverse Problems, Springer, New York, 2005.

[12] Cao, Y.M., Sun, J., Yu, D.H., "Preparation and properties of nano-Al2O3 particles/polyester/epoxy resin ternary composites", Journal of Applied Polymer Science, Vol. 83, No. 1, pp 70–77, 2002.

[13] Lewis, T. e Nielsen, L., "Dynamic mechanical properties of particulate-filled polymers", J. Applied Polymer Science, Vol.14, No.6, pp 1449-1471, 1970.

[14] Evans, W., Prasher, R. Fish, J., Meakin, P., Phelan, P. e Keblinski, P., "Effect of aggregation and interfacial thermal resistance on thermal conductivity of nanocomposites and colloidal nanofluids", Int. J. of Heat and Mass Transfer, Vol. 51, pp 1431-1438, 2008.

A New Control System for Left Ventricular Assist Devices based on the Physiological Demand of the Patient

George Faragallah[1], Yu Wang[1], Eduardo Divo[2], and Marwan Simaan[1]

[1]*Department of Electrical Engineering*
University of Central Florida, Orlando, FL
Email: georgef@knights.ucf.edu
[2]*School of Engineering Technology*
Daytona State College, Daytona Beach, FL
Email: divoe@daytonastate.edu

Abstract

A new feedback control system is developed to automatically adjust the current supplied to the pump motor of rotary left ventricular assist devices (LVAD) based on continuously estimating the physiological demand of the patient as he or she changes activity levels. This is accomplished by inversely solving for the systemic vascular resistance (SVR) of the patient from a coupled LVAD-Cardiovascular model using information from measured LVAD pump flow. Furthermore, the calculated SVR is used in a patient-specific cardiovascular model that assumes a healthy heart to determine the physiological demand in terms of blood flow. Once the physiological demand is established, the current supplied to the pump motor of the LVAD can be adjusted to achieve the desired blood flow through the cardiovascular system. This process can be performed automatically in a real-time basis using information that is readily available and thus rendering a high degree of applicability. Results from simulated patient data shows that the feedback control system is fast and very stable. Future work will feature a gradient-based approach for the estimation of the SVR and adjusting the pump motor current.

Key Words: Feedback Control Systems, Cardiovascular model, LVAD, Physiological Blood Demand.

1. Introduction

The American Heart Association (AHA) estimates that 5.8 million patients above the age of 20 are suffering from Heart Failure (HF), a condition in which the heart cannot pump enough blood into the circulatory system and thus not providing the body with its needs of nutrients and oxygenated blood. The left ventricular assist device (LVAD) is a mechanical pump that can assist the native heart of the patient in performing its function. This pump can provide an alternative way for the blood to flow with a higher rate between the ventricle and the aorta for HF patients.

Currently, the LVAD is being set on a constant pump speed level that matches the lowest venous return [2]. This technique limits the activity of the patients, prevent their return to workforce and many other forms of life that require the blood flow to be from moderate to high. Nowadays, the LVAD is used as destination therapy device and that requires an automatic feedback controller that can sense the need of the patients to more or less blood flow, so it can manipulate the pump motor current, which directly controls the pump flow, to match the physiological need. This controller will allow the patients to leave the hospital and return to a close to normal life style. Such controller needs real-time measurements of the hemodynamic of the patients. However, under the current technology the implantation of long-term sensor inside the human heart is not possible due to the vulnerability to thrombus formation over the sensing diaphragm and being an extra liability on the battery used to power both the pump and the controller [3]. External sensors like the ones used in pacemakers are not highly reliable to be used in conjunction with the LVAD [4]. The pump flow data seems to be a good candidate to be used to control the pump motor current as it can be easily measured by a flow-meter at one of the pump cannulae.

2. The Cardiovascular Model

The cardiovascular system can be represented by a 5th order circuit model and the LVAD pump can be simulated by a 1st order circuit model. Combining both models will result in a 6th order model that has a minimum number of parameters that can offer enough complexity to give an accurate representation of the heart and the LVAD.

Figure 1 shows the 5th order model of the cardiovascular system. This model is adopted from previous work [5] where every resistance, inductance, capacitance and diode used in the model here is well explained and a standard value is provided. There is a need, however, to discuss in more details some important elements in this circuit like: R_S which represents the systemic vascular resistance (can be denoted also as SVR) that can be used to simulate the level of activity experienced by the patient, higher value means that the patient is resting and lower value means that the arteries are offering less resistance to the blood flow because of the high level of activity of the patient (like running, exercising, etc.) The left ventricular compliance $C(t)$ is the inverse of the elastance function of the heart $E(t) = 1/C(t)$ and by changing $E(t)$ we can vary the level of the heart sickness of the patient (i.e. how severe is the heart failure), if $E_{max} = 2$ mmHg/ml , then the left ventricle is healthy. E_{max} can be varied from 1 mmHg/ml to 0.25 mmHg/ml to represent moderate to severe heart failure, respectively. D_A and D_M are the ideal diode representations of the aortic and mitral valves. The opening and closing of these valves are controlled by the pressures across them; hence they can simulate the four phases of the cardiac cycle as mentioned in Table 1.

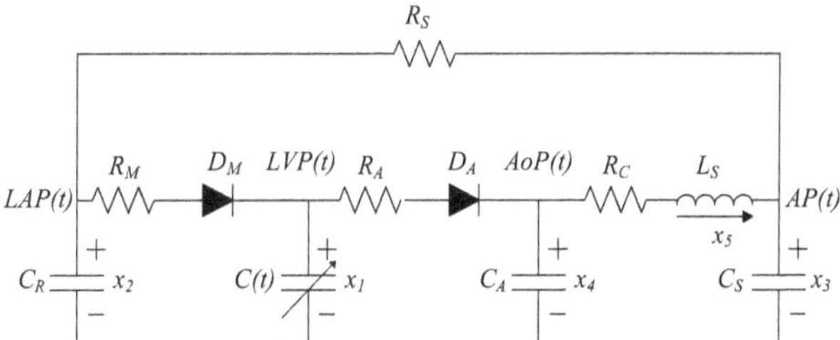

Figure 1: Cardiovascular Circuit model

Table 1: Phases of the cardiac cycle

Modes	Valves		Phases
	Mitral	Aortic	
1	closed	closed	Isovolumic relaxation
2	open	closed	Filling
1	closed	closed	Isovolumic contraction
3	closed	open	Ejection
-	open	open	Not feasible

3. The combined LVAD-Cardiovascular model

When adding the 1st order circuit model representing the LVAD pump to the 5th order model in Fig. 2 we will have a 6th order model that is shown in Fig. 1. The pump functions in parallel to the heart of the patient, hence the parallel connection of the LVAD pump between the left ventricle and the aorta.

Table 2 indicates the six state variables for this circuit model. It is worth mentioning that H represents the pressure difference across the pump [6] and can be calculated by the following equation:

$$H = R^{*} x_6 + L^{*} \frac{dx_6}{dt} - \frac{\gamma}{x_6} i(t) \tag{1}$$

Where $i(t)$ is the pump motor current and the control variable for the model, and R^* and L^* are the summation of resistance and inductances in the pump model, respectively. R_K is a time-varying nonlinear pressure-dependent resistor to simulate the phenomenon of suction (suction happens when the pump tries to draw more blood than available, this may result in the heart muscle collapsing).

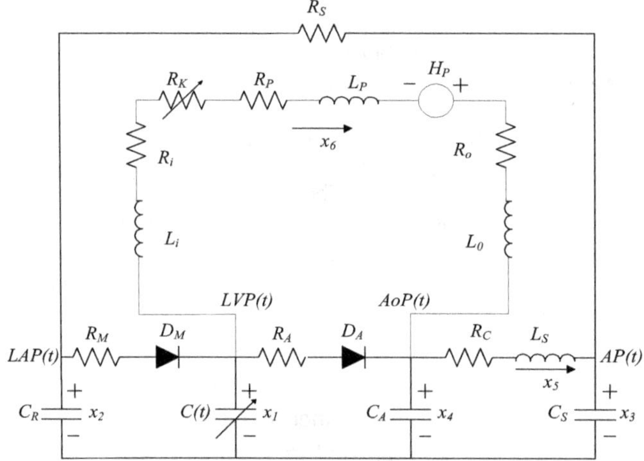

Figure 2: Combined Cardiovascular and LVAD model

The state space representation of the combined model can be written in the following form:

$$\dot{x} = A(t)\,x + P(t)\,p(x) + b\,i(t) \tag{2}$$

Where:

$$A = \begin{bmatrix} \frac{-\dot{C}(t)}{C(t)} & 0 & 0 & 0 & 0 & \frac{-1}{C(t)} \\ 0 & \frac{-1}{R_S C_R} & \frac{1}{R_S C_R} & 0 & 0 & 0 \\ 0 & \frac{1}{R_S C_S} & \frac{-1}{R_S C_S} & 0 & \frac{1}{C_S} & 0 \\ 0 & 0 & 0 & 0 & \frac{-1}{C_A} & \frac{1}{C_A} \\ 0 & 0 & \frac{-1}{L_S} & \frac{1}{L_S} & \frac{-R_C}{L_S} & 0 \\ \frac{1}{L^*} & 0 & 0 & \frac{-1}{L^*} & 0 & \frac{-R^*}{L^*} \end{bmatrix} \tag{3}$$

And:

$$P(t) = \begin{bmatrix} \frac{1}{C(t)} & \frac{-1}{C(t)} \\ \frac{-1}{C_R} & 0 \\ 0 & 0 \\ 0 & \frac{1}{C_A} \\ 0 & 0 \\ 0 & 0 \end{bmatrix}, \quad p(x) = \begin{bmatrix} \frac{1}{R_M} r(x_2 - x_1) \\ \frac{1}{R_A} r(x_1 - x_4) \end{bmatrix}, \quad b = \begin{bmatrix} 0 \\ 0 \\ 0 \\ 0 \\ 0 \\ \frac{\gamma}{L^* x_6} \end{bmatrix} \tag{4}$$

Where the $r(\xi)$ in $p(x)$ expression is defined by:

$$r(\xi)=\begin{cases}\xi & if\ \xi \geq 0\\ 0 & if\ \xi < 0\end{cases} \tag{5}$$

Table 2: State variables in the cardiovascular model

Variables	Name	Physiological meaning (units)
$x_1(t)$	LVP(t)	Left ventricle pressure (mmHg)
$x_2(t)$	LAP(t)	Left atrial pressure
$x_3(t)$	AP(t)	Arterial pressure (mmHg)
$x_4(t)$	AoP(t)	Aortic pressure (mmHg)
$x_5(t)$	Q_T(t)	Total flow (ml/s)
$x_6(t)$	Q_P(t)	Pump flow (ml/s)

4. Development of a Feedback Controller

The LVAD-Cardiovascular model is controlled by the pump motor current which consequently controls the pump motor speed [6] according to the following relation:

$$\omega(t)=\sqrt{\frac{\gamma\, i(t)}{\beta\, Q(t)}} \tag{6}$$

Where γ and β are constants, $i(t)$ is the pump motor current, $\omega(t)$ is the pump speed and $Q(t)$ is the pump flow. When the pump speed is increased the pump flow will increase and vice versa. The aim of the feedback controller is to provide the patient with the required amount of blood flow depending on the level of activity of the patient. The current technology doesn't allow the implantation of sensors inside the human heart for long-term applications; hence there is a need to depend on the pump flow, which is accessible through the installation of a flow-meter sensor inside the pump cannulae, as a feedback variable to automatically adjust and control the pump motor current.

It is shown through Fig. 3 which is obtained by simulation (using the 6th order model presented above) that the mean of the pump flow is inversely proportional to the systemic vascular resistance at a constant pump motor current.

A block diagram for the proposed feedback controller is shown in Fig. 4.The controller consists of three stages before a decision is made to update the pump motor current and hence increase or decrease the mean pump flow.

The first stage, labeled "detect change in pump flow", will keep reading the mean pump flow signal until a change is detected between two consecutive readings. This change is evidence that the activity level of the patient has changed and there is a need to adjust the pump motor current to respond to the new physiological demand. The first stage can be thought of as a gate to the rest of the feedback controller blocks. The controller will only work if the first block sends a signal as a response to change in the mean pump flow. The second stage, labeled "Estimate the SVR using the 6th order model" will estimate the current SVR by running the 6th order model under the same conditions as the heart-LVAD system of the patient (i.e. using the same E_{max}, HR and $i(t)$). Initially the SVR will be determined based on the following non-linear equation:

$$R_{Snew}=\begin{cases}R_{Sold}+\Delta R_S & ,\ if\ Q_{old} > Q_{new}\\ R_{Sold}-\Delta R_S & ,\ if\ Q_{old} < Q_{new}\end{cases} \tag{7}$$

And afterwards a similar criterion will be used:

$$R_S(k+1)=\begin{cases}R_S(k)+\Delta R_S & ,\ if\ Q_k > Q_{k+1}\\ R_S(k)-\Delta R_S & ,\ if\ Q_K < Q_{k+1}\end{cases} \tag{8}$$

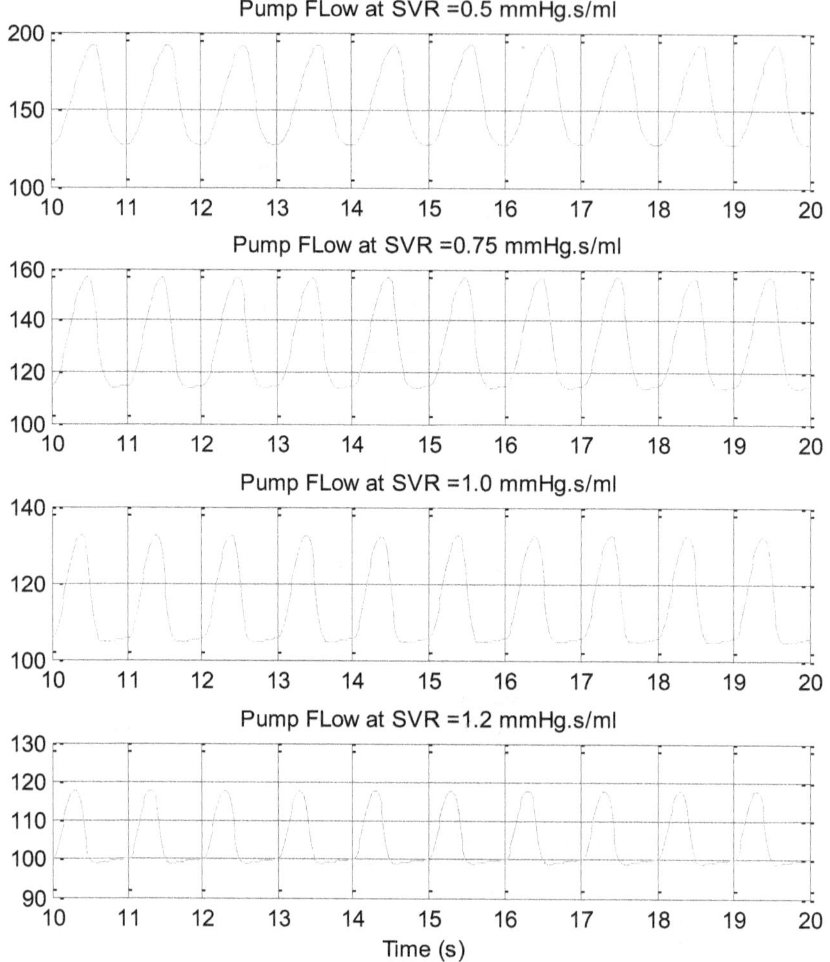

Figure 3: Pump flow signals at $i(t) = 0.18\ amp$ and different SVR values

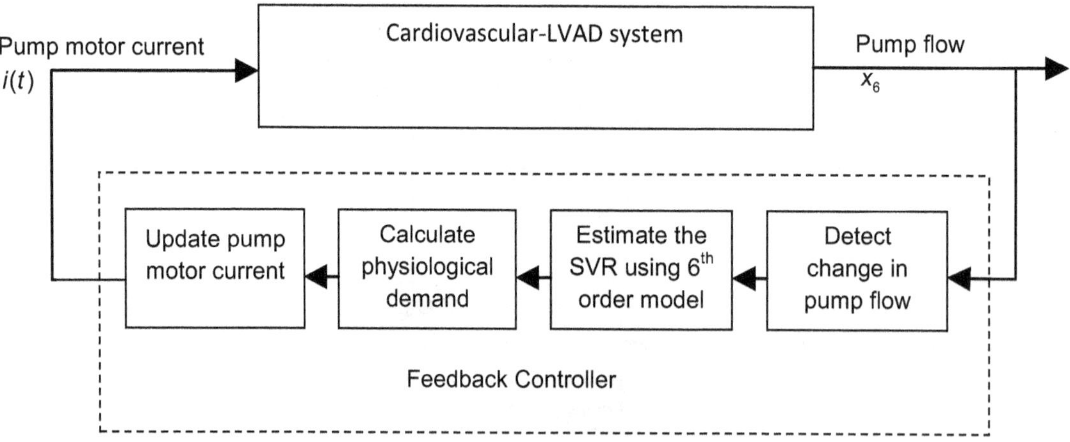

Figure 4: Block diagram for the LVAD feedback controller

The feedback controller will keep trying different values for SVR, according to the rule above, until the mean pump flow generated by the 6th order model is equal to the new measurement of the actual mean pump flow of the patient. At this point the feedback controller determines that the final SVR is the same SVR for the patient. The third stage, labeled "Calculate physiological demand for estimated SVR", will make use of this SVR by plugging it into the 5th order system to determine the required mean pump flow under the current activity level. The E_{max} used in the 5th order model will equal 2 mmHg/ml to represent a healthy heart. This is done since the objective here is to determine what a healthy heart will produce in such circumstances and try to match it by the LVAD. Once this information is obtained, the controller will manipulate the pump motor current until the mean pump flow reaches the desired level.

5. References

[1] Lloyd-Jones D, Adams RJ, Brown TM, Carnethon M, Dai S, De Simone G, Ferguson TB, Ford E, Furie K, Gillespie C, GO A, Greenlund K, Haase N, Hailpern S, Ho PM, Howard V, Kissela B, Kittner S, Lackland D, Lisabeth L, Marelli A, McDermott MM, Meigs J, Mozaffarian D, Mussolino M, Nichol G, Roger VL, Rosamond W, Sacco R, Sorlie P, Stafford R, Thom T, Wasserthiel-Smoller S, Wong ND, Wylie-Rosett J., "Heart disease and stroke statistics – 2010 update: a report from the American Heart Association," Circulation, 2010;121:e46-e215.

[2] Vollkron M., Schima H., Huber L., Benkowski B., Morello G., and Wieselthaler G., "Control of implantable axial blood pumps based on physiological demand," Proceedings of the 2006 American Control Conference, Minneapolis, MN, USA, June 14-16, 2006.

[3] Bertram C D, "Measurement for implantable rotary blood pumps," Physiological Measurements, 26(2005), R99-R117.

[4] Olsen D.B., "The history of continous-flow blood pumps," Artificial Organs 24(6):401-404.

[5] Simaan M.A., Ferreira A., Chen S., Antaki J. F. and Galati, D. G. "A Dynamical State Space Representation and Performance Analysis of a Feedback-Controlled Rotary Left Ventricular Device," IEEE Trans. On Control Systems Technology, Vol. 17, No. 1, 2009, pp 15-28.

[6] Faragallah G., Wang Y., Divo E. and Simaan M.A. "A New Current-Based Control Model of the Combined Cardiovascular and Rotary Left Ventricular Assist Device," Proceeding of 2011 American Control Conference, San Francisco, CA, USA, June 29 – July 1, 2011.

An inverse heat transfer problem for the determination of the cooling system in thermoplastic injection

A.Agazzi[1], *V.Sobotka*[1], *R.Le Goff*[2], *D.Garcia*[2], *Y.Jarny*[1]

[1] *Laboratoire de Thermocinétique de Nantes UMR 6607*
Email: vincent.sobotka@univ-nantes.fr
[2] *Pôle Européen de Plasturgie, Oyonnax, France*
Email: david.garcia@poleplasturgie.com

Abstract

In this paper an optimal design of the cooling system for thermoplastic injection process is proposed. The problem statement is based on a morphological analysis of a 3D part. Heat transfer is modeled in the mould and the plastic part. A semi-crystalline polymer is considered. The heat source released by the crystallization is taken into account through the specific heat term. The heat conduction problem is non-linear. The heat coming from the polymer is dissipated by heat sinks located in the mould. A continuous distribution of heat sinks along the cooling surface is assumed. The inverse heat transfer problem is formulated in order to determine fluid temperature distribution $T_\infty(s)$ along the cooling surface so as to cool the part to a desired level of temperature meanwhile reducing thermal gradients in the part.

Key words: Inverse heat transfer conduction, conjugate gradient, optimal design, injection process

1. Introduction

Injection moulding is the most widely used process in the plastic industry. This cyclic process allows the manufacturing of thermoplastic parts with high production rate. Each cycle is decomposed in four essential stages: filling the cavity, melt packing, solidification of the part and ejection. Due to the low thermal diffusivity of the polymer (~10^{-7} m²/s), heat transfer in the part is slow and more than 60% of the total time of the process is dedicated to the cooling of the part. The cooling phase influences thus the productivity but also the quality [1,2] of the manufactured parts. Indeed many defects such as warpage, shrinkage, thermal residual stresses, sink marks are due to a non efficient cooling. The design of the cooling system is therefore essential. Today, the development of new rapid prototyping processes, like laser sintering, allows the construction of very complex channels shapes. Thanks to these techniques the concept of conformal cooling [3], i.e channels that conform to the surface of the mould cavity, is of primary interest. The use of this concept in designing the channels may improve the efficiency of the cooling in terms of production rate and part quality as compared with conventional design [4]. The use of optimization procedures based on criteria of quality and productivity to design the cooling system is then necessary to reach the best configuration, without conducting any experiment [5-7]. However quality and productivity are two inconsistent objectives. Indeed, productivity can be increased by lowering channels temperatures and bringing them closer to the part surface. Quality is linked to a uniform cooling in the part. It can be achieved by moving channels away from the part surface [5] and by rising the coolant fluid temperature. The optimal cooling system configuration is then a compromise between uniformity of the temperature field and cycle time. The control parameters to achieve these objectives are then the location and shape of the channels, the coolant fluid flow rate and temperature and in a lesser extent the mould material. Two main kinds of methodology are found in the literature. The first one consists in the determination of the optimal location, size of the channels and the fluid flow rate from an initial configuration in order to minimize an objective function [6,7]. The second approach is based on a conformal cooling surface. Lin [5] defines a cooling surface representing the envelope of the part where the cooling channels are located. Optimal conditions (location and size of the channels) are seeked on this cooling surface. The main drawback of these methods is that the number and the shape of channels are fixed before the optimization is performed. We propose then an optimization procedure to determine the distribution of temperature along a cooling surface, which reaches a dual objective related to criteria of quality and productivity. The shape and location of the channels are determined in a last stage, not described in this paper.

2. Heat transfer modeling

Two domains are considered : Ω_1 for the polymer and Ω_2 for the mould. Let us define Γ_3 the surface including all the cooling channels. Their distribution can either be discrete or continuous. A discrete distribution corresponds to classical cooling channels whereas a continuous one corresponds to a cooling surface. This configuration is sketched in Figure 1a. The total cycle time t_c can be divided in two sub time intervals. The first one $]0;t_f[$ corresponds to the cooling of the polymer part. The second $]t_f;t_c[$ corresponds to the interval between the ejection time and the injection of the next part. Several cycles are necessary to reach a periodic steady-state as shown in Figure 1b. Industrial production occurs under this condition. Thus, to be representative, heat transfer analysis and optimization of cooling must be performed after this periodic steady-state is reached.

During the interval $]0;t_f]$ heat transfer can be modeled, in a first approach, as the instantaneous contact between two media at two different temperatures; the filling phase being neglected. The thermal condition at the interface between the mould and the polymer is modeled by introducing a thermal contact resistance ($R_{contact}$) whose value is assumed to remain constant during the whole cooling time, as well as thermophysical properties of the mould. The heat released by the polymer during crystallization is taken into account through the dependence of the specific heat with temperature. The problem is then non-linear. After the ejection of the part, $]t_f;t_c[$, there is no polymer in the moulding cavity. Heat transfer is therefore only modeled in Ω_2 with convective boundary condition on Γ_2.

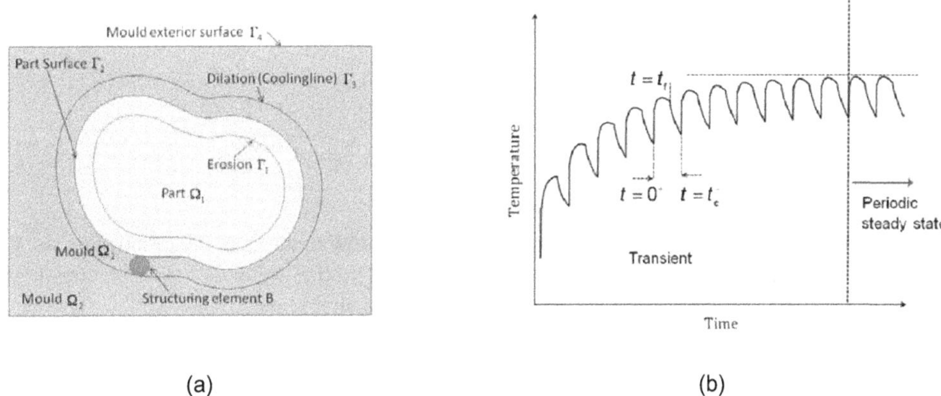

(a) (b)

Figure 1: Schematic of the mould and the polymer part (a), Cycle mould surface temperature (b)

Considering the previous hypotheses, heat transfer during the cooling of injection moulding for a single cycle on $]0;t_c[$ is described by the following set of equations $E(T_1,T_2,T_\infty)=0$.

Polymer part

$$\rho_1(T_1)C_{p_1}(T_1)T_{1,t} = \nabla.(\lambda_1(T_1)\nabla T_1) + S \text{ in } \Omega_1 \times]0;t_f[$$

$$-\lambda_1(T_1)T_{1,n_1}\big|_{\Gamma_2} = \frac{T_1 - T_2}{R_{contact}} \text{ on } \Gamma_2 \times]0;t_f[$$

$$T_1(t = 0^+) = T_{inj} \text{ in } \Omega_1$$

Mould

$$\rho_2 C_{p_2} T_{2,t} = \nabla.(\lambda_2 \nabla T_2) \text{ in } \Omega_2 \times]0;t_c[$$

$$-\lambda_2 T_{2,n_2}\big|_{\Gamma_2} = \begin{cases} \dfrac{T_2 - T_1}{R_{contact}} \text{ on } \Gamma_2 \times]0;t_f[\\ h_{nat\,conv}(T_2 - T_{surrounding}) \text{ on } \Gamma_2 \times]t_f;t_c[\end{cases}$$

$$-\lambda_2 T_{2,n_2}\big|_{\Gamma_3} = h(T_2 - T_\infty) \text{ on } \Gamma_3 \times]0;t_c[\tag{1}$$

$$-\lambda_2 T_{2,n_2}\big|_{\Gamma 4} = h(T_2 - T_{surrounding}) \text{ on } \Gamma_4 \times]0;t_c[$$

$$T_2(t = 0^+) = T_{mould}(t_c^-) \text{ in } \Omega_2$$

The periodic aspect is taken into account through the periodic condition of the mould: $T_2(t = 0^+) = T_{mould}(t_c^-)$. T_{mould} stands for the temperature field in the mould at the end of the previous cycle (Figure 1b). The injection temperature remains the same at each new cycle, $T_1(t = 0^+) = T_{inj}$. Figure 1b represents the evolution of the moulding cavity surface temperature for several cycles. During a cycle, the maximum temperature is obtained at the time ($t=0^+$) corresponding to the contact between the mould and the polymer. The minimum temperature is found at $t=t_c^-$, the end of the cycle. In order to be removed from the mould without damages, a minimum thickness of the polymer part must be solidified. We define the erosion surface Γ_1 issued from the morphological analysis [8] that corresponds to this thickness. One objective of the optimization problem is to reach a prescribed temperature level T_{ejec} on this surface, notably at the ejection time. The dilation surface Γ_3 is also defined from the morphological analysis and corresponds to the cooling surface. A continuous distribution of heat sinks along this surface is chosen. The introduction of a cooling surface is very interesting in this problem statement because it prevents from choosing the number, the size and the shape of cooling channels before optimization is carried out. This represents an important advantage in case of complex parts where the location of channels is not intuitive. Another benefit is that the domain Ω_2' between the cooling surface Γ_3 and the edge of the mould Γ_4 is not represented. Indeed, the surface Γ_3 acts as a thermal barrier; the cooling of the polymer part not being affected by the heat transfer in this domain. Therefore, the mould domain modeled is restricted to Ω_2 . The idea of regulation by fluid thermal barrier is not new and was used through the Stratoconception® process [9]. The main drawback is that only one fluid temperature is used, which is not enough to reduce surface gradients. In this work the variable to be optimized is the fluid temperature distribution T_∞ along the cooling surface Γ_3.

3. Optimization problem

3.1 Objective function to be minimized

The determination of the optimal distribution $T_\infty(s), s \in \Gamma_3$, around the part is formulated as the minimization of an objective function J composed of two terms J_1 and J_2 computed over the subinterval $[t_f - \tau; t_f[$. In practice this interval is reduced to the ejection time, that is $\tau \to 0$.

$$J(T_\infty) = J_1(T_\infty) + J_2(T_\infty) \tag{2}$$

The goal of the first term J_1 is to reach a desired temperature level (T_{ejec}) along the erosion of the part Γ_1 at the ejection time t_f . This term is related to the efficiency of the process. The second term J_2 used in many works [6,7] aims to homogenize the temperature distribution on the surface of the part Γ_2 and therefore reduces the components of thermal gradients both along the surface Γ_2 and through the thickness of the part. This term is linked to the quality of the part. The best solution will be a compromise between quality and efficiency. This compromise is achieved by introducing a weighting parameter ΔT_{ref} [K], between the two terms of the function [10].

$$J(T_\infty) = \underbrace{\int_0^{t_f}\int_{\Omega_1}\left(\frac{T_{ejec}-T_1}{T_{inj}-T_{ejec}}\right)^2 \sigma(t)\,\eta\,\mathrm{d}\Omega\,\mathrm{dt}}_{J_1} + \underbrace{\int_0^{t_f}\int_{\Gamma_2}\left(\frac{\overline{T_1}-T_1}{\Delta T_{ref}}\right)^2 \sigma(t)\,\mathrm{d}\Gamma\,\mathrm{dt}}_{J_2} \quad (3)$$

With, T_1 solution of the model equations $E(T_1, T_2, T_\infty) = 0$ and with $\overline{T_1} = \frac{1}{\Gamma_2}\int_{\Gamma_2} T_1\,\mathrm{d}\Gamma$, $\sigma(t) = \begin{cases} 0 \text{ if } t \in \left]0; t_f - \tau\right[\\ 1 \text{ if } t \in \left[t_f - \tau; t_f\right[\end{cases}$ and $\eta = \begin{cases} 1 \text{ on } \Gamma_1 \\ 0 \text{ elsewhere} \end{cases}$

3.2 Numerical resolution

Numerous numerical methods can be used to minimize the objective function J. Tang *et al.* [6] use the Powell's conjugate direction search method. Mathey *et al.* [7] use the Sequential Quadratic Programming which is a method based on gradients. It can be found not only deterministic methods but also evolutionary methods. Huang *et al.* [11] use a genetic algorithm to reach the solution. This last kind of algorithm is very time consuming because it tries a lot of range of solution. In practice time spent for mould design must be minimized hence a deterministic method which reaches an acceptable local solution more rapidly is preferred. In this work the conjugate gradient method is used to find the optimal value of $T_\infty^*(s)$ on Γ_3 which minimizes the criterion J [12].

3.2.1 The conjugate gradient algorithm

This iterative algorithm consists at each iteration k+1, in correcting the previous estimate $T_\infty^{\ k}$ according to $T_\infty^{k+1} = T_\infty^k + \rho^k w^k$ in order to obtain $J(T_\infty^{k+1}) < J(T_\infty^k)$. This computation is iterated while $J > \varepsilon$, the stopping criterion. In this expression w^k is the search direction and ρ^k the descent length. By naming ∇J the vector gradient of the functional J, the vector w^k is determined in obedience to the gradient equations: $w^{(1)} = -\nabla J^{(1)}$ at the first iteration and $w^{(k)} = -\nabla J^{(k)} + \frac{\|\nabla J^{(k)}\|^2}{\|\nabla J^{(k-1)}\|^2} w^{(k-1)}$.

3.2.2 Determination of the descent length

The descent length ρ^k is computed to minimize the following scalar function $\phi(r)$

$$\phi(r) = J(T_\infty^k + rw^k)$$
$$\rho^k = Argmin(\phi(r)) \quad (4)$$

The determination of ρ^k requires the knowledge of δT_1 and δT_2, the sensitivities of the temperature fields according to the variations of parameters δT_∞. They are given by solving the sensitivity equations:

$$\sum_i \frac{\partial E}{\partial T_i}\delta T_i + \frac{\partial E}{\partial T_\infty}\delta T_\infty = 0 \quad (5)$$

where E is the equation set of the direct problem and T_∞ is the distribution to be estimated:

In the polymer domain Ω_1:

$$\frac{\partial \rho_1(T_1)\, C_{p_1}(T_1)\delta T_1}{\partial t} = \Delta(\lambda_1(T_1)\delta T_1) \text{ in } \Omega_1 \times \left]0; \mathrm{t_f}\right[$$
$$-\left.\frac{\partial \lambda_1(T_1)\delta T_1}{\partial n_1}\right|_{\Gamma_2} = \frac{\delta T_1 - \delta T_2}{R_{\text{contact}}} \text{ on } \Gamma_2 \times \left]0; \mathrm{t_f}\right[$$
$$\delta T_1 = 0 \text{ in } \Omega_1 \text{ at t=0}$$

In the mould domain Ω_2:

$$\rho_2\, C_{p_2}\frac{\partial \delta T_2}{\partial t} = \nabla.(\lambda_2 \nabla T_2) \text{ in } \Omega_2 \times \left]0; \mathrm{t_f}\right[$$
$$-\lambda_2\left.\frac{\partial \delta T_2}{\partial n_2}\right|_{\Gamma_2} = \frac{\delta T_2 - \delta T_1}{R_{\text{contact}}} \text{ on } \Gamma_2 \times \left]0; \mathrm{t_f}\right[\quad (6)$$
$$-\lambda_2\left.\frac{\partial \delta T_2}{\partial n_2}\right|_{\Gamma_3} = h(\delta T_2 - \delta T_\infty) \text{ on } \Gamma_3 \times \left]0; \mathrm{t_f}\right[$$
$$\delta T_2 = 0 \text{ in } \Omega_2 \text{ at t=0}$$

3.2.3 Determination of the gradient components

As there is no explicit relation between the objective function and the parameters to be estimated, a Lagrangian technique is developped by introducing an adjoint variable Ψ.

$$L(T_i, T_\infty, \Psi_i) = J(T_\infty) - \sum_i \int_0^{t_f} \langle \rho_i\, C_i\, \frac{\partial T_i}{\partial t} - \nabla.(\lambda_i\, \nabla T_i), \Psi_i \rangle_{\Omega_i} \mathrm{dt} \tag{7}$$

The main idea is to fix Ψ so as $\delta L(T, Z, \Psi)$ can be written $\delta L(T, T_\infty, \Psi) = \langle J'(T, T_\infty), \delta T_\infty \rangle$. If Ψ is fixed, the derivative of the Lagrangian can be written :

$$\mathrm{d}L = \sum_i \frac{\partial L}{\partial T_i} \delta T_i + \frac{\partial L}{\partial T_\infty} \delta T_\infty \tag{8}$$

Ψ is chosen to verify: $\sum_i \int_0^{t_f} \langle \frac{\partial L}{\partial T_i}, \delta T_i \rangle_{\Omega_i} dt = 0 \quad \forall\, \delta T_i$. This condition leads to the set of adjoint equations.

In the polymer domain Ω_1:

$$\rho_1(T_1) C_{p_1}(T_1) \frac{\partial \Psi_1}{\partial t} + \lambda_1(T_1) \Delta \Psi_1 = 2\left(\frac{T_{ejec} - T_1}{\left(T_{inj} - T_{ejec}\right)^2}\right) \sigma(t) \eta(r,z) \text{ in } \Omega_1 \times]0; t_f[$$

$$-\lambda_1(T_1) \left.\frac{\partial \Psi_1}{\partial n_1}\right|_{\Gamma_2} = \frac{\Psi_1 - \Psi_2}{R_{\text{contact}}} + 2\left(\frac{\overline{T_1} - T_1}{\Delta T_{1max}^2}\right) \sigma(t) \text{ on } \Gamma_2 \times]0; t_f[$$

$$\Psi_1 = 0 \text{ in } \Omega_1 \text{ at } t_f$$

In the mould domain Ω_2:

$$\rho_2\, C_{p_2} \frac{\partial \Psi_2}{\partial t} + \lambda_2 \Delta \Psi_2 = 0 \text{ in } \Omega_2 \times]0; t_f[$$

$$-\lambda_2 \left.\frac{\partial \Psi_2}{\partial n_2}\right|_{\Gamma_2} = \frac{\Psi_2 - \Psi_1}{R_{\text{contact}}} \text{ on } \Gamma_2 \times]0; t_f[$$

$$-\lambda_2 \left.\frac{\partial \Psi_2}{\partial n_2}\right|_{\Gamma_3} = h\Psi_2 \text{ on } \Gamma_3 \times]0; t_f[$$

$$T_2 = 0 \text{ in } \Omega_2 \text{ at } t_f \tag{9}$$

When Ψ is solution of the adjoint equations (9) and T solution of the direct problem (1), the Lagrangian is equal to the cost function hence :

$$\delta L = \delta J \tag{10}$$

An explicit relation can be found to express the gradient components as a function of the unknown parameters:

$$\nabla J_i = \int_0^{t_f} \int_{\Gamma_3} h \Psi_2\, \omega_i(s) d\Gamma dt \quad, s \epsilon \Gamma_3, 1 \le i \le N_P \tag{11}$$

In which $T_\infty(s)$ is expressed by $T_\infty(s) = \sum_{i=1}^{N_P} \xi_i \omega_i(s)$, $s \in \Gamma_3$ with $\{\omega_i\}_{i=1,N_p}$ a given set of N_p basis functions on Γ_3.

4. Numerical results

The cooling of the box in 3D is studied. Because of the symmetry, numerical results are performed on only one quarter of the box (Figure 2). The optimization problem is implemented in Comsol® multiphysics v. 3.5a. Finite elements analysis with quadratic elements is used. 7900 parameters (N_p) have been estimated. Time spent for one iteration of optimization is about 2.10^4 s on a quad core 64 bits Intel Xeon CPU W3550 (3.07GHz) 12Go RAM. Parameters required for the simulation are listed in the table below (Table 1). In this paper, ρ, λ for the polymer are assumed to be constant.

Polymer

$\rho = 978\ kg/m^3$

$\lambda = 0.24\ W/m.K$

$T\,[K]$	328	488	502	508	563
$C(T)[J/kg.K]$	1536	3105	10180	3410	2635

Mould

$\rho = 7800\ kg/m^3$

$\lambda = 15\ W/m.K$

$C = 420\ J/kg.K$

Parameters values:

$T_{ejec} = 488\ K$

$T_{inj} = 563\ K$

$\Delta T_{ref} = 100\ K$

$t_f = 7s$

$t_C = 12s$

Table 1: Thermophysical parameters and data values for the simulation

The minimum of the cost function is obtained from the 31th iteration. Evolution of the two terms of the criterion is plotted in Figure 3. During the first four iterations J_1 and J_2 decrease. Then J_2 remains constant whereas J_1 keeps on decreasing. Globally the cost function J diminishes. There is one order of magnitude between the first term J_1 and the second term J_2. This ratio is due to the choice of ΔT_{ref}. With $\Delta T_{ref} = 100K$ priority is given to reach the level of temperature (T_{ejec}) in the part, but the second term is large enough for introducing a bias. At the 31th iteration, the average optimal temperature obtained with T_∞^* at $t = t_f$, along the erosion of the part is 390K for a desired ejection temperature of 388K. The standard deviation calculated along the erosion with the mean temperature fixed as the ejection temperature is 4.7K and the standard deviation on the part surface is 5.15K. In Figure 4, optimal fluid temperatures T_∞^* are plotted on a cross-section of the studied domain. Results show that to get a uniform cooling in the part, fluid temperature is not uniform. Channels will be built with respect to this temperature profile in a next step.

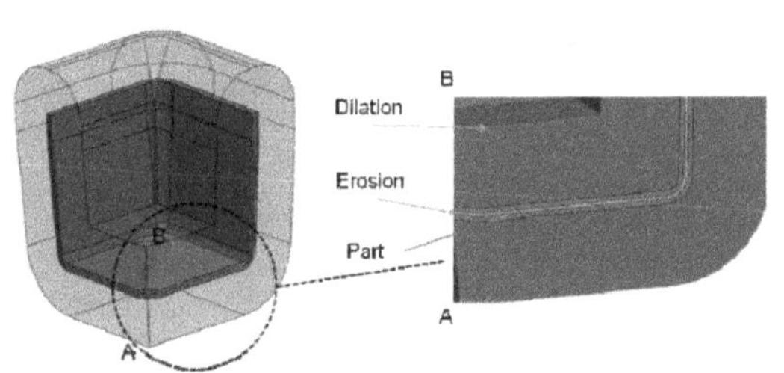

Figure 2: CAD of the box

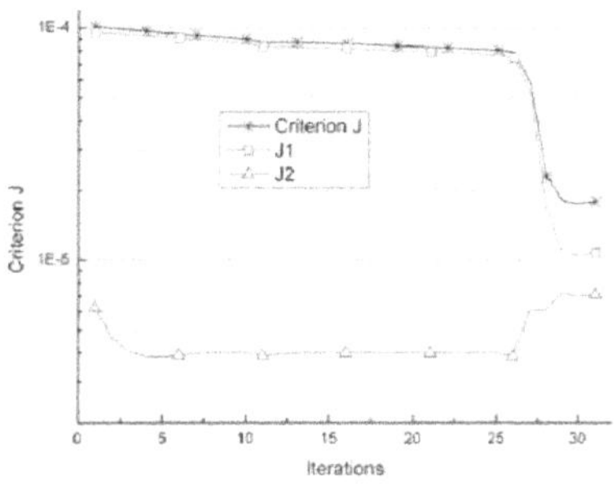

Figure 3: Criterion vs. iterations

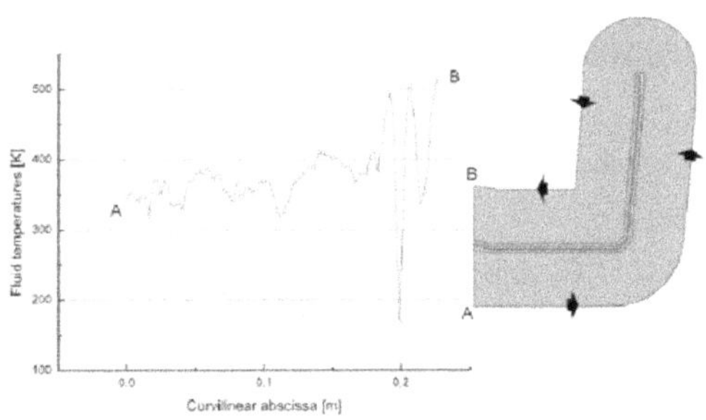

Figure 4: T_∞^* versus curvilinear abscissa along [AB]

5. Conclusion

The design of the cooling system of a mould used in thermoplastic injection process has been formulated as a non linear inverse heat conduction problem. The criterion to be minimized involves two antagonistic terms: one related to the productivity (temperature level reached in a fixed cycle time) and the other to the quality of the part (minimal thermal gradient at the ejection time). The control variable is the fluid temperature distribution $T_\infty(s)$ located on a conformal surface. The numerical resolution is achieved by the classical conjugate gradient algorithm and by solving a set of adjoint equations. Numerical results show the feasibility of the method for the design of a 3D cooling system. Based on these results, the manufacturing of an experimental mould is in progress.

6. References

[1] Pichon, J F., Injection des matières plastiques, Dunod, France 2001.

[2] Zöllner, O., Optimized Mould Temperature Control, Technical report, Bayer AG, Leverkusen, Germany, 1997. http://search.bayermaterialsciencenafta.com/inquiragw/ui.jsp

[3] Dimitrov, D and Moammer, A, "Investigation towards the impact of conformal cooling on ther performance of injection moulds for the packaging industry", RAPDASA's proceedings, 8th annual International conference on transportation weight Reduction, pp. 67-74, November 07-09, 2007, Pilanesburg, South Africa.

[4] Xu, X, Sachs, E and Allen S., "The design of conformal cooling channels in injection moulding tooling", Polymer engineering and science, Vol. 41, No. 7, pp. 1265-1279, 2001.

[5] Lin, J.C., " Optimum cooling system design of a free form injection mould using an abductive network ", Journal of Mat. Proc. Tech., Vol. 120,No 1-3, pp. 226-236, 2002

[6] Tang, L.Q, Pochiraju, K, Chassapis, C, Manoochehri, S, " A computer-aided optimization approach fpr the design of injection mould cooling systems ", Journal of mechanical design, Vol 120, No 2, pp. 165-174, 1998

[7] Mathey, E. , Penazzi, L., Schmidt, F. M., Rondé-Oustau, F., " Automatic optimization of the cooling of injection mould based on the boundary element method", Materials processing and design: Modeling, Simulation and Applications - NUMIFORM 2004 - Proceedings of the 8th International Conference on Numerical Methods in Industrial Forming Processes, Vol 712, No 1, pp. 222-227, 2004

[8] Serra, J., Image analysis and mathematical morphology, Academic press, London, 1982.

[9] Pelaingre, C., Abel, C., Thabourey, J., Barlier , C., "Moule de fonderie sous pression réalisé par Stratoconception", 10èmes Assises Européennes du Prototypage Rapide, Maison de la Mécanique, September 14th, 2004, Paris, France.

[10] Agazzi, A., Sobotka, V., Le Goff, R., Garcia, D., Jarny, Y., " A methodology for the design of effective cooling system in injection moulding", International journal of material forming, Vol 3, Suppl 1, pp.13-16, 2010.

[11] Huang, J., Fadel, G.M., " Bi-objective optimization design of heterogeneous injection mould cooling systems", Journal of mechanical design, Vol 123, No 2, pp. 226-239, 2001

[12] Agazzi, A., Sobotka, V., Le Goff, R., Garcia, D., Jarny, Y., " The cooling phase of the thermoplastic injection process: an inverse heat ransfer control problem", ECCM IV Proceedings, May 16-21, 2010, Paris, France.

Determination of the Optimum Fluid Temperature History for Reducing Transient Stress Intensity Factor of Crack in Pipe during Start-up and Shut-down by the Multiphysics Inverse Analysis Method

Mitsumasa MAEKAWA[1], *Shiro KUBO* [1,2] *and Seiji IOKA* [1]

[1]*Department of mechanical Engineering, Graduate School of Engineering*
Osaka University, 2-1, Yamadaoka, Suita, Osaka, 565-0871, Japan
[2]*Email: kubo@mech.eng.osaka-u.ac.jp*

Abstract

It is important to reduce the transient stress intensity factors for managing and extending the lives of pipes in plants. In this problem, heat conduction, elastic deformation, heat transfer, liquid flow should be considered, and therefore the problem is of a multidisciplinary nature. In this study, An inverse method is proposed for determining the optimum temperature history, which reduces transient stress intensity factor considering the multidisciplinary physics. The multidisciplinary complex problem is decomposed into a heat conduction problem, a heat transfer problem, and a thermal stress problem. An analytical solution of the temperature distribution in the thickness direction, thermal stress distribution and the stress intensity factor is obtained. The maximum stress intensity factor is minimized for the case where inner surface temperature $T_s(t)$ is expressed in terms of the 4th order polynomial function of time t. Finally, from the temperature distributions, the optimum fluid temperature history is obtained for reducing the stress intensity factors.

Key Words: stress intensity factor, crack, pipe, multiphysics, thermal stress

1. Introduction

Thermal stress problem is a serious problem to be considered for the safety of structures and their components. For example, when high temperature fluid flows into a pipe, a transient temperature distribution in the pipe induces thermal stresses. When the thermal stress is high, it causes thermal fatigue fracture in many cases. In addition, when there is a crack in the pipe, high stress field develops near the crack tip, and the stress intensity factor is high. Therefore, the transient stress intensity factor of a crack in the pipe is important for evaluating the integrity and lives of pipes and plants. It is requested to reduce the transient stress intensity factor for extending the lives of pipes and plants. In this problem, heat conduction, elastic deformation, heat transfer, liquid flow should be considered. Therefore this problem is of a multidisciplinary nature.

A number of research works have been carried out on inverse problems of thermomechanical physics. Shao [1] studied the solution of temperature, displacement, and thermal/mechanical stresses in a functionally graded circular hollow cylinder by using a multi-layered approach based on the theory of laminated composites. A.K.Tikhe *et al.* [2] studied an inverse thermoelastic problem of transient heat conduction in a thin finite circular plate with the given temperature distribution on the interior surface of a thin circular plate and determined the thermal deflection on the outer curved surface of a thin circular plate. However, few studies are available in the literature, which deals with the multiphysics problem consisting of high temperature fluid inflow, heat transfer, heat conduction, and reduction of transient thermal stress.

Ishizaka *et al.* [3,4] and Uchida *et al.* [5,6] proposed an inverse method for determining the optimum thermal load history which reduces transient thermal stress considering the multiphysics. When a pipe has a crack, the stress intensity factor is necessary for evaluating the integrity of the equipment.

In this paper, an effective and multidisciplinary inverse analysis method is proposed for determining the optimum temperature history, which reduces the transient stress intensity factor. The transient stress intensity factor is reduced by controlling fluid temperature history.

The corresponding traditional direct problem deals with the determination of transient temperature distribution and thermal deformation and thermal stress and stress intensity factor from given high temperature fluid inflow. The present problem which determines the optimum fluid temperature history minimizing the maximum transient stress intensity factor is an inverse problem [7]. For solving the inverse problem, it is necessary to obtain the relationship between the fluid temperature, the temperature distribution in pipe, thermal stress in pipe, and stress intensity factor of crack.

When high temperature fluid inflow and internal wall temperature rises, then transient thermal stress in pipe arises and a crack has transient stress intensity factor due to non-steady heat conduction. Then we decompose the multidisciplinary complex problem into the following problems.

- Heat transfer problem for analyzing fluid temperature history
- Heat conduction problem for analyzing internal temperature distribution of a pipe
- Thermoelastic deformation problem for analyzing internal thermal stress distribution of a pipe

In the first place of an inverse problem analysis, we define a certain inner surface temperature of a pipe as a boundary condition. In the second place, we analyze temperature distribution in pipe, thermal stress distribution in a pipe and stress intensity factor of crack. In the third place, the inner surface temperature minimizing the maximum value of transient stress intensity factor is determined. Finally, we obtain the optimum fluid temperature history from the inner surface temperature.

We define three inner temperature history functions of time, which ensure the continuity of the temperature increasing rate of inner surface and are monotonically increasing functions. Then, by superposing the results for these functions, we obtain a polynomial function of inner surface temperature, which minimizes the maximum stress intensity factors of the crack in the pipe.

2. Analysis Procedure

We consider a thin-walled pipe that has crack on outer surface or on inner surface as shown in Fig. 1. The inner radius r_i is much larger than the thickness l. As a typical and simple situation of the pipe, it is presumed that the inner surface is axi-symmetrically heated, and the outer surface is insulated. We take r'-axis from the inner surface in the radius direction and define the temperature distribution by $u(r',t)$, while x-axis is taken from outer surface to the crack tip according to the crack whose length is a. The pipe is heated from the initial temperature T_m (t<0) to $T_m+T_s(t)$.

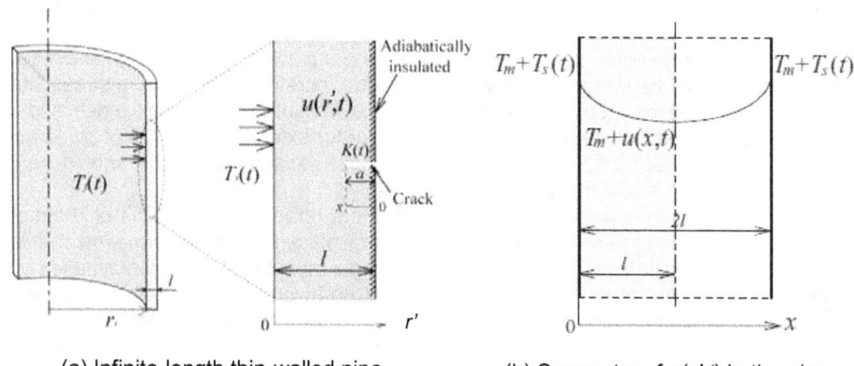

(a) Infinite-length thin-walled pipe (b) Symmetry of $u(r',t)$ in the pipe

Figure 1: Analysis model

In the first place, we analyze the internal temperature distribution $u(r',t)$ of a pipe from the inner surface temperature $T_s(t)$ using the eigenfunction expansion method. This heat transfer problem can be replaced by the problem in which both ends of flat plate of thickness $2l$ is heated to $T_m+T_s(t)$ from the symmetry as shown in Fig. 1(b). The one-dimensional heat equation is written as follows.

$$\frac{\partial u(r',t)}{\partial t} = \kappa \frac{\partial^2 u(r',t)}{\partial r'^2} \tag{1}$$

where κ is a constant independent of temperature, and $u(r',t)$ is subjected to the following boundary conditions.

$$u(r',0) = 0 \qquad (0 \le r' \le 2l) \tag{2}$$

$$u(0,t) = T_s(t) \qquad (0 < t \le \infty) \tag{3}$$

$$u(2l,t) = T_s(t) \qquad (0 < t \le \infty) \tag{4}$$

Temperature distribution is obtained from these boundary conditions as follows.

$$u(r',t) = T_s(t) - 2\sum_{n=0}^{\infty} \frac{\sin\left(\phi_n \frac{r'}{l}\right)}{\phi_n} e^{-\frac{\kappa\phi_n^2}{l^2}t} \int_0^t \frac{dT_s(\tau)}{d\tau} e^{\frac{\kappa\phi_n^2}{l^2}\tau} d\tau \tag{5}$$

Here ϕ_n is defined as,

$$\phi_n = \frac{(2n+1)\pi}{2} \tag{6}$$

In the second place, we obtain the thermal stress in a thin-walled pipe with a given temperature distribution in the pipe. When the pipe is axi-symmetrically heated, radial temperature distribution $u(r)$ induces axial stress $\sigma_z(r)$. The axial strain $\varepsilon_z(r,t)$ is given by,

$$\varepsilon_z(r,t) = \frac{\sigma_z(r,t)}{E} + \alpha u(r,t) \tag{7}$$

This problem is replaced by a flat plate problem because we consider a thin-walled pipe. So, $\varepsilon_z(r,t)$ takes a constant value, and we obtain,

$$\varepsilon_z(t) = \frac{\alpha \int_{r_i}^{r_o} u(r,t)dr}{r_b - r_a} = \frac{\alpha \int_0^l u(r',t)dr}{l} \tag{8}$$

From Eqs. (7) and (8), we obtain the following equation for the axial stress distribution $\sigma_z(r',t)$.

$$\sigma_z(r',t) = \alpha E \left\{ \frac{\sin\left(\phi_n \frac{r'}{l}\right)}{\phi_n} - \frac{1}{\phi_n} \right\} \times e^{-\frac{\kappa\phi_n^2}{l^2}t} \int_0^t \frac{dT_s(\tau)}{d\tau} e^{\frac{\kappa\phi_n^2}{l^2}\tau} d\tau \tag{9}$$

In the third place, the axial stress of a pipe can be taken as a residual stress. Then, the stress intensity factor of the crack $K(t)$ is expressed as follows,

$$K(t) = \int_{x=0}^{x=a} \frac{2F}{\sqrt{\pi a}\sqrt{1 - x^2/a^2}} \sigma_z(x,t)dx \tag{10}$$

where F is a correction factor.

Finally, we can obtain the fluid temperature $T_f(t)$ from the inner surface temperature $T_s(t)$ of the pipe. The heat flux $q(t)$ is written as follows.

$$q(t) = h(T_f(t) - T_s(t)). \tag{11}$$

$q(t)$ is written from Fourier's law as follows.

$$q(t) = \lambda \left. \frac{\partial u(r',t)}{\partial r'} \right|_{r'=0} . \tag{12}$$

From Eqs. (11) and (12), we can obtain the fluid temperature history $T_f(t)$ as follows.

$$T_f(t) = T_s(t) + \frac{2\lambda}{hl} \sum_{n=0}^{\infty} e^{-\frac{\kappa\phi_n^2}{l^2}t} \int_0^t \frac{dT_s(\tau)}{d\tau} e^{\frac{\kappa\phi_n^2}{l^2}t} d\tau . \tag{13}$$

If we know the inner surface temperature $T_s(t)$ reducing transient stress intensity factor, we can obtain the optimum temperature history of fluid flowing in a pipe, even when h has nonlinear temperature-dependence.

3. Minimization of the Maximum Stress Intensity Factor

We consider a case where the inner surface temperature of a pipe T_s rises from 0 [K] at t=0 [s] to T_e [K] at t=t_e [s], keeping $T_s = T_e$ [K] after t=t_e [s]. We minimize the maximum stress intensity factor of a crack by superposing the solutions for several functions of inner surface temperature $T_{si}(t)$ (i=1,2,3...). Firstly, $T_{si}(t)$ is defined by the (i+1)-th order function of time t as follows.

$$T_{si}(t) = \begin{cases} T_e - T_e\left(\frac{t_e - t}{t_e}\right)^{i+1} & (0 \le t \le t_e) \\ T_e & (t_e \le t) \end{cases} \qquad (i = 1,2,3,..j) \tag{14}$$

$T_{si}(t)$ is defined by the (i+1)-th order function (i=1,2,3...) because the 1st order function induces discontinuity of the inner surface temperature history. For the inner surface temperature $T_{si}(t)$, temperature distribution $u_i(r',t)$, thermal stress distribution $\sigma_i(r',t)$ in the pipe and stress intensity factor $K_i(t)$ are analyzed by substituting Eq. (14) into Eqs. (5), (9) and (10).

The temperature $T_s(t)$ minimizing the stress intensity factor of the crack is given by choosing the optimum value of ξ_i as follows.

$$T_s(t) = \sum_{i=1}^{j} \xi_i T_{si}(t) \tag{15}$$

where ξ_i is a parameter, which satisfies the following condition.

$$\sum_{i=1}^{j} \xi_i = 1 . \tag{16}$$

Under the condition of Eq. (16), $T_s(t)$ satisfies the conditions that T_s=0 [K] at t=0 [s] and T_s=T_e [K] at t_e [s]. Then, using the superposition principle, stress intensity factor of the crack $K(t)$ is expressed as follows.

$$K(t) = \sum_{i=1}^{j} \xi_i K_i(t) . \tag{17}$$

We perform minimization of the maximum value of stress intensity factor of the crack on the outer surface and on the inner surface.

4. Numerical Example

We consider the case where the inner surface temperature of a pipe rises from 0 [K] at t=0 [s] to 100 [K] at t=100 [s], and keeps 100 [K] from 100 [s] to 200 [s] (i.e. T_e=100 and t_e=100). The thermal and mechanical properties we used are shown in Table 1. Then we define $T_s(t)$ is given as the 4th order polynomial function (i.e. j=3). So Eq. (14) is expressed as follows.

$$T_s(t) = \xi_1 T_{s1}(t) + \xi_2 T_{s2}(t) + \xi_3 T_{s3}(t) \tag{18}$$

If we get the ξ_i (i=1,2,3) which minimizes the maximum value of $K(t)$, the optimum fluid temperature history $T_f(t)$ is obtained by Eq. (13).

Table1: Thermal and mechanical properties

Property	Symbol	Unit	Value
Radial thickness	l	[m]	0.03
Thermal diffusivity	κ	[m^2/s]	4.0485×10^{-6}
Coefficient of thermal expansion	α	[°C^{-1}]	17.3×10^{-6}
Yong's modulus	E	[MPa]	197000
Thermal conductivity	λ	[W/(m·°C)]	16.0
Coefficient of heat transfer	h	[W/(m^2·°C)]	5500
Crack length	a	[m]	0.003

The minimization of the maximum value of stress intensity factor of a crack on the outer surface of a pipe was made. We calculated ξ_i from Eqs. (16) and (17) to minimize the maximum stress intensity factor occurring at the crack tip. The optimum values minimizing the stress intensity factor are ξ_1=5.48, ξ_2=-12.49, ξ_3=8.01. The maximum value of the optimum stress intensity factor $K(t)$ is 40.3[MPa] occurring at t=37.6[s] and t=98.1[s]. History of $K(t)$ is shown in Fig. 2(a). This figure shows that $K(t)$ has two extremal values and the maximum stress intensity factor is reduced compared with $K_i(t)$ for i=1,2 and 3. Then, the optimum inner surface temperature $T_s(t)$ which minimizes the maximum stress intensity factor of the crack is shown in Fig. 3(a). The optimum fluid temperature history $T_f(t)$ is shown in Fig. 4(a). If we know the inner surface temperature $T_s(t)$ minimizing the maximum stress intensity factor of the crack, we can obtain the optimum temperature history of fluid flowing in the pipe.

The minimization of the maximum absolute value of stress intensity factor of a crack on the inner surface of a pipe was made. The optimum values minimizing the stress intensity factor are ξ_1=4.18, ξ_2=-9.00, ξ_3=5.82. The maximum value of the optimum stress intensity factor of the crack $K(t)$ is 59.7[MPa] occurring at t=32.6[s] and t=91.5[s]. History of $K(t)$ is shown in Fig. 2(b). This figure shows that $K(t)$ has two extremal values and the maximum stress intensity factor is reduced compared with $K_i(t)$ for i=1,2 and 3. Then, the optimum inner surface temperature $T_s(t)$ is shown in Fig.3(b). The optimum fluid temperature history of the crack $T_f(t)$ is also shown in Fig. 4(b).

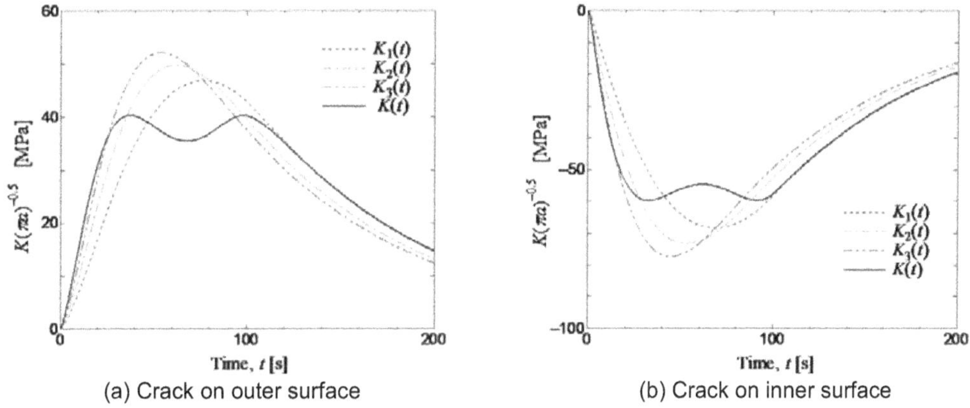

(a) Crack on outer surface (b) Crack on inner surface

Figure 2: History of stress intensity factor $K(t)$ and $K_i(t)$ (i=1,2,3)

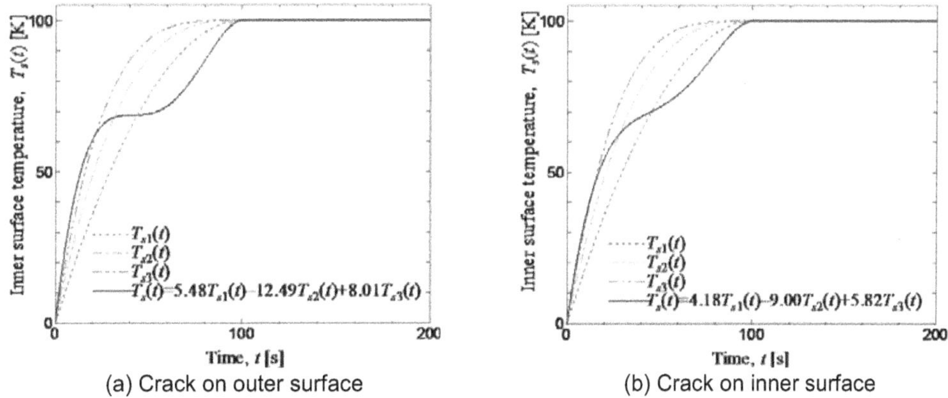

(a) Crack on outer surface (b) Crack on inner surface

Figure 3: History of inner surface temperature $T_s(t)$ and $T_{si}(t)$ (i=1,2,3)

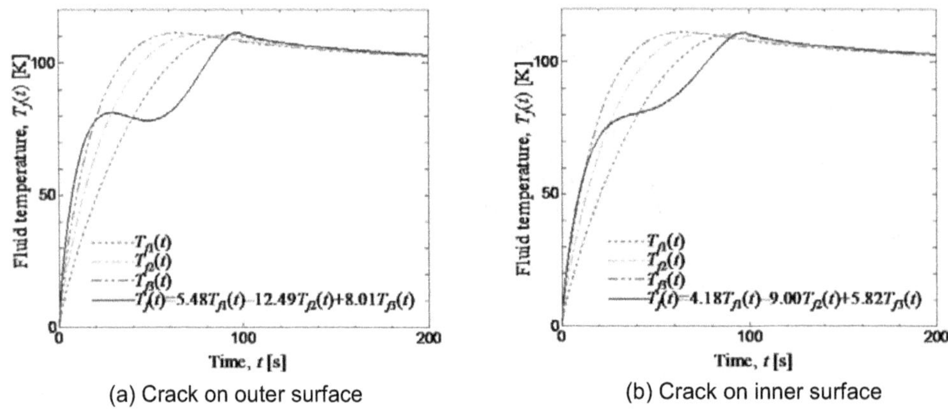

(a) Crack on outer surface (b) Crack on inner surface

Figure 4: History of fluid temperature $T_f(t)$ and $T_{fi}(t)$ (i=1,2,3)

5. Conclusions

When high temperature fluid flows in a thin-walled pipe which has crack, transient thermal stress occurs. Therefore, it is important to evaluate the transient stress intensity factor for evaluating integrity of the structure. In this study, the optimum fluid temperature history was obtained using the inverse analysis method based on the problem decomposition.

The analytical solutions of internal temperature distribution, internal transient thermal stress distribution, stress intensity factor of crack and fluid temperature are written using the inner surface temperature. We define three inner temperature histories of the functions of time, which ensure the continuity of the temperature increasing rate of inner surface and are monotonically increasing functions. Then, by superposing the results for these functions, we obtain the 4th order polynomial function of inner surface temperature which minimizes the maximum stress intensity factor of the crack on outer surface or on inner surface. The reduction of transient stress intensity factor is made. It is found that the inverse analysis method based on problem decomposition is effective.

6. References

[1] Shao, Z.S. Mechanical and thermal stresses of a functionally graded circular hollow cylinder with finite length, *International Journal of Pressure Vessels and Piping*, 82 pp. 155-163, 2005

[2] Tikhe, A.K. and Deshmukh, K.C. Inverse heat condition problem in a thin circular plate and its thermal deflection, Applied Mathematical Modelling 30, pp. 554-560, 2006

[3] Ishizaka, T. Kubo, S. and Ioka, S. Proceedings of PVP2006-ICPVT-11, paper # PVP2006-ICPVT-11-93618, ASME, pp.1-8, 2006

[4] Isizaka, T. Kubo, S. and Ioka, S. Multiphysics Inverse Analysis Method for Reducing Transient Thermal Stress in a Thick-Walled Pipe, ASME, Proceedings of PVP2009, Paper # PVP2009-77588, pp.1-7, 2009

[5] Kubo, S. Uchida, K. Ishizaka, T. and Ioka, S. Determination of the Optimum Temperature History of Inlet Water for Minimizing Thermal Stress in a Pipe by the Multiphysics Inverse Analysis, Journal of Physics: Conference Series (6th International Conference on Inverse Problems in Engineering: Theory and Practice (ICIPE 2008)) Vol. 35, pp.1-9, 2008

[6] Uchida, K. Reduction of Transient Thermal Stress and Estimate Heat Transfer Coefficient by the Multiphysics Inverse Analysis Method, Master Thesis Osaka University, 2008

[7] Kubo, S. Inverse Problems Related to the Mechanics and Fracture of Solids and Structures, JSME Int.J., Ser.I, Vol.31, No.2, pp.157-166, 1988

Uniqueness of the solution to inverse scattering problem with non-overdetermined data

A G Ramm

Department of Mathematics
Kansas State University, Manhattan, KS 66506-2602, USA
ramm@math.ksu.edu

Abstract

Let $q(x)$ be a real-valued compactly supported sufficiently smooth function. It is proved that the scattering data $A(-\beta, \beta, k)$ $\forall \beta \in S^2$, $\forall k > 0$, determine q uniquely. Under the same assumptions on $q(x)$ it is proved that the scattering data $A(\beta, \alpha_0, k)$ $\forall \beta \in S^2$, $\forall k > 0$, and a fixed α_0, the direction of the incident plane wave, determine q uniquely. The above scattering data are non-overdetermined in the sense that the scattering data depends on the same number of variables as the potential $q(x)$, that is, on three variables. earlier there were no uniqueness results for the solution of three-dimensional inverse scattering problems with non-overdetermined scattering data.

MSC: 35P25, 35R30, 81Q05;
Key words: inverse scattering, non-overdetermined data, backscattering.

1 Introduction

The scattering solution $u(x, \alpha, k)$ solves the scattering problem:

$$[\nabla^2 + k^2 - q(x)]u = 0 \quad in \quad \mathbb{R}^3, \tag{1}$$

$$u = e^{ik\alpha\cdot x} + A(\beta, \alpha, k)\frac{e^{ikr}}{r} + o\left(\frac{1}{r}\right), \quad r := |x| \to \infty, \ \beta := \frac{x}{r}. \tag{2}$$

Here $\alpha, \beta \in S^2$ are the unit vectors, S^2 is the unit sphere, the coefficient $A(\beta, \alpha, k)$ is called the scattering amplitude, $q(x)$ is a real-valued compactly supported sufficiently smooth function. We want to determine $q(x)$ given the backscattering data $A(-\beta, \beta, k)$, or the data $A(\beta, \alpha_0, k)$ $\forall \beta \in S^2$, $\forall k > 0$, $\alpha_0 \in S^2$ is fixed. These are 3D inverse scattering problems with non-overdetermined data: the data depend on the same number of variables as the $q(x)$, i.e., on three variables. The first uniqueness results for such problems were obtained in [2], [3].

Assumption A): We assume that q is compactly supported, i.e., $q(x) = 0$ for $|x| > a$, where $a > 0$ is an arbitrary large fixed number; $q(x)$ is real-valued, i.e., $q = \overline{q}$; and $q(x) \in H_0^\ell(B_a)$, $\ell > 3$.

Here B_a is the ball centered at the origin and of radius a, and $H_0^\ell(B_a)$ is the closure of $C_0^\infty(B_a)$ in the norm of the Sobolev space $H^\ell(B_a)$ of functions whose derivatives up to the order ℓ belong to $L^2(B_a)$.

It was proved in [5] that if $q = \overline{q}$ and $q \in L^2(B_a)$ is compactly supported, then the resolvent kernel $G(x, y, k)$ of the Schrödinger operator $-\nabla^2 + q(x) - k^2$ is a meromorphic function of k on the whole complex plane k, analytic in $\text{Im}k \geq 0$, except, possibly, of a finitely many simple poles at the points ik_j, $k_j > 0$, $1 \leq j \leq n$, where $-k_j^2$ are negative eigenvalues of the selfadjoint operator $-\nabla^2 + q(x)$ in $L^2(\mathbb{R}^3)$. Consequently, the scattering amplitude $A(\beta, \alpha, k)$, corresponding to the above q, is a restriction to the positive semiaxis $k \in [0, \infty)$ of a meromorphic on the whole complex k-plane function.

It was proved by the author ([6]), that the *fixed-energy scattering data* $A(\beta, \alpha) := A(\beta, \alpha, k_0)$, $k_0 = const > 0$, $\forall \beta \in S_1^2$, $\forall \alpha \in S_2^2$, determine real-valued compactly supported $q \in L^2(B_a)$ uniquely. Here S_j^2, $j = 1, 2$, are arbitrary small open subsets of S^2 (solid angles).

In [9] (see also monograph [10], Chapter 5, and [7]) an analytical formula is derived for the reconstruction of the potential q from exact fixed-energy scattering data, and from noisy fixed-energy scattering data, and stability estimates and error estimates for the reconstruction method are obtained. To the author's knowledge, these are the only known until now theoretical error estimates for the recovery of the potential from noisy fixed-energy scattering data in the three-dimensional inverse scattering problem.

In [8] stability results are obtained for the inverse scattering problem for obstacles.

The scattering data $A(\beta, \alpha)$ depend on four variables (two unit vectors), while the unknown $q(x)$ depends on three variables. In this sense the inverse scattering problem, which consists of finding q from the fixed-energy scattering data $A(\beta, \alpha)$, is overdetermined.

Historical remark. In the beginning of the forties of the last century physicists raised the the following question: is it possible to recover the Hamiltonian of a quantum-mechanical system from the observed quantities, such as S-matrix? In the non-relativistic quantum mechanics the simplest Hamiltonian $\mathbf{H} = -\nabla^2 + q(x)$ can be uniquely determined if one knows the potential $q(x)$. The S-matrix in this case is in one-to-one correspondence with the scattering amplitude A: $S = I - \frac{k}{2\pi i} A$, where I is the identity operator in $L^2(S^2)$, A is an integral operator in $L^2(S^2)$ with the kernel $A(\beta, \alpha, k)$, and $k^2 > 0$ is energy. Therefore, the question, raised by the physicists, is reduced to an inverse scattering problem: can one determine the potential $q(x)$ from the knowledge of the scattering amplitude. We have briefly discussed this problem above.

Since the above question was raised, there were no uniqueness theorems for three-dimensional inverse scattering problems with non-overdetermined data. The goal of this paper is to outline a proof of such theorems. The results are formulated in Theorem 1.1:

Theorem 1.1 *If Assumption A) holds, then the data* $A(-\beta,\beta,k)$ $\forall\beta\in S^2$, $\forall k>0$, *determine* q *uniquely. Under the same assumptions, the data* $A(\beta,\alpha_0,k)$ $\forall\beta\in S^2$, *a fixed* $\alpha=\alpha_0\in S^2$, *and* $\forall k>0$, *determine* q *uniquely.*

Remark 1. The conclusion of Theorem 1.1 remains valid if the scattering data $A(-\beta,\beta,k)$, or the scattering data $A(\beta,\alpha_0,k)$ are known $\forall\beta\in S_1^2$ and $k\in(k_0,k_1)$, where $(k_0,k_1)\subset[0,\infty)$ is an arbitrary small interval, $k_1>k_0$, and S_1^2 is an arbitrary small open subset of S^2. The assumption $\ell>3$ can be relaxed to $\ell>2$.

In Section 2 we formulate some known auxiliary results and introduce some notations. In Section 3 an outline of the proof of Theorem 1.1 is given.

The results, presented in this paper, were reported in [2], [3]. We follow closely the outline of the ideas from these papers.

2 Auxiliary results

Let

$$F(g):=\tilde g(\xi)=\int_{\mathbb{R}^3} g(x)e^{i\xi\cdot x}dx,\quad g(x)=\frac{1}{(2\pi)^3}\int_{\mathbb{R}^3} e^{-i\xi\cdot x}\tilde g(\xi)d\xi. \tag{3}$$

If $f*g:=\int_{\mathbb{R}^3} f(x-y)g(y)dy$, then

$$F(f*g)=\tilde f(\xi)\tilde g(\xi),\quad F(f(x)g(x))=\frac{1}{(2\pi)^3}\tilde f*\tilde g. \tag{4}$$

If

$$G(x-y,k):=\frac{e^{ik[|x-y|-\beta\cdot(x-y)]}}{4\pi|x-y|}, \tag{5}$$

then

$$F(G(x,k))=\frac{1}{\xi^2-2k\beta\cdot\xi},\qquad \xi^2:=\xi\cdot\xi. \tag{6}$$

The scattering solution $u=u(x,\alpha,k)$ solves (uniquely) the integral equation

$$u(x,\alpha,k)=e^{ik\alpha\cdot x}-\int_{B_a} g(x,y,k)q(y)u(y,\alpha,k)dy, \tag{7}$$

where

$$g(x,y,k):=\frac{e^{ik|x-y|}}{4\pi|x-y|}. \tag{8}$$

If

$$v=e^{-ik\alpha\cdot x}u(x,\alpha,k), \tag{9}$$

then

$$v=1-\int_{B_a} G(x-y,k)q(y)v(y,\alpha,k)dy, \tag{10}$$

where G is defined in (5). Define ϵ by the formula

$$v = 1 + \epsilon. \tag{11}$$

Then (10) can be rewritten as

$$\epsilon(x, \alpha, k) = -\int_{\mathbb{R}^3} G(x - y, k) q(y) dy - T\epsilon, \tag{12}$$

where $T\epsilon := \int_{B_a} G(x - y, k) q(y) \epsilon(y, \alpha, k) dy$. Fourier transform of (12) yields (see (4),(6)):

$$\tilde{\epsilon}(\xi, \alpha, k) = -\frac{\tilde{q}(\xi)}{\xi^2 - 2k\alpha \cdot \xi} - \frac{1}{(2\pi)^3} \frac{1}{\xi^2 - 2k\alpha \cdot \xi} \tilde{q} * \tilde{\epsilon}. \tag{13}$$

An essential ingredient of our proof in Section 3 is the following lemma, proved by the author in [10], p.262, and in [9].

Lemma 2.1 *If $A_j(\beta, \alpha, k)$ is the scattering amplitude corresponding to potential q_j, $j = 1, 2$, then*

$$-4\pi[A_1(\beta, \alpha, k) - A_2(\beta, \alpha, k)] = \int_{B_1} [q_1(x) - q_2(x)] u_1(x, \alpha, k) u_2(x, -\beta, k) dx, \tag{14}$$

where u_j is the scattering solution corresponding to q_j.

Consider an algebraic variety $\mathcal{M}$ in $\mathbb{C}^3$ defined by the equation

$$\mathcal{M} := \{\theta \cdot \theta = 1, \quad \theta \cdot \theta := \theta_1^2 + \theta_2^2 + \theta_3^2, \quad \theta_j \in \mathbb{C}, \ 1 \leq j \leq 3.\} \tag{15}$$

This is a non-compact variety, intersecting $\mathbb{R}^3$ over the unit sphere S^2.

Let $R_+ = [0, \infty)$. The following result is proved in [11], p.62.

Lemma 2.2 *If Assumption A) holds, then the scattering amplitude $A(\beta, \alpha, k)$ is a restriction to $S^2 \times S^2 \times R_+$ of a function $A(\theta', \theta, k)$ on $\mathcal{M} \times \mathcal{M} \times \mathbb{C}$, analytic on $\mathcal{M} \times \mathcal{M}$ and meromorphic on $\mathbb{C}$, θ', $\theta \in \mathcal{M}$, $k \in \mathbb{C}$.*

The scattering solution $u(x, \alpha, k)$ is a meromorphic function of k in $\mathbb{C}$, analytic in $\mathrm{Im}k \geq 0$, except, possibly, at the points $k = ik_j$, $1 \leq j \leq n$, $k_j > 0$, where $-k_j^2$ are negative eigenvalues of the selfadjoint Schrödinger operator, defined by the potential q in $L^2(\mathbb{R}^3)$. These eigenvalues can be absent, for example, if $q \geq 0$.

We need the notion of the Radon transform: $\hat{f}(\beta, \lambda) := \int_{\beta \cdot x = \lambda} f(x) d\sigma$, where $d\sigma$ is the element of the area of the plane $\beta \cdot x = \lambda$, $\beta \in S^2$, λ is a real number. The following properties of the Radon transfor will be used: $\int_{B_a} f(x) dx = \int_{-a}^{a} \hat{f}(\beta, \lambda) d\lambda, \int_{B_a} e^{ik\beta \cdot x} f(x) dx = \int_{-a}^{a} e^{ik\lambda} \hat{f}(\beta, \lambda) d\lambda, \hat{f}(\beta, \lambda) = \hat{f}(-\beta, -\lambda)$.These properties are proved, e.g., in [12], pp. 12, 15. We also need the following Phragmen-Lindelöf lemma, which is proved, e.g., in [1], p.69.

Lemma 2.3 *Let $f(z)$ be holomorphic inside an angle $\mathcal{A}$ of opening $< \pi$; $|f(z)| \leq c_1 e^{c_2|z|}$, $z \in \mathcal{A}$, $c_1, c_2 > 0$ are constants; $|f(z)| \leq M$ on the boundary of $\mathcal{A}$; and f is continuous up to the boundary of $\mathcal{A}$. Then $|f(z)| \leq M$, $\forall z \in \mathcal{A}$.*

3 Outline of Proof of Theorem 1.1

The scattering data in Remark 1 determine uniquely the scattering data in Theorem 1.1 by Lemma 2.2. Assume that potentials q_j, $j = 1, 2$, generate the same scattering data: $A_1(-\beta, \beta, k) = A_2(-\beta, \beta, k) \quad \forall \beta \in S^2, \quad \forall k > 0$, and let $p(x) := q_1(x) - q_2(x)$. Then by Lemma 2.1, see equation (14), one gets

$$0 = \int_{B_a} p(x)u_1(x, \beta, k)u_2(x, \beta, k)dx, \qquad \forall \beta \in S^2, \ \forall k > 0. \tag{16}$$

By (9) and (11) one can rewrite (16) as

$$\int_{B_a} e^{2ik\beta\cdot x}[1 + \epsilon(x, k)]p(x)dx = 0 \quad \forall \beta \in S^2, \ \forall k > 0, \tag{17}$$

where $\epsilon(x, k) := \epsilon := \epsilon_1(x, k) + \epsilon_2(x, k) + \epsilon_1(x, k)\epsilon_2(x, k)$. By Lemma 2.2 the relations (16) and (17) hold for complex k, $k = \frac{\kappa+i\eta}{2}, \quad \kappa+i\eta \neq 2ik_j, \quad \eta \geq 0$. Using formulas (3)-(4), one derives from (17) the relation

$$\tilde{p}((\kappa + i\eta)\beta) + \frac{1}{(2\pi)^3}(\tilde{\epsilon} * \tilde{p})((\kappa + i\eta)\beta) = 0 \qquad \forall \beta \in S^2, \ \forall \kappa \in \mathbb{R}, \tag{18}$$

where the notation $(f * g)(z)$ means that the convolution $f * g$ is calculated at the argument $z = (\kappa + i\eta)\beta$.

One has

$$\sup_{\beta \in S^2} |\tilde{\epsilon} * \tilde{p}| := \sup_{\beta \in S^2} |\int_{\mathbb{R}^3} \tilde{\epsilon}((\kappa + i\eta)\beta - s)\tilde{p}(s)ds| \leq \nu(\kappa, \eta) \sup_{s \in \mathbb{R}^3} |\tilde{p}(s)|, \tag{19}$$

where $\nu(\kappa, \eta) := \sup_{\beta \in S^2} \int_{\mathbb{R}^3} |\tilde{\epsilon}((\kappa + i\eta)\beta - s)|ds$. We prove that if $\eta = \eta(\kappa) = O(\ln \kappa)$ is suitably chosen (see [3]), then the following inequality holds:

$$0 < \nu(\kappa, \eta(\kappa)) < 1, \qquad \kappa \to \infty. \tag{20}$$

We also prove that

$$\sup_{\beta \in S^2} |\tilde{p}((\kappa + i\eta(\kappa))\beta)| \geq \sup_{s \in \mathbb{R}^3} |\tilde{p}(s)|, \quad \kappa \to \infty, \tag{21}$$

and then it follows from (18)-(21) that $\tilde{p}(s) = 0$, so $p(x) = 0$, and Theorem 1.1 is proved. Indeed, it follows from (18) and (21) that, for sufficiently large κ and a suitable $\eta(k) = O(\ln k)$, one has $\sup_{s \in \mathbb{R}^3} |\tilde{p}(s)| \leq \frac{1}{(2\pi)^3}\nu(\kappa, \eta(\kappa)) \sup_{s \in \mathbb{R}^3} |\tilde{p}(s)|$. If (20) holds, then the above equation implies that $\tilde{p} = 0$. This and the injectivity of the Fourier transform imply that $p = 0$. A detailed proof of estimates (20) and (21), that completes the proof, is given in [3]. *This completes the outline of the proof of Theorem 1.1.*

References

[1] B. Levin, Distribution of zeros of entire functions, AMS, Providence, RI, 1980.

[2] A.G.Ramm, "Uniqueness theorems for inverse scattering problem with non-overdetermined data", J.Phys A., pp. 112001, 2010.

[3] A.G.Ramm, "Uniqueness theorem for inverse scattering with backscattering data", Eurasian Jourm. Math, 1, No. 3, pp. 82-95, 2010.

[4] A.G.Ramm, On the analytic continuation of the solution of the Schrödinger equation in the spectral parameter and the behavior of the solution to the nonstationary problem as $t \to \infty$", Uspechi Mat. Nauk, 19, 192-194, 1964.

[5] A.G.Ramm, "Some theorems on analytic continuation of the Schrödinger operator resolvent kernel in the spectral parameter", Izvestiya Acad. Nauk Armyan. SSR, Mathematics, 3, pp. 443-464, 1968.

[6] A.G.Ramm, "Recovery of the potential from fixed energy scattering data", Inverse Problems, 4, pp. 877-886, 1988.

[7] A.G.Ramm, "Stability estimates in inverse scattering", Acta Appl. Math., 28, No. 1, pp. 1-42, 1992.

[8] A.G.Ramm, "Stability of the solution to inverse obstacle scattering problem", J.Inverse and Ill-Posed Problems, 2, No. 3, pp. 269-275, 1994.

[9] A.G.Ramm, "Stability of solutions to inverse scattering problems with fixed-energy data", Milan Journ of Math., 70, pp. 97-161, 2002.

[10] A.G.Ramm, Inverse problems, Springer, New York, 2005.

[11] A.G.Ramm, Scattering by obstacles, D.Reidel, Dordrecht, 1986.

[12] A.G.Ramm, A.I.Katsevich, The Radon transform and local tomography, CRC Press, Boca Raton 1996.

A Fourier Series Method for Solving Ordinary Differential Equations with Non-Constant Coefficients Arising in Inverse Shape Design

Daniel P. Baker[1] and George S. Dulikravich[2]
[1]P.O. Box 124, Lemont, PA 16851
Email: baker.daniel@verizon.net
[2]Mechanical and Materials Engineering Department
Florida International University, Miami, FL 33174
Email: dulikrav@fiu.edu

Abstract

An analytical method of integrating ordinary differential equations with non-constant coefficients arising from an elastic membrane concept for inverse shape design is presented utilizing Fourier series formulation. The non-homogeneous ordinary differential equation with non-constant coefficients mimics forced oscillations of a system of mass-damper-spring elements linked in parallel where coefficients of mass, damper and spring are non-constant. This elastic membrane concept for inverse shape design requires knowledge only of the surface field variables distribution on the body to perform a shape update. Thus, it can be implemented without modifying an existing field analysis code. The proposed formulation allows each segment of an evolving shape to move at its own optimal speed thus potentially significantly reducing the required number of shape updates until it matches the specified surface field data.

1. Introduction

The elastic membrane approach to aerodynamic inverse shape design was first proposed by Garabedian and McFadden [1]. This concept treats the surface of an aerodynamic body as a membrane that deforms under aerodynamic loads until it achieves specified (target) surface pressure distribution. The original model [1] for the evolution of an airfoil shape to create a specified (target) surface pressure distribution was given by a simple mass-damper-spring equation for forced vibrations

$$\beta_0 \Delta n + \beta_1 \frac{d\Delta n}{dx} + \beta_2 \frac{d^2 \Delta n}{dx^2} = C_p^{\text{target}} - C_p^{\text{actual}} \tag{1}$$

Here, Δn's are defined as shape corrections along outward normal vectors, and $\beta_{0\text{-}2}$ are user supplied constants coefficients that control the rate of convergence of the shape evolution process, while C_p^{target} and C_p^{actual} are the specified (target) and the actual (computed) local surface pressure coefficients, respectively. This technique was modified by Malone *et al.* [2], giving

$$\beta_0 \Delta y + \beta_1 \frac{d\Delta y}{dx} + \beta_2 \frac{d^2 \Delta y}{dx^2} = C_p^{\text{target}} - C_p^{\text{actual}} \tag{2}$$

so that shape modifications are in the y-direction only, thus preventing the chord length from changing. Equation (2) is traditionally solved for Δy shape corrections using a finite difference approach by discretizing along the airfoil contour. This approach has slow convergence especially with the flow-field analysis codes of increasing non-linearity [3-6] because of the truncation errors resulting from numerical differentiation. The iterative process of evolving solution of equation (2) has been significantly accelerated by using Fourier series formulation for a *de facto* analytical integration of equation (2) as presented by Baker and Dulikravich [3-7]. However, even this can be improved upon if it could be possible to have each segment of a body surface move with its own values of coefficients of the mass, damper and spring, that is, by allowing these coefficients to be arbitrary functions of the x-coordinate.

This paper offers an attempt to derive an analytical framework for integrating such ordinary differential equations with non-constant coefficients by utilizing Fourier series.

2. Analytical Formulation

Let us represent two functions, Yu(x) and Yk(x), in terms of complete Fourier series so that

$$Y_u = \sum_{n=0}^{n_{max}} [a_n \cos(nx) + b_n \sin(nx)] \quad (3)$$

$$Y_k = \sum_{n=0}^{n_{max}} [e_n \cos(nx) + f_n \sin(nx)] \quad (4)$$

These two Fourier series (with the same number of terms) multiply together to make another Fourier series that will have an equal number of terms.

$$Y_k Y_u = \sum_{n=0}^{n_{max}} [a_n \cos(nx) + b_n \sin(nx)] \sum_{m=0}^{n_{max}} [e_m \cos(mx) + f_m \sin(mx)] = \\ \sum_{n=0}^{n_{max}} \sum_{m=0}^{n_{max}} \{a_n [e_m \cos(mx) + f_m \sin(mx)] \cos(nx) + b_n [e_m \cos(mx) + f_m \sin(mx)] \sin(nx)\} \quad (5)$$

The following four trigonometric identities will simplify this.

$$\begin{aligned} \cos(nx)\cos(mx) &= \frac{1}{2}[\cos(n+m)x + \cos(n-m)x] \\ \cos(nx)\sin(mx) &= \frac{1}{2}[\sin(n+m)x - \sin(n-m)x] \\ \sin(nx)\cos(mx) &= \frac{1}{2}[\sin(n+m)x + \sin(n-m)x] \\ \sin(nx)\sin(mx) &= \frac{1}{2}[-\cos(n+m)x + \cos(n-m)x] \end{aligned} \quad (6)$$

Hence,

$$Y_k Y_u = \frac{1}{2} \sum_{n=0}^{n_{max}} \sum_{m=0}^{n_{max}} [(e_m a_n + f_m b_n) \cos(n-m)x + (e_m b_n - f_m a_n) \sin(n-m)x] \\ + \sum_{n=0}^{n_{max}} \sum_{m=0}^{n_{max}} [(e_m a_n - f_m b_n) \cos(n+m)x + (e_m b_n + f_m a_n) \sin(n+m)x] \quad (7)$$

or

$$Y_k Y_u = \frac{1}{2} \sum_{n=0}^{n_{max}} \sum_{m=0}^{n} [(e_m a_n + f_m b_n) \cos(n-m)x + (e_m b_n - f_m a_n) \sin(n-m)x] \\ + \frac{1}{2} \sum_{n=0}^{n_{max}} \sum_{m=0}^{n_{max}-n} [(e_m a_n - f_m b_n) \cos(n+m)x + (e_m b_n + f_m a_n) \sin(n+m)x] \\ \frac{1}{2} \sum_{n=0}^{n_{max}} \sum_{m=n+1}^{n_{max}} [(e_m a_n + f_m b_n) \cos(n-m)x + (e_m b_n - f_m a_n) \sin(n-m)x] \\ + \frac{1}{2} \sum_{n=0}^{n_{max}} \sum_{m=n_{max}-n+1}^{n_{max}} [(e_m a_n - f_m b_n) \cos(n+m)x + (e_m b_n + f_m a_n) \sin(n+m)x] \quad (8)$$

The first two lines of the above sum produce harmonics with wavenumber inclusively between zero and nmax. The third line produces wavenumbers less than zero. The fourth line produces wavenumbers greater than n_{max}. The third line is corrected using the following identities.

$$\begin{aligned} \cos(-A) &= \cos(A) \\ \sin(-A) &= -\sin(A) \end{aligned} \tag{9}$$

This results in

$$\begin{aligned} Y_k Y_u &= \frac{1}{2}\sum_{n=0}^{n_{max}}\sum_{m=0}^{n}[(e_m a_n + f_m b_n)\cos(n-m)x + (e_m b_n - f_m a_n)\sin(n-m)x] \\ &+ \frac{1}{2}\sum_{n=0}^{n_{max}}\sum_{m=0}^{n_{max}-n}[(e_m a_n - f_m b_n)\cos(n+m)x + (e_m b_n + f_m a_n)\sin(n+m)x] \\ &\frac{1}{2}\sum_{n=0}^{n_{max}}\sum_{m=n+1}^{n_{max}}[(e_m a_n + f_m b_n)\cos(m-n)x - (e_m b_n - f_m a_n)\sin(m-n)x] \\ &+ \frac{1}{2}\sum_{n=0}^{n_{max}}\sum_{m=n_{max}-n+1}^{n_{max}}[(e_m a_n - f_m b_n)\cos(n+m)x + (e_m b_n + f_m a_n)\sin(n+m)x] \end{aligned} \tag{10}$$

Wavenumbers greater than n_{max} will be aliased back into the available wavenumber spectrum by an FFT, thus modifying the fourth line as follows

$$\begin{aligned} Y_k Y_u &= \frac{1}{2}\sum_{n=0}^{n_{max}}\sum_{m=0}^{n}[(e_m a_n + f_m b_n)\cos(n-m)x + (e_m b_n - f_m a_n)\sin(n-m)x] \\ &+ \frac{1}{2}\sum_{n=0}^{n_{max}}\sum_{m=0}^{n_{max}-n}[(e_m a_n - f_m b_n)\cos(n+m)x + (e_m b_n + f_m a_n)\sin(n+m)x] \\ &\frac{1}{2}\sum_{n=0}^{n_{max}}\sum_{m=n+1}^{n_{max}}[(e_m a_n + f_m b_n)\cos(m-n)x - (e_m b_n - f_m a_n)\sin(m-n)x] \\ &+ \frac{1}{2}\sum_{n=0}^{n_{max}}\sum_{m=n_{max}-n+1}^{n_{max}}[(e_m a_n - f_m b_n)\cos(2n_{max}-n-m)x - (e_m b_n + f_m a_n)\sin(2n_{max}-n-m)x] \end{aligned} \tag{11}$$

This expression can be rewritten in a more convenient form by changing the subscripts so that m = n – t in the first line, m = t – n in the second line, m = t + n in the third line, and m = 2nmax – n – t in the fourth.

$$\begin{aligned} Y_k Y_u &= \frac{1}{2}\sum_{n=0}^{n_{max}}\sum_{t=0}^{n}[(e_{n-t} a_n + f_{n-t} b_n)\cos(tx) + (e_{n-t} b_n - f_{n-t} a_n)\sin(tx)] \\ &+ \frac{1}{2}\sum_{n=0}^{n_{max}}\sum_{t=n}^{n_{max}}[(e_{t-n} a_n - f_{t-n} b_n)\cos(tx) + (e_{t-n} b_n + f_{t-n} a_n)\sin(tx)] \\ &\frac{1}{2}\sum_{n=0}^{n_{max}}\sum_{t=1}^{n_{max}-n}[(e_{t+n} a_n + f_{t+n} b_n)\cos(tx) - (e_{t+n} b_n - f_{t+n} a_n)\sin(tx)] \\ &+ \frac{1}{2}\sum_{n=0}^{n_{max}}\sum_{t=n_{max}-n}^{n_{max}-1}[(e_{2n_{max}-n-t} a_n - f_{2n_{max}-n-t} b_n)\cos(tx) - (e_{2n_{max}-n-t} b_n + f_{2n_{max}-n-t} a_n)\sin(tx)] \end{aligned} \tag{12}$$

Thus, the effect of the nth mode of the first multiplier on the tth mode of the product is

$$\begin{aligned}[Y_k Y_u]_{t,n} &= \frac{1}{2}[(e_{n-t}a_n + f_{n-t}b_n)\cos(tx) + (e_{n-t}b_n - f_{n-t}a_n)\sin(tx)] + \\ &\frac{1}{2}[(e_{t-n}a_n - f_{t-n}b_n)\cos(tx) + (e_{t-n}b_n + f_{t-n}a_n)\sin(tx)] + \\ &\frac{1}{2}[(e_{t+n}a_n + f_{t+n}b_n)\cos(tx) - (e_{t+n}b_n - f_{t+n}a_n)\sin(tx)] + \\ &\frac{1}{2}[(e_{2n_{max}-n-t}a_n - f_{2n_{max}-n-t}b_n)\cos(tx) - (e_{2n_{max}-n-t}b_n + f_{2n_{max}-n-t}a_n)\sin(tx)]\end{aligned} \tag{13}$$

where Fourier components e_m and f_m with m out of the range $[0,n_{max}]$ are equal to zero. This reduces to

$$\begin{aligned}[Y_k Y_u]_{t,n} &= \left[\left(\frac{e_{n-t}+e_{t-n}+e_{t+n}+e_{2n_{max}-n-t}}{2}\right)a_n + \left(\frac{f_{n-t}-f_{t-n}+f_{t+n}-f_{2n_{max}-n-t}}{2}\right)b_n\right]\cos(tx) \\ &+\left[\left(\frac{-f_{n-t}+f_{t-n}+f_{t+n}-f_{2n_{max}-n-t}}{2}\right)a_n + \left(\frac{e_{n-t}+e_{t-n}-e_{t+n}-e_{2n_{max}-t-n}}{2}\right)b_n\right]\sin(tx)\end{aligned} \tag{14}$$

Similarly,

$$\begin{aligned}\left[Y_k \frac{dY_u}{dx}\right]_{t,n} &= \left[n\left(\frac{f_{n-t}-f_{t-n}+f_{t+n}-f_{2n_{max}-n-t}}{2}\right)a_n - n\left(\frac{e_{n-t}+e_{t-n}+e_{t+n}+e_{2n_{max}-n-t}}{2}\right)b_n\right]\cos(tx) \\ &+\left[n\left(\frac{e_{n-t}+e_{t-n}-e_{t+n}-e_{2n_{max}-t-n}}{2}\right)a_n - n\left(\frac{-f_{n-t}+f_{t-n}+f_{t+n}-f_{2n_{max}-n-t}}{2}\right)b_n\right]\sin(tx)\end{aligned} \tag{15}$$

$$\begin{aligned}\left[Y_k \frac{d^2Y_u}{dx^2}\right]_{t,n} &= \left[-n^2\left(\frac{e_{n-t}+e_{t-n}+e_{t+n}+e_{2n_{max}-n-t}}{2}\right)a_n - n^2\left(\frac{f_{n-t}-f_{t-n}+f_{t+n}-f_{2n_{max}-n-t}}{2}\right)b_n\right]\cos(tx) \\ &+\left[-n^2\left(\frac{-f_{n-t}+f_{t-n}+f_{t+n}-f_{2n_{max}-n-t}}{2}\right)a_n - n^2\left(\frac{e_{n-t}+e_{t-n}-e_{t+n}-e_{2n_{max}-t-n}}{2}\right)b_n\right]\sin(tx)\end{aligned} \tag{16}$$

For example, consider the 2nd order linear ordinary differential equation with a forcing function D(x)

$$A(x)\frac{d^2y}{dx^2} + B(x)\frac{dy}{dx} + C(x)y = D(x) \tag{17}$$

We wish to solve for the Fourier components that would be generated by an FFT for the solution y based on the Fourier components generated by an FFT for A(x), B(x), C(x), and the forcing function, D(x) where

$$A(x) = \sum_{n=0}^{n_{max}}[Ac_n\cos(nx) + As_n\sin(nx)] \quad \text{(known)} \tag{18}$$

$$B(x)=\sum_{n=0}^{n_{max}}\left[Bc_n\cos(nx)+Bs_n\sin(nx)\right] \quad \text{(known)} \tag{19}$$

$$C(x)=\sum_{n=0}^{n_{max}}\left[Cc_n\cos(nx)+Cs_n\sin(nx)\right] \quad \text{(known)} \tag{20}$$

$$D(x)=\sum_{n=0}^{n_{max}}\left[Dc_n\cos(nx)+Ds_n\sin(nx)\right] \quad \text{(known)} \tag{21}$$

$$y(x)=\sum_{n=0}^{n_{max}}\left[yc_n\cos(nx)+ys_n\sin(nx)\right] \quad \text{(unknown)} \tag{22}$$

Substituting these five equations in the example equation (17) leads to the fact that the tth cosine mode of the LHS must equal the tth cosine mode of the RHS in the governing equation (17). Hence,

$$\sum_{n=0}^{n_{max}}\left\{\begin{array}{l}\left[\begin{array}{l}Cc_{n-t}+Cc_{t-n}+Cc_{t+n}+Cc_{2n_{max}-n-t}\\ +n\left(Bs_{n-t}-Bs_{t-n}+Bs_{t+n}-Bs_{2n_{max}-n-t}\right)\\ -n^2\left(Ac_{n-t}+Ac_{t-n}+Ac_{t+n}+Ac_{2n_{max}-n-t}\right)\end{array}\right]yc_n\\ +\left[\begin{array}{l}Cs_{n-t}-Cs_{t-n}+Cs_{t+n}-Cs_{2n_{max}-n-t}\\ -n\left(Bc_{n-t}+Bc_{t-n}+Bc_{t+n}+Bc_{2n_{max}-n-t}\right)\\ -n^2\left(As_{n-t}-As_{t-n}+As_{t+n}-As_{2n_{max}-n-t}\right)\end{array}\right]ys_n\end{array}\right\}=2Dc_t \tag{23}$$

Similarly, the tth sine mode of the LHS must equal the tth sine mode of the RHS in equation (17). Hence,

$$\sum_{n=0}^{n_{max}}\left\{\begin{array}{l}\left[\begin{array}{l}-Cs_{n-t}+Cs_{t-n}+Cs_{t+n}-Cs_{2n_{max}-n-t}\\ +n\left(Bc_{n-t}+Bc_{t-n}-Bc_{t+n}-Bc_{2n_{max}-n-t}\right)\\ -n^2\left(-As_{n-t}+As_{t-n}+As_{t+n}-As_{2n_{max}-n-t}\right)\end{array}\right]yc_n\\ +\left[\begin{array}{l}Cc_{n-t}+Cc_{t-n}-Cc_{t+n}-Cc_{2n_{max}-n-t}\\ -n\left(-Bs_{n-t}+Bs_{t-n}+Bs_{t+n}-Bs_{2n_{max}-n-t}\right)\\ -n^2\left(Ac_{n-t}+Ac_{t-n}-Ac_{t+n}-Ac_{2n_{max}-n-t}\right)\end{array}\right]ys_n\end{array}\right\}=2Ds_t \tag{24}$$

Thus, $2*n_{max}+1$ linear equations can be solved for $2*n_{max}+1$ unknown variables yc_0, yc_n, and ys_n that make up the analytical solution for y. Notice that no explicit boundary conditions are included in the system. Implicit boundary conditions arise from all basis functions being periodic. Thus, the final solution y will be periodic. In many cases, no additional boundary conditions need be applied to the system

3. Example of a Need for Boundary Conditions

However, consider the equation

$$y''=0 \tag{25}$$

solved using this method. We can get the analytic solution as

$$\begin{aligned} & y' = \text{const1} \\ & y = (\text{const1})x + \text{const2} \end{aligned} \tag{26}$$

As the basis fuctions guarantee a periodic solution y,

$$\begin{aligned} & y(0) = \text{const2} \\ & y(L) = (\text{const1})L + \text{const2} \\ & y(0) = y(L) \\ & \text{thus} \\ & \text{const1} = 0 \end{aligned} \tag{27}$$

However, for const2 to be determined uniquely, another boundary condition must be explicitly specified (perhaps y(0)). This equation deficiency will manifest itself in the set of equations as either a 0=0 equation or a set of linearly dependant equations. In either case, the system will not have sufficient rank, and will not be invertible. Therefore, care must be taken when examining the resulting system of equations (23) and (24) to determine when auxilliary boundary condition equations are needed.

One final note: this procedure can be likewise extended to higher order linear equations. However, the process results in increasingly stiff matrices of equations, due to the presence of powers of n in derivative terms. Thus, while n-squared appears in the above 2nd order equation, n-cubed would occur in a 3rd order equation, and so forth.

4. Summary

An analytical formulation was derived for integration of inhomogeneous ordinary differential equations with non-constant coefficients by utilizing a Fourier series method. Such equations can be used to model a distributed mass-damper-spring system oscillating and dampening while driven by an arbitrary distributed forcing function. This model is used for a robust inverse design of shapes of objects. The presented analytical formulation offer an enhanced convergence rate of the iterative determination of shapes when compared to the current model that uses constant mass-damper-spring coefficients.

5. References

[1] Garabedian, P. and McFadden, G., "Design of Supercritical Swept Wings," AIAA Journal, Vol. 20, No. 3, pp. 289-291, 1982.

[2] Malone, J.B., Vadyak, J. and Sankar, L.N., "A Technique for the Inverse Aerodynamic Design of Nacelles and Wing Configurations," AIAA Journal of Aircraft, Vol. 24, No. 1, pp. 8-9, 1987.

[3] Dulikravich, G.S. and Baker, D.P., "Fourier Series Analytical Solution for Inverse Design of Aerodynamic Shapes," Inverse Problems in Engineering Mechanics - ISIP'98, eds. Tanaka M, Dulikravich G.S., Elsevier Science, Ltd., U.K., pp. 427-436, 1998.

[4] Dulikravich, G.S. and Baker, D.P., "Using Existing Flow-Field Analysis Codes for Inverse Design of Three-dimensional Aerodynamic Shapes," Recent Development of Aerodynamic Design Methodologies - Inverse Design and Optimization, eds. Fujii. K. and Dulikravich G.S., Vieweg Series on Notes on Numerical Fluid Mechanics, vol. 68, Springer, pp. 89-112, 1999.

[5] Dulikravich, G.S. and Baker, D.P., "Aerodynamic Shape Inverse Design Using a Fourier Series Method", AIAA paper 99-0185, AIAA Aerospace Sciences Meeting, Reno, NV, Jan. 11-14, 1999.

[6] Baker, D.P., "A Fourier Series Approach to the Elastic Membrane Inverse Shape Design Problem in Aerodynamics," M.Sc. thesis, Dept. of Aerospace Eng., The Pennsylvania State University, University Park, PA, U.S.A., 1999.

[7] Baker, D.P., Dulikravich, G.S., Dennis, B.H. and Martin, T.J., "Inverse Determination of Eroded Smelter Wall Thickness Variation Using an Elastic Membrane Concept", ASME Journal of Heat Transfer, Vol. 132, Issue 5, pp. 052101-1/052101-8, 2010

Author Index

www.ingramcontent.com/pod-product-compliance
Lightning Source LLC
Chambersburg PA
CBHW080719220326
41520CB00056B/7145

* 9 7 8 0 6 1 5 4 7 1 0 0 6 *